The science of genetics has undergone a period of very rapid and significant development in recent years, and the area of poultry genetics has been no exception. This book provides a balanced and up-to-date account of the major areas of genetics and evolution in the domestic fowl, including transmission genetics, cytogenetics, molecular genetics and quantitative genetics. In each chapter, a brief explanation of the genetic principles is followed by a full discussion illustrated by key examples. This book will be of great interest to postgraduate students and researchers in the fields of genetics, agriculture and veterinary medicine, as well as to poultry breeders (both commercial and non-commercial).

***Genetics and evolution of the domestic fowl***

# Genetics and evolution of the domestic fowl

LEWIS STEVENS
*Department of Biological and Molecular Sciences*
*University of Stirling*

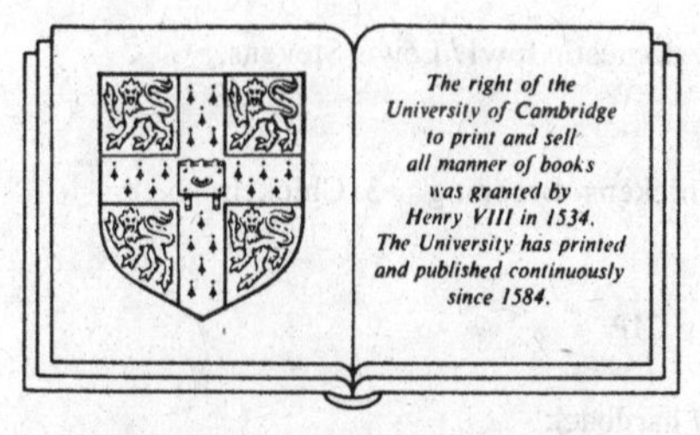


CAMBRIDGE UNIVERSITY PRESS
*Cambridge*
*New York Port Chester Melbourne Sydney*

CAMBRIDGE UNIVERSITY PRESS
Cambridge, New York, Melbourne, Madrid, Cape Town, Singapore, São Paulo

Cambridge University Press
The Edinburgh Building, Cambridge CB2 2RU, UK

Published in the United States of America by Cambridge University Press, New York

www.cambridge.org
Information on this title: www.cambridge.org/9780521403177



First published 1991
This digitally printed first paperback version 2005

*A catalogue record for this publication is available from the British Library*

*Library of Congress Cataloguing in Publication data*

Stevens, Lewis.
Genetics and evolution of the domestic fowl / Lewis Stevens.
p. cm.
Includes index.
1. Chickens–Genetics. 2. Chickens–Breeding. 3. Chickens–Evolution. I. Title.
SF492.S74 1991
636.5′0821–dc20 91–9211 CIP

ISBN-13 978-0-521-40317-7 hardback
ISBN-10 0-521-40317-0 hardback

ISBN-13 978-0-521-01757-2 paperback
ISBN-10 0-521-01757-2 paperback

*To the memory of my father*
*for whom the fortunes of war denied so much*

# *Contents*

# *Preface*

It is over forty years since the publication of Hutt's classic text *The Genetics of the Fowl.* During that period there have been enormous advances in the understanding of genetics in general. With the advent of gene cloning these advances are likely to continue at an increased pace for at least the next decade. On a different level there have also been sweeping changes in the commercial rearing of the domestic fowl as broilers and layers. These changes have been in part attributable to the new strains that have been introduced as a result of research into breeding. An area that has changed least is the way in which the poultry fancier breeds for 'perfection' as he or she sees it. It is with these factors in mind that this book has been written. Whilst not aiming to produce an encyclopaedic work like that of Hutt's and the new multiauthor reference work *Poultry Breeding and Genetics*, edited by R.D. Crawford, it is intended that this book will provide both an introduction to, and a useful survey of, the recent developments in the genetics and evolution of the domestic fowl. There are some areas covered in Hutt's book that have changed little since 1950, particularly the 'classical' genetics of many of the physiological and anatomical characters. Although the inheritance of several additional characters has been studied, and in some cases the genes controlling them have been mapped, the molecular bases of most of these are still unknown. A large number of biochemical characters have been studied in the past twenty years including enzymes, immunoglobulins, histocompatibility antigens, blood group substances and egg proteins, and several genes have been cloned and sequenced.

The classical methods of gene mapping are very time consuming since they involve breeding large numbers of birds. The long generation time is, of course, one reason why the fowl fell out of favour with the geneticist. At the turn of the century, when Bateson and Punnett, working in Cambridge, studied the inherited characters in the fowl, it could be regarded as at the frontiers of the newly emerging subject of genetics. However, once simpler organisms with shorter generation times and fewer chromosomes had been studied, e.g. *Drosophila*, the fowl moved into the background as far as most geneticists were concerned.

One difficulty for a number of years had been to establish the chromosome number. This proved difficult because of the presence of the minute microchromosomes. It was only with the improved cytochemical techniques of the 1960s and 1970s that the diploid number was finally established as 78, and that detailed cytogenetic studies have been possible.

Within the last decade new methods of gene mapping have been developed involving cell hybridisation and complementary DNA. The use of DNA fingerprinting in domestic fowl is now under way. These methods have already made an enormous impact on human genetics and are starting to make their presence felt in poultry genetics. We can thus expect to see genes mapped at a much higher rate in the future, while the last decade has seen the complete sequence of a number of genes of the fowl resolved. Work is also under way on the introduction of foreign genes into the domestic fowl. Thus, the genetics of the domestic fowl is entering a new phase. The advances in molecular biology have also made an impact on the study of the evolution of the domestic fowl. It is particularly with these developments in mind that the author felt the time was ripe for an overall résumé.

I have tried in this book to cater for people interested in the domestic fowl having different backgrounds. For this reason some of the basic principles both of molecular and classical genetics are discussed, with examples taken from the domestic fowl. However, there is obviously a limit to how far one can go without simply writing a general introduction to genetics which, in any case, would not be of interest to a geneticist wishing to know the current position with regard to the fowl. A compromise thus had to be reached. I have endeavoured to give fuller background information on the newer developments such as cell hybridisation, gene cloning and sequencing, whereas with the more established areas I have referred the reader to more general textbooks.

Chapters 3, 4, 5, 6 and 8 are concerned mainly with transmission genetics, illustrated with examples from the domestic fowl, and require little molecular background knowledge. Chapters 1 and 10 are concerned primarily with evolution. Chapters 11 (immunogenetics) and 12 (gene cloning and sequencing) are the most molecularly orientated, and some basic background material for these chapters is given in the first half of Chapter 2. The second half of Chapter 2 is concerned with the fowl karyotype. Chapter 9 is an introduction to quantitative genetics as applied to the domestic fowl; this is particularly important for understanding breeding programmes. Chapter 7 deals with the genetics of three types of tissues, namely muscle, nervous tissue and the skeleton and it includes both transmission genetics and molecular genetics. Genetics and molecular biology has a vocabulary of its own, and so a glossary is included for those unfamiliar with some of the terminology.

A number of people have helped at various stages in the production of this book. My interest in poultry genetics arose from two happenstances. My daughters, whilst at primary school, brought home two two-week-old Leghorn chicks that needed a home, and then two years later an opportunity to give some elementary genetics lectures arose in the department in which I teach.

After I had written a draft of the first six chapters, Mrs Alison Jones kindly read them and provided some useful feedback as to how accessible they might be to poultry fanciers. My daughter Dr Rowena Stevens read Chapter 11 and made some helpful comments on the immunogenetics. Professor Trevor R. Morris, School of Agriculture, University of Reading read the whole draft manuscript and offered a number of useful suggestions, which I have incorporated in the final manuscript. I would like to thank Dr Sara Trevitt of Cambridge University Press, who, from the time she first received the draft manuscript, has been helpful and efficient in piloting it through to the final stages of production.

To all those I have mentioned I am grateful for the help they have given, but most of all I would like to thank my wife Dr Evelyn Stevens for her help and encouragement throughout the whole period in which I was writing the book, and for the care and effort which she has spent on reading the whole manuscript, and for the advice that she has given.

Sheriffmuir L.S.
July 1990

# 1

# *The history and evolution of the domestic fowl*

## 1.1 Introduction

It is convenient to divide the evolutionary history of the present day domestic fowl into three phases: the first is the evolution of the genus *Gallus*, the second is the emergence of the domestic fowl from its progenitor(s) within the genus *Gallus*, and the third is the appearance of the large number of present day breeds and varieties. These three phases occupy very different time spans. The origin of life on earth is estimated to have occurred about 3000 million years ago, whereas the genus *Gallus* probably dates from about 8 million years (Helm-Bychowski & Wilson, 1986). The domestication of the fowl in the region of the Indus valley is believed to have occurred by 2000 BC (Zeuner, 1963), but more recent archaeological evidence shows that a much earlier domestication occurred in China *c*. 6000 BC (West & Zhou, 1989). The origin of most of the present day breeds and varieties dates from the last century, although a few are considerably older. Figure 1.1 illustrates the phases of evolution up to that of the genus *Gallus*, in relation to other significant events. In this chapter these three phases are considered in turn giving some of the supporting evidence for each stage in evolution. This will be preceded by a brief outline of the probable mechanism of evolution which is intended to provide a minimum of background information for the later sections of the chapter (for a more detailed account, see Callow, 1983).

## 1.2 Mechanism of evolution

The mechanism of evolution as proposed by Darwin is that individuals of a species within a population show considerable but continuous variation in form and physiology, and that this variation arises randomly and is heritable. A process of natural selection occurs. Since resources are limited those individuals best adapted to the environment will survive, hence the notion of 'Survival of the Fittest'. Darwin was unable to give a clear indication as to how variation arose or how it was inherited. Since Darwin's time much detailed knowledge has been acquired about living organisms particularly at the cellular and molecular level. Two important breakthroughs that have had a bearing on understanding the mechanism of evolution have been (i) Mendel's Laws of Inheritance and (ii) the elucidation of the structure of the genetic material (DNA) followed by the understanding of the genetic code. The discovery at the beginning of this century of the work which Mendel had carried out nearly 40 years previously provided the background for biologists to consider the mechanism by which variation occurs. Both Fisher and Haldane considered these problems in their respective books *The Causes of Evolution* (1932) and *Genetic Theory of Evolution* (1930). Important conclusions reached were (i) that the rate of increase in 'fitness' of a population was related to the genetic variance within that population, i.e. a population having a high genetic variance would adapt more quickly, and (ii) natural selection was a more potent force than mutational events in bringing about this change.

**Fig. 1.1** Chart of the time of emergence of vertebrate classes, including the genus *Gallus*, in relation to geological eras.

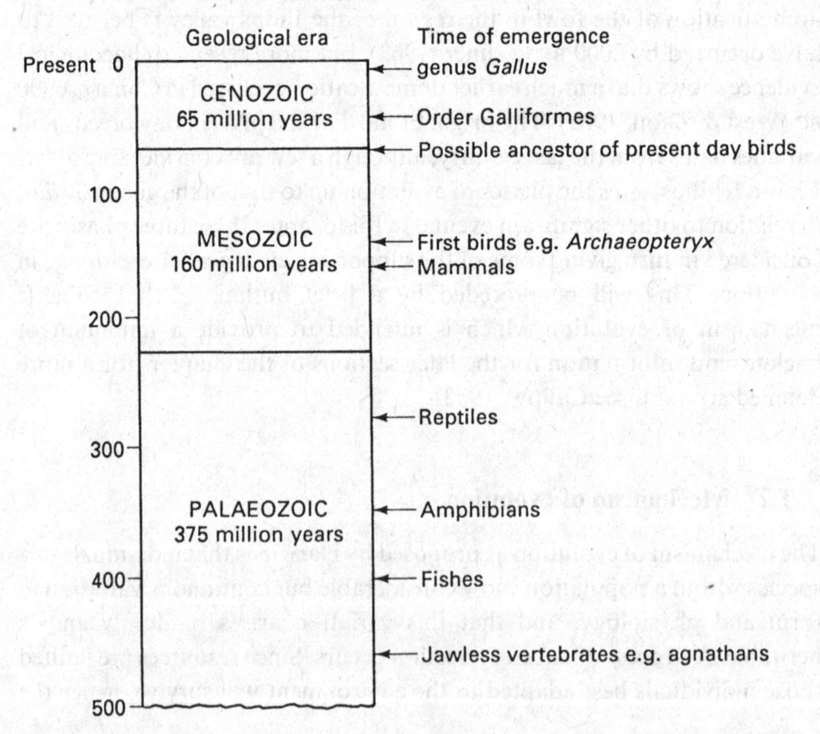

Since the elucidation of the genetic code in the 1960s and the sequencing of many proteins and nucleic acids it is now possible to understand the molecular nature of mutational events and the precise differences in the nucleotide sequences (see Chapter 2, section 2.1) within genes. These differences in sequences are responsible for the variation. There is still much debate about many aspects of the mechanism of evolution, in particular the importance of **positive selection** and the so called **'neutralist hypothesis'**. The 'neutralists' propose that many of the mutations which occur are neutral, in the sense that most of the amino acids replaced by others in proteins have little effect on the functioning of those proteins. Thus neutral changes will not be acted upon by the process of selection but will become randomly fixed within a population and changes will occur gradually by a process of random 'genetic drift'. The positive selectionists, on the other hand, take the view that most mutations will be detrimental but that these will be eliminated from the population by natural selection, whereas the rare advantageous mutations will be retained. Clearly both processes could be, and probably are, operating, but the relative importance of the two is difficult to assess.

## 1.3 The evolution of the genus *Gallus*

The evidence for the path of evolution in general comes from a variety of lines of investigation that include (i) palaeontology, (ii) geographical distribution, (iii) embryology and (iv) the structure and sequence of proteins and nucleic acids. Darwin's evidence for his theory of evolution came largely from the first three, but the evidence from the sequences of proteins and nucleic acids which has become increasingly available since the 1960s has very largely supported and greatly strengthened conclusions arrived at from the first three. There are, however, some awkward exceptions where differences in the sequences of homologous proteins from different species do not match the palaeontological evidence (Schwabe, 1986). During the late nineteenth and first half of the twentieth century it was possible to determine the relative ages of fossils only by their positions in sedimentary rocks; it is now possible to date them reasonably accurately using radiometric dating methods (Miller & Orgel, 1974). These methods involve measuring the proportions of particular radioactive isotopes in fossils. If the rate at which a radioisotope decays is known, then in principle the age of the fossil can be determined.

The age of the earth is generally reckoned to be about 4500 million years with the first forms of life appearing at about 3000 million years, the first eukaryotes at 1300 million years, the first mammals at 200 million years and the first birds at 150 million years. The present Class Aves comprises 27 orders and 170 families. There is still much debate about the methods of classification of living organisms, particularly the divisions into families, genera and species. The system used at the present time is built on the Linnaeus classification of 1758. Since Darwin's time the importance of evolutionary relationships has had a major influence on the method of classification. Evolution provides two kinds of event that can be taken into account in classifications: (i) the order of descent from ancestors, and (ii) the extent of divergence from ancestors. The former emphasises similarities and is the basis of the **cladistic method**, formulated by Hennig (1979); the latter emphasises differences and is known as the **phenetic method** (Sneath & Sokal, 1973). The cladistic method is concerned with genealogy, i.e. branching sequences, and thus with evolutionary relationships. In the phenetic method it is assumed to be impossible in many cases to classify according to genealogy, because it is often not possible to distinguish convergent and divergent characters. This method focuses on the similarities and differences between species. A large number of characters have to be considered and the differences are given similarity coefficients. These are related in a quantitative manner which gives numerical values for the differences between species. With certain groups of organism the one method may be advantageous over the other and vice versa. These two systems are discussed in detail by Ridley (1986) who advocates strongly the importance of evolution as the basis of classification.

The class Aves has for many years been thought of as displaying many more uniform anatomical features than other classes of vertebrates. It has been suggested that the structural requirements for flight have imposed constraints on the anatomical changes that could be tolerated. A recent study by Wyles, Kunkel & Wilson (1983) has refuted this view. They have made quantitative comparisons between two birds, the hummingbird and the albatross, and between two mammals, the seal and the cat. The four species were chosen as being two pairs of extremes within the two classes. Using an index 'H' (the Manhattan index) they found that the birds showed greater anatomical differences than the mammals. Also Wyles *et al.* suggest that the evolution of modern birds has occurred over a shorter period than was once supposed. The traditional view was that modern birds had ancestral roots in fossil birds with teeth dating from 135 to 165 million years ago. None of these fossils can be convincingly related to modern birds and they suggest that modern birds stem from one of the ancient bird species that lived about 65 million years ago. Geological evidence of a catastrophe about 65 million years ago allows for the possibility of two rounds of bird evolution (Alvarez, 1983), starting before and after the catastrophe. This is also consistent with the modest accumulation of point mutations causing amino acid substitutions in proteins of modern birds. Recent comparisons by Patton & Avise (1986) of the electrophoretic patterns of proteins from the orders Anseriformes (mainly waterfowl) and Passeriformes (a large order of perching birds including e.g. finches, warblers, pipits, thrushes and crows) show smaller genetic differences than expected by comparison with non-avian vertebrates. However, Patton & Avise (1986) suggest this may possibly be attributable to a deceleration in the rate of protein evolution. Schwabe (1986) doubts the justification for using this type of *ad hoc* argument. There is thus still debate concerning the period over which modern birds have evolved and whether the class Aves is 'oversplit' relative to other groups of vertebrates.

The standard classification of the relevant section of the Galliformes which includes the domestic fowl is given below (Fig. 1.2). According to this classification the turkey (*Meleagris*) is more distantly related to the genus *Gallus* than pheasants (*Phasianus*), partridges (*Perdix*) and Old World quails (*Coturnix*). Recent evidence suggests that this is not so. When the amino acid sequences of the enzyme lysozyme from seven Phasianoid birds are compared (Jolles *et al.*, 1979) and the average minimal mutation distances are calculated, the pheasant appears the most distantly related to

**Fig. 1.2** Standard classification of the Order Galliformes.

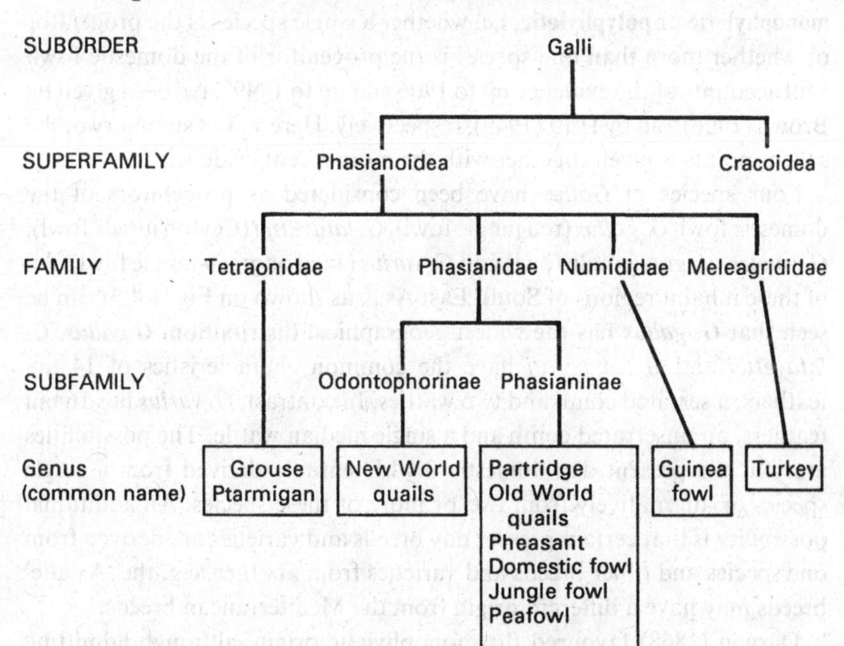

the genus *Gallus* of the Phasianoid birds; the turkey, on the other hand, is more closely related. The **minimum mutational distance** is the minimum number of nucleotide substitutions required by the genetic code (see Appendix V) to account for the observed number of amino acid sequence differences in two proteins. This is also supported by immunological evidence. Further, Schnell & Wood (1976) made morphological comparisons between *Gallus*, *Phasianus* and *Meleagris*, measuring 51 skeletal dimensions. Their phenetic analysis also suggests that turkeys are as closely related to *Gallus* as pheasants.

Recently Helm-Bychowski & Wilson(1986) have compared restriction maps between seven of the Phasianoid birds. **Restriction mapping** entails fragmentation of the DNA isolated from the species concerned using enzymes called restriction endonucleases which cause splitting at regions where there is a specific base sequence (see Chapter 12, section 12.2a). Changes in the base sequences through mutations may change the number of restriction sites and hence the restriction map. From their comparison they suggest a revised cladistic (or phylogenetic) relationship between the species with possible times of divergence (Fig. 1.3).

## 1.4 The origin of the domestic fowl

The main issue concerning the origin of the domestic fowl is whether it is **monophyletic** or **polyphyletic**, i.e. whether a single species is the progenitor or whether more than one species is the progenitor of the domestic fowl. Full accounts of the evidence up to 1906 and up to 1949 have been given by Brown (1906) and by Hutt (1949), respectively. Here a brief summary of the salient points is given together with the more recent evidence.

Four species of *Gallus* have been considered as progenitors of the domestic fowl: *G. gallus* (red jungle fowl), *G. lafayettei* (Ceylon jungle fowl), *G. sonnerati* (grey jungle fowl) and *G. varius* (Java or green jungle fowl). All of these inhabit regions of South East Asia, as shown on Fig. 1.4. It can be seen that *G. gallus* has the widest geographical distribution. *G.gallus*, *G. lafayettei* and *G. sonnerati* have the common characteristics of 14 tail feathers, a serrated comb and two wattles. In contrast, *G. varius* has 16 tail feathers, an unserrated comb and a single median wattle. The possibilities are that the present day domestic fowl is entirely derived from a single species or alternatively from two or more of these species. An additional possibility is that certain present day breeds and varieties are derived from one species and other breeds and varieties from another, e.g. the 'Asiatic' breeds may have a different origin from the Mediterranean breeds.

Darwin (1868) favoured the monophyletic origin, although admitting that the evidence was not as strong as was the case for the origin of domesticated pigeons. The main points which led him to this view were: (i) the domestic fowl mated more freely with *G. gallus* and the progeny showed a higher degree of fertility, (ii) the voice of *G. gallus* most closely resembled that of the domestic fowl, (iii) breeding with the domestic fowl often resulted in offspring with the black–red plumage similar to that of *G. gallus*. He believed this to be a form of reversion to ancestral type, whereas it is now known to be due to the interaction of complementary genes from the parents. The voices of *G. lafayettei* and *G. sonnerati* are different from that of *G. gallus*. This has been examined more recently by Collias & Collias (1967). Spectrograms of the crowing of *G. gallus*, *G. sonnerati* and *G. lafayettei* show that the crow of the domestic cock most closely resembles that of *G. gallus*.

The counter-arguments in favour of the polyphyletic origin of the domestic fowl were initiated by Darwin's colleague, Tegetmeier, who felt

**Fig. 1.3** Proposed evolutionary tree for Galliformes (Helm-Bychowski & Wilson, 1986).

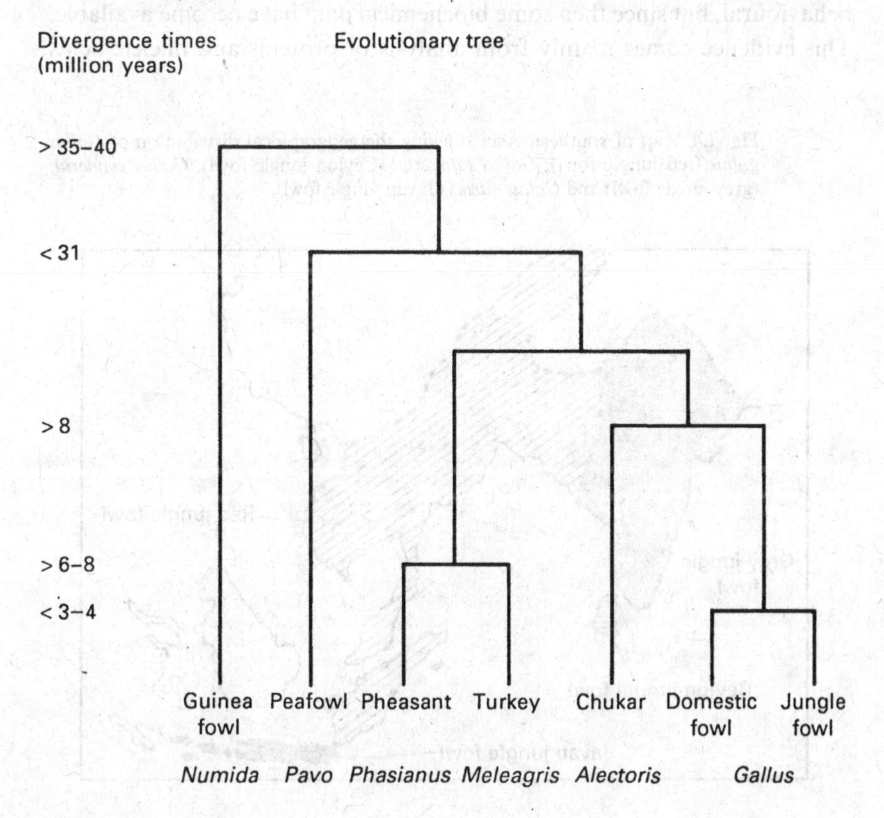

that the Asiatic breeds, particularly the Cochin, had a different origin from the Mediterranean breeds. The Asiatic breeds have short wings and a wide drooping tail and are much less 'flighty'. Also, an important structural difference is that in the Cochin the long axis of the occipital foramen (the aperture through which the spinal cord emerges from the skull) is vertical whereas in *G. gallus* it is horizontal (Fig. 1.5). These differences, it was felt, were unlikely to have been bred in. Also, Tegetmeier was more successful in interbreeding the wild *Gallus* species. Another piece of evidence against *G. gallus* being the sole progenitor of the domestic fowl is that it lacks the silver gene (*S*). This is a sex-linked dominant gene (see Chapter 4, section 4.5). A pair of alleles *S* and *s* exist in many domestic breeds, *S* being the dominant gene and giving rise to silver plumage and *s* the recessive gene giving rise to gold plumage. *G. gallus* apparently has only the *s* gene, since silver variants have not been described. All specimens of *G. sonnerati*, on the other hand, are grey or silver and thus have the *S* gene. By a similar line of reasoning the extended black gene *E* (see Chapter 6, section 6.5e) can be attributed to *G. varius*, since it is lacking in *G. gallus*.

Almost all of the evidence until the 1950s was morphological or behavioural, but since then some biochemical data have become available. This evidence comes mainly from analysis of proteins and nucleic acids

**Fig. 1.4** Map of southern Asia showing the geographical distribution of *Gallus gallus* (red jungle fowl), *Gallus lafayettei* (Ceylon jungle fowl), *Gallus sonnerati* (grey jungle fowl) and *Gallus varius* (Javan jungle fowl).

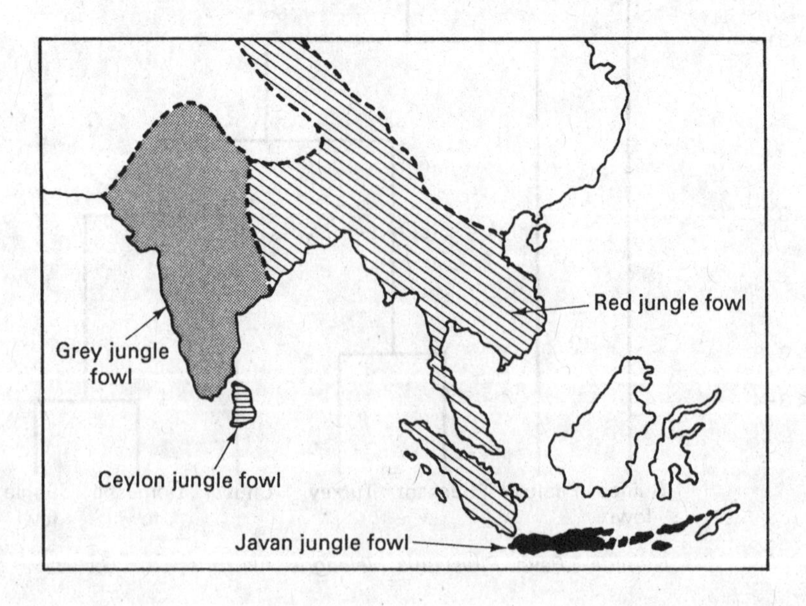

from the species concerned. The proteins most readily available in the case of the fowl are either egg proteins or blood proteins. Baker *et al.* (1971) and Baker & Manwell (1972) made extensive studies of the egg-white proteins, separating them by starch gel electrophoresis; there are about ten main groups of protein in egg white. Those that show greatest polymorphisms on the basis of electrophoretic behaviour are the ovoglobulins, and it is in this group that differences can be found between the different species of *Gallus*. The domestic fowl and *G. gallus* differ consistently from *G. sonnerati* in their $G_2$ and $G_3$ ovoglobulins and in the enzyme glutamyl peptidase (Baker & Manwell, 1972). *G. lafayettei* differs from *G. gallus* only in the $G_2$ ovoglobulin (Stratil, Hála & Hasek, 1969). Baker *et al.* (1971) conclude that the domestic fowl is derived largely if not entirely from *G. gallus*. Ovoglobulins are discussed more extensively in Chapter 10, section 10.5.

The particular proteins which Baker *et al.* (1971) studied, namely ovoglobulins, are at present one of the least well characterised egg-white proteins. Not all of them have been purified to homogeneity (Stevens & Duncan, 1988) and thus cannot yet be sequenced. Electrophoretic methods will separate proteins if they have different net charges, but a number of polymorphic forms may have the same net charge and so will go undetected. A more precise way to examine the differences between the different

**Fig. 1.5** The occipital foramen of *A*, the Cochin breed of domestic fowl, and *B*, *Gallus gallus*. *C*, skull of the fowl showing the position of the occipital foramen (Darwin, 1868).

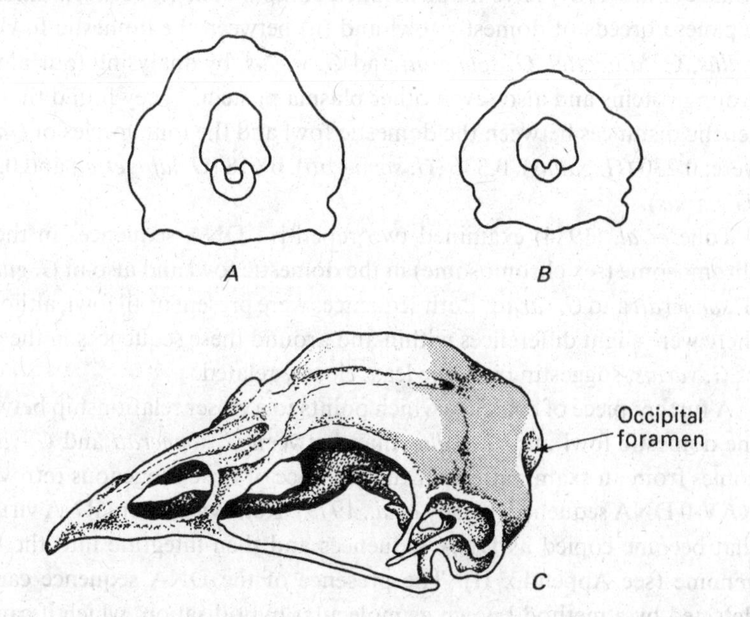

*Gallus* species might be to examine the amino acid sequence of individual well-characterised proteins. This is a good method to resolve evolutionary relationships between families, orders or classes, i.e. long term evolution, but it is much less successful when considering relationships between species and subspecies. The reason is that the number of differences between individual proteins from related species is likely to be very low. Recently the complete sequences of the third domain of ovomucoid from domestic fowl, *G. lafayettei* and *G. sonnerati*, have been determined (Laskowski *et al.*, 1987) and all three are identical. It generally requires the sequencing of a large number of different proteins to find differences between related species and subspecies. This is very time consuming and at present such data are not available.

Since it is possible to sequence DNA much more quickly than to sequence proteins, it is likely that a range of gene sequences will become available for comparison before that of proteins (for details see Chapter 12). However, at present less precise methods may be used which enable a comparison of a large number of proteins relatively easily, the most popular of which is the separation of proteins by polyacrylamide gel electrophoresis, mentioned above. Nei (1975, 1987) has developed statistical methods for estimating the number of gene differences. The parameter which is calculated is termed the '**genetic distance**' and it is a measure of the accumulated allelic differences per locus. The genetic distance is related to the divergence time and can be used to examine relationships between species and subspecies. Okada *et al.* (1984) have made detailed comparisons (i) between different Japanese breeds of domestic fowl and (ii) between the domestic fowl, *G. gallus*, *G. lafayettei*, *G. sonnerati* and *G. varius*, by analysing four blood-group proteins and also seven other plasma proteins. They found that the genetic distances between the domestic fowl and the four species of *Gallus* were: 0.230 (*G. gallus*), 0.336 (*G. sonnerati*), 0.658 (*G. lafayettei*) and 0.681 (*G. varius*).

Tone *et al.* (1984) examined two repetitive DNA sequences in the W chromosome (sex chromosome) in the domestic fowl and also in *G. gallus*, *G. sonnerati* and *G. varius*. Both sequences were present in all fowl, although there were slight differences within and around these sequences in the case of *G. varius*, suggesting it is the least closely related.

A further piece of evidence which points to a closer relationship between the domestic fowl and *G. gallus* than between *G. sonnerati* and *G. varius* comes from an examination for the presence of the endogenous retrovirus RAV-0 DNA sequence (Frisby *et al.*, 1979). Retroviruses are RNA viruses, that become copied as DNA sequences and then integrate into the host genome (see Appendix II). The presence of the DNA sequence can be detected by a method known as molecular hybridisation, which if carried out under stringent conditions detects only accurate copies of the RNA. When hybridisation is carried out under stringent conditions, the RAV-0 DNA sequence can only be detected in the domestic fowl and in *G. gallus*, but not in the other two jungle fowl. Frisby *et al.* (1979) suggest that RAV-0 first infected *G. gallus* after speciation but before domestication, since the DNA sequence is not found in Japanese quail or Ring-necked pheasant.

On balance the evidence supports the idea that *G. gallus* is the species most closely related to the domestic fowl, although it should be noted that the domestic fowl being compared in the work of Okada *et al.* is the 'average' of the Japanese native breeds, and slightly different results might be obtained using European domestic stock. It seems, therefore, that *G. gallus* is the main progenitor, but that genes from other species may have been introduced by hybridisation. Also it is quite possible that the Asiatic breeds might have a larger contribution from the other species. Throughout this section the term 'domestic fowl' has been used. This is sometimes designated *Gallus domesticus* and sometimes *Gallus gallus domesticus*, depending on whether it is regarded as a species or subspecies.

## 1.5 Evolution of the present day breeds and varieties of domestic poultry

Unlike the first two phases of evolution of the domestic fowl, the evidence for the most recent phase is based almost entirely on historical accounts rather than scientific evidence. The terms **breed** and **variety** are rather ill-defined, though breeds in the case of domestic poultry are generally distinguished by their shape and form, whereas varieties within a breed usually differ in the colour of their plumage. The original domestication of the fowl was assumed to have occurred in India and the earliest records suggest that the initial purpose was for cock fighting rather than egg laying or 'table quality'. Cock fighting is known to have taken place in India over 3000 years ago. More recent archaeological evidence referred to in section 1.1 (West & Zhou, 1989) suggests that domestication occurred in China by *c.* 6000 BC, but whether this occurred independently of that in India, or whether it spread thence to India is not clear. From India the domestic fowl spread westwards to Europe. The possible routes are indicated in Fig. 1.6. Thus there appear to be at least two routes by which the fowl reached Europe, and this occurred by the sixth century BC. The spread to America is believed by some to have occurred with the Spanish conquest, followed later by a further influx from English, French and Dutch colonisers. Other evidence suggests there may have been an earlier introduction via the Pacific. A detailed historical account is given by Crawford (1990).

Until recently there were relatively few breeds as such, although undoubtedly the fowl would have been selected for cock fighting and for culinary purposes. Several characters, such as silkiness, frizzling, rumplessness and polydactyly, were undoubtedly present in domestic fowl for several centuries, but selective breeding for these characters and the 'creation' of the present day European and American breeds came into vogue only within the last 200 years.

Expansion was at its peak in the Victorian era. The exhibition of poultry in Britain began in the first half of the nineteenth century and between 1814 and 1864 the number of recognised breeds increased from 12 to 34. Currently there are 65 breeds recognised by the British Poultry Club. A similar expansion occurred in America and several European countries. The early nineteenth century saw an influx of Asiatic breeds such as Cochins, Brahmas, Langshams, Silkies and Aseels brought in by the sea routes. There is evidence that a few breeds were in existence before the nineteenth century. The Dorking, or at least its ancestor, is believed to have been in Britain in Roman times. It has a fairly distinguishing feature in having five toes on each foot. The evolution of native breeds in Japan has been quite distinct from those in Europe. There are at least eleven different breeds and they are thought to have originated in the Edo era (1600–1860); they are concentrated in different regions of Japan (Okada *et al.*, 1984). The history of poultry breeding in Britain is described by Easom Smith (1976) and in Britain, Europe and America by Brown (1906), and a general review covering early history to the present day is given by Crawford (1984, 1990).

**Fig. 1.6** Possible routes by which the domestic fowl became dispersed from their earliest known location, the latter shown as a cross-hatched area.

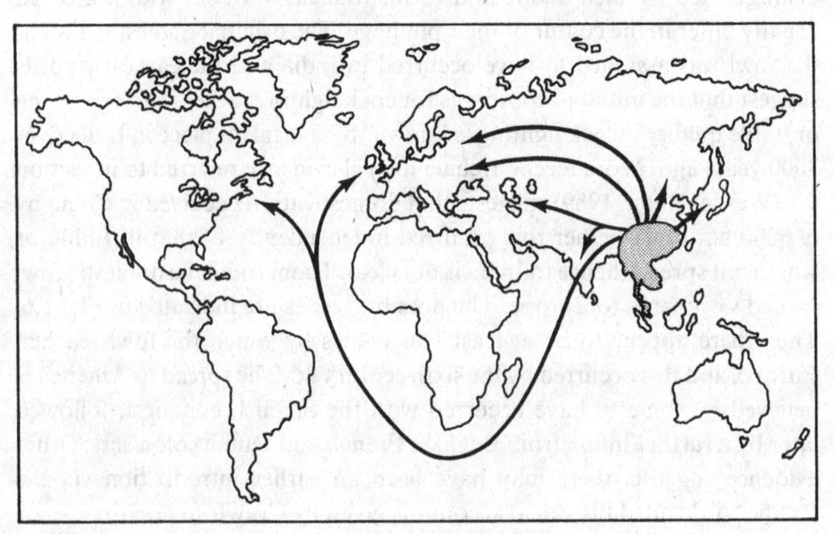

As was mentioned in section 1.4 analysis of the egg-white proteins has been used as evidence for the progenitor of the domestic fowl. Baker (1968) has also studied the egg-white proteins of 37 different breeds of fowl. Again the ovoglobulins showed the most useful patterns. She found that fowl of English and Northern European origin, e.g. Old English Game, Dorking, Hamburgh, Leghorn, Orpington, Houdan, Breda and Polish possessed only one type of $G_3$ ovoglobulin, namely $G_3^A$ whereas the ovoglobulin $G_3^B$ was most frequently found in the domestic fowl of Asiatic origin. This supports the hypothesis that current Asiatic breeds may have evolved separately from most of the 'European' breeds.

## References

Alvarez, L. W. (1983). Experimental evidence that an asteroid impact led to the extinction of many species 65 million years ago. *Proceedings of the National Academy of Sciences of USA*, **80**, 627–42.

Baker, C. M. A. (1968). Molecular genetics of avian proteins. IX. Interspecific and intraspecific variation of egg white proteins of the genus *Gallus*. *Genetics*, **58**, 211–26.

Baker, C. M. A. & Manwell, C. (1972). Molecular genetics of avian proteins. XI. Egg proteins of *Gallus gallus*, *G. sonnerati* and hybrids. *Animal Blood Groups and Biochemical Genetics*, **3**, 101–7.

Baker, C. M. A., Manwell, C., Jayaprakash, N. & Francis, N. (1971). Molecular genetics of avian proteins. X. Egg white protein polymorphism of indigenous Indian chickens. *Comparative Biochemistry and Physiology*, **40B**, 147–53.

Brown, E. (1906). *Races of Domestic Poultry*. Reprinted 1985. Liss, UK: Nimrod Books.

Callow, P. (1983). *Evolutionary Principles*. Glasgow: Blackie.

Collias, N. E. & Collias, E. C. (1967). A field study of the red jungle fowl in north-central India. *Condor*, **69**, 360–86.

Crawford, R. D. (1984). Domestic fowl. In *Evolution of Domesticated Animals*, ed. I. L. Masin, pp. 298–311. New York: Longman.

Crawford, R. D. (1990). Origin and history of poultry species. In *Poultry Breeding and Genetics*, ed. R.D. Crawford, pp. 1–42. Amsterdam: Elsevier.

Darwin, C. (1868). *Variation of Animals and Plants*, pp. 273–335. London: Murray.

Easom Smith, H. (1976). *Modern Poultry Development*. Liss, UK: Spur Publications.

Frisby, D. P., Weiss, R. A., Roussel, M. & Stehelin, D. (1979). The distribution of endogenous chicken retrovirus sequences in the DNA of galliform birds does not coincide with avian phylogenetic relationships. *Cell*, **17**, 623–34.

Helm-Bychowski, K. M. & Wilson, A. C. (1986). Rates of nuclear DNA evolution in pheasant-like birds: Evidence from restriction maps. *Proceedings of the National Academy of Sciences of USA*, **83**, 688–92.

Hennig, W. (1979). *Phylogenetic Systematics*, 2nd edn. Urbana: University of Illinois Press.

Hutt, F. B. (1949). *Genetics of the Fowl*. New York: McGraw Hill.
Jolles, J., Ibrahimi, I. M., Prager, E. M., Schoentgen, F., Jolles, P. & Wilson, A. C. (1979). Amino acid sequence of pheasant lysozyme. Evolutionary change affecting processing of prelysozyme. *Biochemistry*, **18**, 2744–52.
Laskowski, M., Kato, I., Ardelt, W., Cook, J., Denton, A., Empie, M. W., Kohr, W. J., Park, S. J., Parks, K., Schatzley, B. L., Schoenberger, O. L., Tashiro, M., Vichot, G., Whatley, H. E., Wieczorek, A. & Wieczorek, M. (1987). Ovomucoid third domains from 100 avian species: Isolation, sequences and hypervariability of enzyme–inhibitor contact residues. *Biochemistry*, **26**, 202–21.
Miller, S. L. & Orgel, L. E. (1974). *The Origins of Life on the Earth*. New York: Prentice Hall.
Nei, M. (1975). *Molecular Population Genetics and Evolution*. Amsterdam: North Holland.
Nei, M. (1987). *Molecular Evolutionary Genetics*. New York: Columbia University Press.
Okada, I., Yamamoto, Y., Hashiguchi, T. & Ito, S. (1984). Phylogenetic studies on the Japanese native breeds of chickens. *Japanese Poultry Science*, **21**, 318–29.
Patton, J. C. & Avise, J. C. (1986). Evolutionary genetics of birds IV: Rates of protein divergence in waterfowl (Anatidae). *Genetica*, **68**, 129–43.
Ridley, M. (1986). *Evolution and Classification*. London: Longman.
Schnell, G. D. & Wood, D. S. (1976). More on chicken–turkey–pheasant resemblances. *Condor*, **78**, 550–3.
Schwabe, C. (1986). On the validity of molecular evolution. *Trends in Biochemical Sciences*, **11**, 280–2.
Sneath, P. H. A. & Sokal, R. R. (1973). *Numerical Taxonomy*. San Francisco: W.H. Freeman.
Stevens, L. & Duncan, D. (1988). Peptide mapping of ovoglobulins G2A and G2B in the domestic fowl. *British Poultry Science*, **29**, 665–9.
Stratil, A., Hála, K. & Hasek, M. (1969). Syngenic lines of chickens IV. Genetic polymorphism in the proteins of four lines of chickens. *Folia Biologica*, **15**, 306–8.
Tone, M., Sakaki, Y., Hashiguchi, T. & Mizuno, S. (1984). Genus specificity and extensive methylation of the W chromosome-specific repetitive DNA sequences from the domestic fowl, *Gallus gallus domesticus*. *Chromosoma (Berlin)*, **89**, 228–37.
West, B. & Zhou, B.-X. (1989). Did chickens go north? New evidence for domestication. *World's Poultry Science Journal*, **45**, 205–18.
Wyles, J. S., Kunkel, J. G. & Wilson, A. C. (1983). Birds, behaviour and anatomical evolution. *Proceedings of the National Academy of Sciences of USA*, **80**, 4394–7.
Zeuner, F. E. (1963). *A History of Domesticated Animals*. London: Hutchinson.

# 2

# *The cellular organisation of genetic material*

The first sections of this chapter (2.1–2.3) form a résumé on the nature and functioning of genes, how they are organised and how they replicate. The second half (sections 2.4 and 2.5) then describes specifically the chromosomes in the domestic fowl, including chromosomal abnormalities.

## 2.1 The nature of the gene and its organisation on the chromosome

Since the turn of the century it has been apparent that genes are the units of inheritance and, since Watson and Crick's proposal in 1953 for the double helical structure of DNA, their molecular nature has been clear. However, there is still much to be learned about their organisation and also the details of replication and expression. Genes are comprised of DNA, a macromolecule having a double helical structure in which its two strands are arranged in antiparallel fashion. The information content of DNA arises from the specific order of the structural units attached to the backbone of the macromolecule; these structural units are known as **bases**. There are four different bases in DNA: adenine, thymine, guanine and cytosine, usually abbreviated to their first letters A, T, G, and C. Each gene may have as many as 1000 or more bases arranged in a unique sequence along the macromolecule. DNA can be replicated and this is achieved by specific base pairing arrangements between the two strands. Adenine on one strand pairs with thymine on the second strand and likewise guanine pairs with cytosine. This ensures that the two strands are complementary in the sequence of bases throughout the two chains. The genes are thus pieces of DNA; their length is related to the amount of coded information they contain.

Each acts as a unit of inheritance since the order of the bases in the DNA is copied as a sequence of similar bases (**transcribed**) into messenger RNA using the enzyme RNA polymerase. Messenger RNAs are a group of macromolecules with a similar arrangement of bases to DNA (one difference is that the base thymine is replaced by a structurally similar base, uracil [U]) and are synthesised using one of the DNA strands as a template. The messenger RNAs are decoded using ribosomes into particular sequences of amino acids (**translated**), which comprise the proteins (Fig. 2.1). The process is known as translation since each sequence of three bases on messenger RNA codes for one of the twenty amino acids which occur in proteins. The relationship between the base sequence and that of the amino acids into which they are translated is known as the **Genetic Code** (see Appendix V). The full complement of proteins in a cell or organism ultimately determines its phenotypic character.

The number of different genes in an organism is related to its complexity. In a eukaryote (see Glossary for a definition) the genes are organised on a number of chromosomes and there is a large number of genes on each chromosome. It is difficult to assess the total number of genes an organism contains. Although the amount of DNA per cell can be measured accurately not all of it codes for particular proteins. Some may have a structural role and some may not be expressed but may be important in the evolution of new genes. The fruit-fly (*Drosophila*), which has four pairs of chromosomes in each somatic cell, is known to have over 5000 genes whereas man has 23 pairs of chromosomes which are estimated to have a total of between 50 000 and 100 000 genes. The domestic fowl, although having 39 pairs of chromosomes, has less than half the amount of DNA per cell than man and therefore probably has between 10 000 and 30 000 genes. The arrangement of genes on the chromosomes is often likened to that of beads on a string,

**Fig. 2.1** Transcription and translation.

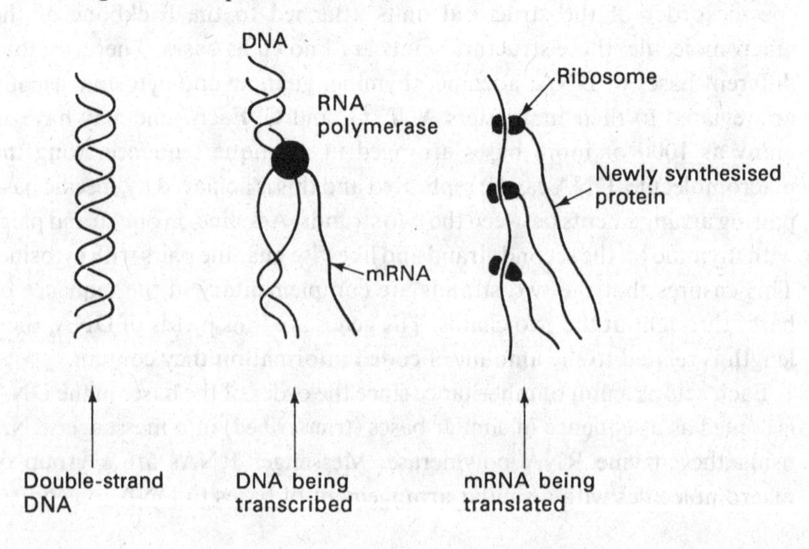

thus conveying the notion that each gene is a discrete unit, and that all of the units are arranged in a linear array. However, since 1977 this view has had to be substantially modified. It has been shown that the individual genes, at least in eukaryotes, do not comprise a single discrete length of DNA but consist of several bits of DNA (**exons**) separated by **intervening sequences** or **introns** (Fig. 2.2).

The DNA is transcribed and the resultant RNA is modified to remove the introns. Thus it is only transcribed copies of the exons that **ex**it the nucleus, hence the name '**ex**on'.The structures of some fowl genes showing the introns and exons are described in detail in Chapter 12. The 'beads on a string' analogy is also an oversimplification when one examines the way in which the double helical thread runs through the chromosome. The appearance of chromosomes under the light microscope depends on the state of the cell cycle in which they are examined. The **cell cycle** is the cyclic series of events which cells undergo each time they divide (see section 2.3). During interphase, between cell divisions, the chromosomes appear in the nucleus mainly as diffusely staining material known as **euchromatin** and individual strands are not discerned. There is also, however, some densely staining material known as **heterochromatin**. The heterochromatin is highly condensed and genes within these regions are not being transcribed. On the other hand, transcription occurs in the diffuse regions of euchromatin. During mitosis and meiosis the chromosomes become highly condensed rather like the heterochromatin and individual chromosomes are readily discerned. Transcription of the genes is temporarily suspended during mitosis and meiosis.

**Fig. 2.2** Transcription of the $\beta$-globin gene, showing the primary transcript and its subsequent processing to form mRNA. The gene is transcribed and the primary transcript is then processed in the nucleus to remove introns. The resulting mRNA, consisting of the **ex**ons only, then **ex**its the nucleus to be translated by the ribosomes in the cytoplasm.

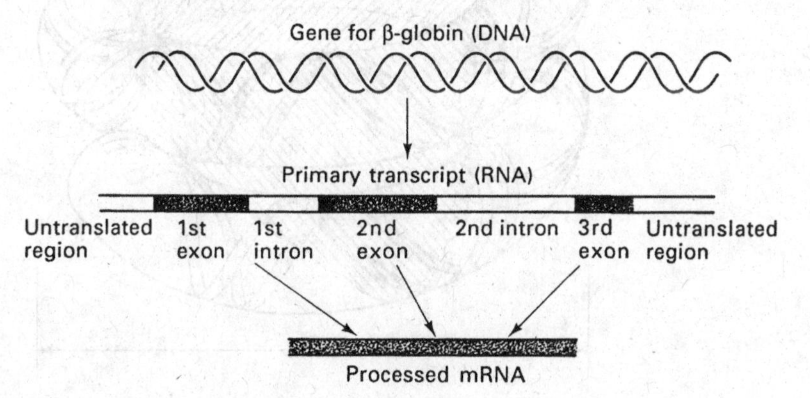

The chromosomes are not composed entirely of DNA but also contain proteins. The main proteins of the chromosomes are a group of basic proteins known as histones. The DNA and histones are arranged in units called **nucleosomes**. These comprise a disc-like core of eight histone molecules and, wrapped around this core, is the DNA forming two complete turns of supercoiling per nucleosome. The nucleosomes lie adjacent to one another and are linked by the continuous thread of DNA, which has a solenoid-like structure (Fig. 2.3), and are also held closely together by a further histone molecule, H1. This supercoiling of DNA leads to a compact arrangement of the genes as in euchromatin and is further compacted in heterochromatin.

## 2.2 Gene replication and expression

Gene replication and expression are described in detail in many standard textbooks (e.g. Lewin, 1990) and thus only the principles will be outlined

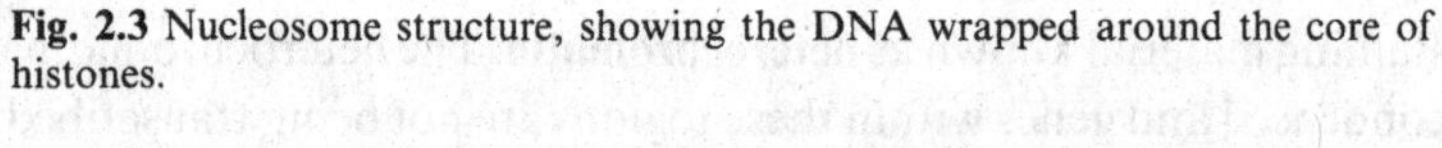

**Fig. 2.3** Nucleosome structure, showing the DNA wrapped around the core of histones.

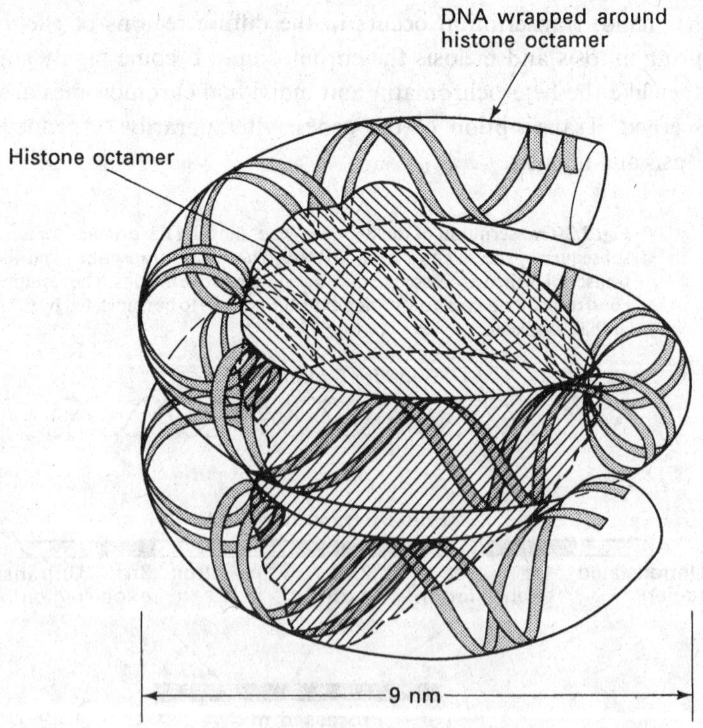

here. Genes are replicated in both mitosis and meiosis by what is termed a **semiconservative** process. In essence, the two complementary strands of DNA separate and two new daughter strands are replicated on each strand, making use of the base pairing arrangement. Thus each daughter cell contains genes comprising two strands of DNA, one of which was an original strand from the parent cell and the other is a newly synthesised strand (Fig. 2.4). Although at first sight this mechanism appears very elegant and simple, a closer examination of the topology shows that this is far from the case. DNA can be likened to a two-stranded rope which is then extensively supercoiled to form a compact arrangement in the chromosomes. For this type of structure to be replicated it has to be in some way uncoiled. If the strands of the parent DNA remained intact throughout replication this would involve very considerable untwisting of the DNA within the close confines of the nucleus. Much of the problem of untwisting is overcome by the presence of a group of enzymes called **topoisomerases** which are capable of catalysing first, a break in one of the strands followed by untwisting and second, rejoining the strands.

The replication of DNA occurs throughout all the chromosomes during a restricted phase of the cell cycle. It is initiated at numerous specific sites along each chromosome. The number of replication origins within the mammalian chromosomes is believed to be in the region of 20000 to 30000. In a simpler organism such as the fruit-fly the number is about 3500.

**Fig. 2.4** Semiconservative replication of DNA. ——, DNA from the parental cell; ---, replicated DNA.

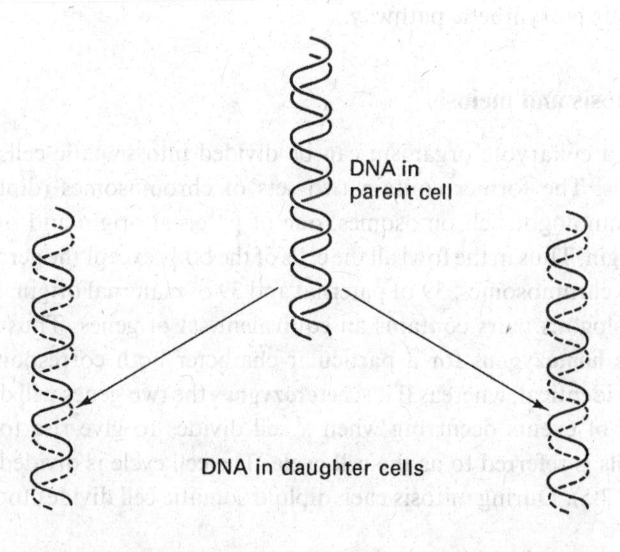

Although the number has not been determined in the fowl it is presumably closer to that of the mammal.

For **gene expression** to occur, one of the strands, known as the coding strand of DNA, has to be copied to form messenger RNA by a process known as transcription. This process has a number of parallels with that of replication. The polymer synthesised comprises four different bases (adenine, uracil, guanine and cytosine) attached to a backbone, but the deoxyribose component is replaced by ribose. Normally only the coding strand is copied. A further difference is that, whereas in replication all the genes of the cell are replicated, in transcription only the genes that are active or being expressed are transcribed. A complex set of controls determines which are transcribed. As was mentioned in section 2.1, the genes in eukaryotes comprise pieces of DNA coding for proteins, interspersed with intervening sequences (introns). The whole section of DNA is transcribed and the resultant RNA is subsequently processed to excise the non-coding regions, leaving intact the message for the gene to be expressed. The messenger RNA is made in the cell nucleus but passes out into the cytoplasm, where it is translated by the ribosomes in a process which involves decoding the messenger RNA to form a sequence of linked amino acids. Thus each gene expressed results in the synthesis of a particular protein. The phenotypic character of a cell, or indeed the whole organism, depends on the particular complement of proteins that are produced. The majority of proteins produced are enzymes, which in turn regulate the amount of other substances produced by the cell. Thus if, for example, a fowl has black feathers this is due to the synthesis of a black pigment, melanin, which requires a particular sequence of enzymes to catalyse the reactions of its biosynthetic pathway.

## 2.3 Mitosis and meiosis

The cells of a eukaryote organism can be divided into **somatic** cells and **germline** cells. The former contain two sets of chromosomes (diploid), known as homologous chromosomes, one of paternal origin and one of maternal origin. Thus in the fowl all the cells of the body except the germline cells have 78 chromosomes, 39 of paternal and 39 of maternal origin. Each of the homologous pairs contains an equivalent set of genes. Thus if an individual is **homozygous** for a particular character both corresponding genes will be identical, whereas if it is **heterozygous** the two genes will differ.

The cycle of events occurring when a cell divides to give rise to two daughter cells is referred to as the **cell cycle**. The cell cycle is divided into phases (Fig. 2.5). During mitosis each diploid somatic cell divides to form two diploid daughter cells. Mitosis occupies a fairly short part of the cell cycle. For most of the time cells are in interphase which is subdivided into three phases, G1, S and G2. During S phase (**synthetic phase**) DNA is replicated so that each chromosome is duplicated. The G1 and G2 phases are known as **presynthetic** and **postsynthetic** gaps. Mitosis can be subdivided into four phases: prophase, metaphase, anaphase and telophase. At the start of mitosis, known as prophase, the chromosomes condense and become clearly visible under the light microscope. Each chromosome that replicated during the S phase into a pair of chromatids can now be seen as a paired unit with the division running lengthwise and joined by a centromere. During the next stage (**metaphase**) the paired chromatids orientate themselves about the centre of the spindle fibres which arise from the two spindle pole bodies. The spindle fibres are attached to the centromeres, and so are able to draw the chromatids to opposite poles. In the subsequent phases the chromatids migrate to opposite poles and eventually the cell divides to give two daughter cells, each with a diploid set of chromosomes.

Mitosis occurs within most tissues of the body, either because the number of cells is increasing during growth, or because dead cells are being replaced. Meiosis occurs only during the formation of the germ cells. **Meiosis** is more complex but has many steps in common with mitosis. There is a synthetic

**Fig. 2.5** Diagram showing the phases of a cell cycle in higher organisms. The times spent in each period varies with the cell type, but the division phase (mitosis) is generally shortest and the S phase, in which the DNA content doubles, is generally the longest. In cells which have ceased to divide, or which need a particular stimulus (e.g. a hormone) to reinitiate division, the $G_1$ period is referred to as the $G_0$ period. $G_0$ and $G_1$ differ in that in the latter the cell is preparing to divide, e.g. an increase in the concentrations of enzymes required for DNA synthesis can be detected.

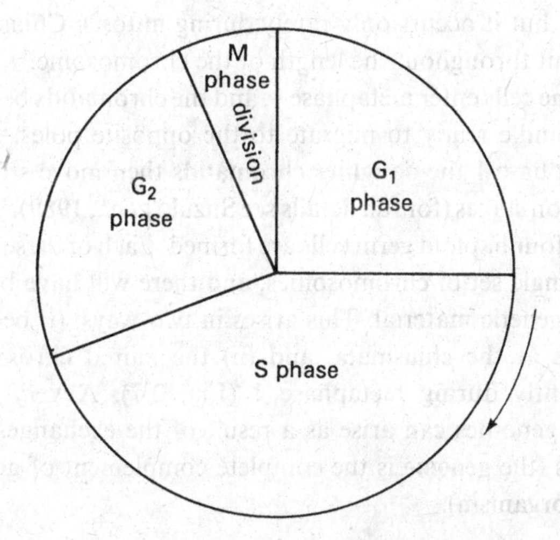

phase prior to meiosis in which the DNA doubles but this is followed by two divisions giving rise to four haploid daughter cells, each cell having only one of each of the homologous chromosomes (haploid)

Mitosis $2n \longrightarrow 4n \longrightarrow 2n + 2n$

Meiosis $2n \longrightarrow 4n \longrightarrow 2n + 2n \longrightarrow n + n + n + n$

n = the haploid number of chromosomes.

Apart from the transition from the diploid to the haploid number of chromosomes another important aspect of meiosis is the **crossing over** which occurs in one of the early stages of meiosis. Meiosis consists of the stages: prophase 1, metaphase 1, anaphase 1, telophase 1, prophase 2, metaphase 2, anaphase 2 and telophase 2. Prophase 1 is further subdivided into sub-phases (for further details of meiosis see Darnell, Lodish & Baltimore, 1986; Suzuki *et al.*, 1989). It is during prophase 1 that the most significant events occur. At this stage the homologous chromosomes are seen to pair up side by side. The chromatids may be in direct physical contact at a number of points during the pairing. A contact point is called a **chiasma** (plural, chiasmata). At these points crossing over occurs; a break occurs in a chromatid and the ends subsequently reattach to the homologous chromatid (Fig. 2.6). Thus a genetic exchange between chromosomes occurs. This may result in altered patterns of inheritance for linked genes (see Chapter 3, section 3.5). Crossing over is a relatively frequent event during meiosis but it occurs only rarely during mitosis. Chiasmata may form at any point throughout the length of the chromosome.

After prophase 1 the cells enter metaphase 1, and the chromatids become aligned along the spindle ready to migrate to the opposite poles. After anaphase 1 and telophase 1 the daughter chromatids then move straight into the second division stages (for full details see Suzuki *et al.*, 1989). Thus, as a result of meiosis four haploid germ cells are formed. Each of these germ cells will possess a single set of chromosomes, and there will have been a reassortment of the genetic material. This arises in two ways: (i) because crossing over occurs at the chiasmata, and (ii) the paired chromatids segregate independently during metaphase 1 (Fig. 2.7). A very large number of different genomes can arise as a result of the exchanges that occur during meiosis (the **genome** is the complete complement of genetic material in a cell or organism).

## 2.4 The domestic fowl karyotype

In sections 2.1 and 2.2 the molecular structure of the gene and its transcription were outlined. The general organisation of the chromosomes seen at a microscopic rather than a molecular level is considered next; the subject is known as cytogenetics. In order to understand fully the inheritance mechanism in any species it is necessary to know how the genes are organised on the chromosomes since, as will be explained in Chapter 5, genes that are on the same chromosome are linked. In other words, the physical linkage of particular genes on a chromosome means that the characters for the genes concerned are not inherited independently. The extent to which characters are inherited in a linked fashion depends on how close together two genes are on a chromosome. The closer they are, the less probable it is that a chiasma will form between them. Therefore, as one of the approaches to understanding inheritance in a species, it is important to know how many chromosomes there are and how many genes they contain. The cytogenetic map of the chromosomes is called the **karyotype**. The shape

**Fig. 2.6** Chiasma formation. The diagram shows a homologous pair of chromosomes, each comprising one pair of chromatids forming a chiasma during prophase 1 of meiosis. The process leads to an exchange of genetic material between the two chromatids (crossing over).

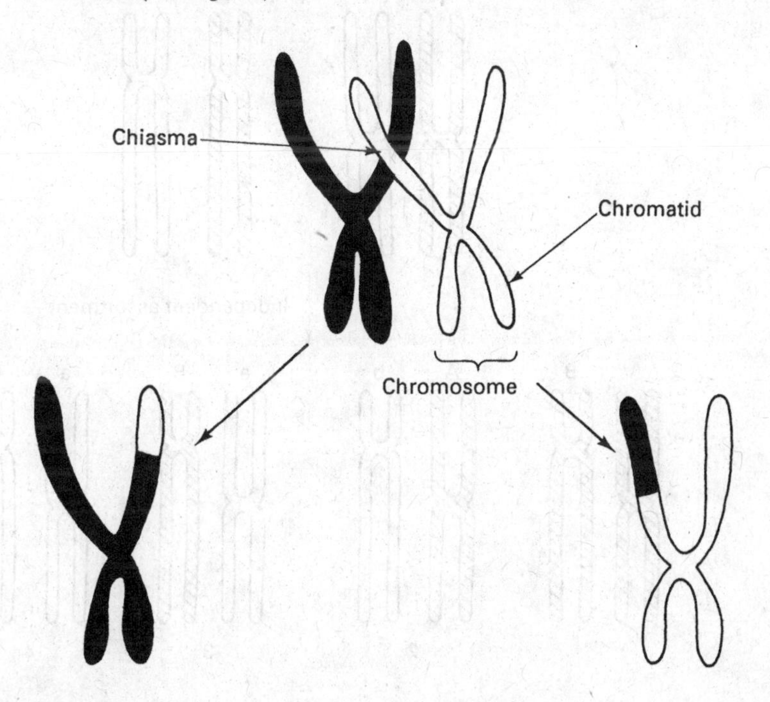

and appearance of the chromosomes is studied after specific staining procedures have been applied to the chromosomes. Since the individual chromosomes are most easily discerned during metaphase of mitosis it is at this stage that they are usually examined, often after inhibiting the later stages of mitosis by using the drug colchicine. The number of paired chromosomes is counted and individual chromosomes are identified by their size, the position of the centromere (Fig. 2.8) and any other constrictions visible. The centromere is the region of contact between two chromatids, to which the spindle fibres attach during mitosis and meiosis.

A set numbering procedure is usually adopted starting from the largest chromosome and proceeding downwards. It is not always easy to determine the precise number of chromosomes in some organisms because the chromosomes often become entangled with one another and sometimes larger ones completely obscure smaller ones. Determination has proved straightforward in the case of *Drosophila* (4 pairs) and *Zea mays* (10 pairs). In the case of the human karyotype, the accepted number was established as 23 pairs only in 1956 by Tjio & Levan. The chromosome number in birds

**Fig. 2.7** Segregation and independent assortment of two pairs of homologous chromosomes (A,a and B,b) during the first division of meiosis, giving rise to four possible arrangements in the daughter cells.

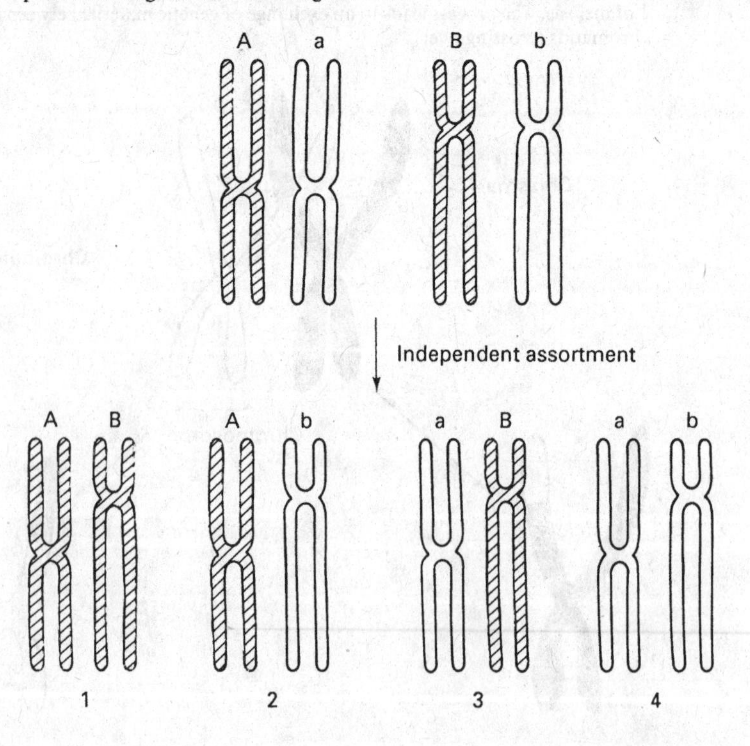

has proved particularly difficult to establish (see Pollock & Fechheimer, 1976). This is because there is such a large size variation from the largest to the smallest chromosome and, added to that, the total number is also high. In fact, the chromosomes are often subdivided into macrochromosomes and microchromosomes.

In the fowl each somatic cell contains 2.5 pg ($2.5 \times 10^{-12}$ g) of DNA (Olofsson & Bernardi, 1983) and this may be estimated to code for in the region of 50 000 genes. The number cannot be determined precisely since the sizes of the individual genes vary and also, more importantly, because there are large regions of DNA that are not genes but are probably structural features concerned with the overall packaging of the genome. These regions are often recognised because they contain highly repetitive sequences of DNA not corresponding to any coded information. DNA sequences can be divided into three categories: (i) unique sequences, (ii) middle repetitive sequences, and (iii) highly repetitive sequences. The unique sequences comprise most of the genes and are transcribed and translated to give the large variety of proteins that occur in the cell. The division between the middle repetitive sequences and the highly repetitive sequences is somewhat arbitrary in terms of copy number and is probably best separated on functional grounds. The middle repetitive sequences include a number of genes that are amplified to give multiple copies such as those for ribosomal RNA and histones. Ribosomal RNA and histones are required to be synthesised rapidly during the early stages of rapid growth in some organisms, and one way to achieve this is to have multiple copies of the genes concerned. The domestic fowl has 10–20 copies of the histone genes (Crawford *et al.*, 1979; Sugarman, Dodgson & Engel, 1983) and 100–145 copies of the ribosomal RNA genes per haploid genome (Muscarella, Vogt & Bloom, 1985).

The highly repetitive sequences account for less than 20% of the total DNA (Olofsson & Bernardi, 1983; Arthur & Straus, 1983). Of the total DNA, 34% of the unique sequences are flanked by repetitive DNA in units up to 7000 base pairs long, and a further 38% of the unique sequences

**Fig. 2.8** The karyotype of the domestic fowl, showing only the largest 9 chromosome pairs.

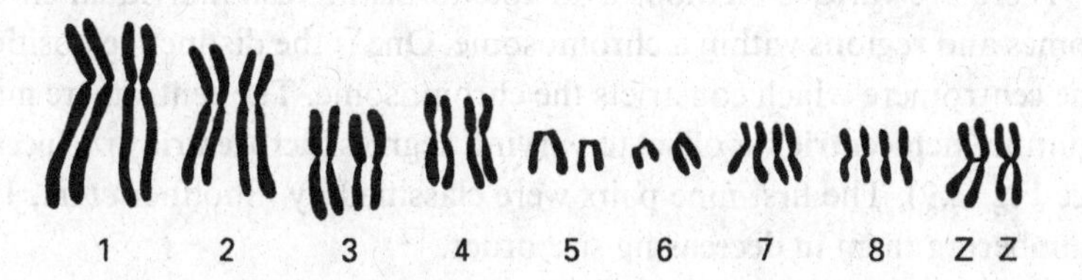

occurs in blocks of 22 000 base pairs or more (Arthur & Straus, 1983). The significance of this arrangement is not yet clear. The highly repetitive regions of DNA usually appear as densely staining heterochromatic regions of DNA.

The 39 pairs of chromosomes present in the somatic cells of the fowl can be subdivided on the basis of size into the largest chromosomes (**macrochromosomes**) (Fig. 2.8) that include the first five pairs, an intermediate set of five pairs, and the smallest 29 pairs (**microchromosomes**). Some workers include the ten largest pairs in the category of macrochromosomes, whereas others include only the largest five. The occurrence of macrochromosomes and microchromosomes is common to all birds that have so far been karyotyped (over 200 species). In mammals the sex chromosomes are designated XY in males and XX in females, the males being the **heterogametic sex**. In birds, including the fowl, the female is the heterogametic sex. The two sex chromosomes are the Z and W chromosomes. The male has two Z chromosomes, and the female one Z and one much smaller W. The Z chromosomes are the fifth largest. The W chromosome corresponds in size with the smallest of the intermediate chromosomes and was not positively identified until 1961 (Ohno, 1961). However, it was found to stain densely like heterochromatin or by use of the C-banding method (see below), and is most easily visible in late prophase. Part of the Z chromosome also stains heterochromatically. Because in birds the female is the heterogametic sex, there are differences in the inheritance of sex-linked genes when compared with mammals; these are discussed in Chapter 5.

For some time after the initial discovery of the microchromosomes there were doubts as to whether they were 'normal' functioning chromosomes. However, more recently they have been shown to have normal centromeres, and to divide in the normal fashion. Also a number of genes have been located on the microchromosomes, and there seems little doubt that they function much as do the macrochromosomes (Bitgood & Shoffner, 1990).

From an examination of the size and the morphology of the chromosomes it is possible to distinguish readily the five pairs of macrochromosomes and, with good preparations, chromosomes 6–10, but the remaining 29 pairs are difficult to distinguish individually and are simply placed in size order (Bloom, 1981).

There are various methods available to distinguish individual chromosomes and regions within a chromosome. One is the distinctive position of the centromere which constricts the chromosome. The centromere may be central (**metacentric**) or offset to varying degrees (**acrocentric** or **telocentric**: see Fig. 2.9). The first nine pairs were classified by (Shoffner *et al.*, 1967), numbering them in decreasing size order.

| | |
|---|---|
| 1 and 2 | metacentric |
| 3 | acrocentric |
| 4 | submetacentric |
| Z | metacentric |
| W | metacentric |
| 6 and 7 | acrocentric |
| 8 and 9 | metacentric |

Methods have been developed largely since 1970 to stain the chromosomes so as to produce specific and unique banding patterns, and this has enabled smaller chromosomes to be classified, and earlier classifications of some to be revised (Bitgood & Shoffner, 1990). These not only enable two chromosomes of similar size and morphology to be distinguished, but also make it possible to pick out specific regions on the chromosome. This is invaluable in the detection of abnormalities. These staining methods were developed for use on human chromosomes, but are now also being used on those of the domestic fowl.

The four main methods are referred to as **Q-banding**, **G-banding**, **R-banding** and **C-banding** (Bitgood & Shoffner, 1990). The quinacrine fluorescence method (Q-banding) was the first to be developed by Caspersson in 1968. He demonstrated that quinacrine dyes produced characteristic and reproducible patterns of fluorescence on the chromosomes when activated by ultraviolet (UV) light. Other methods were soon developed to avoid the need for expensive fluorescent equipment. The G-, R- and C-banding techniques evolved empirically out of an observation of Pardue & Gall (1970), who were attempting to hybridise RNA *in situ* to mouse chromosomes. They noticed that an area near the centromere stained more deeply with Giemsa stain than the rest of the chromosomes. The areas that stained

**Fig. 2.9** Chromosomes showing possible positions for the centromere.

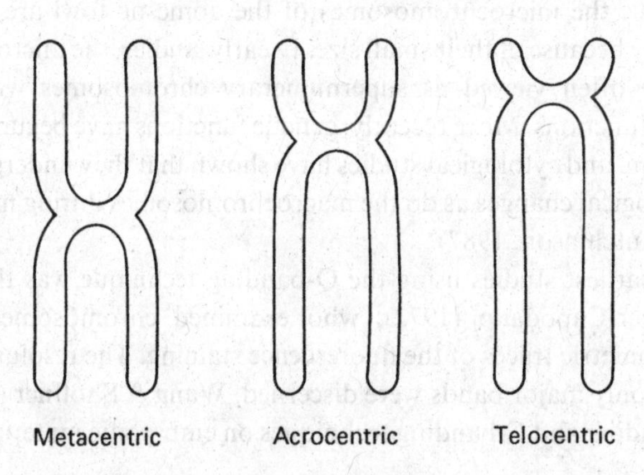

most darkly were those regions where the DNA renatured most rapidly; this occurred in the more highly repetitive sequences. The C-banding (=constitutive) is generated after denaturation with alkali, and this technique distinguishes constitutive heterochromatin. The G-banding technique involves staining with Giemsa stain, usually after a pretreatment with a proteinase, and it gives a banding pattern similar to that of Q-banding. If, on the other hand, the chromosomes are first partially denatured and then stained with Giemsa stain, a banding pattern the reverse of Q-banding is obtained, referred to as, R-banding. The theoretical basis of the staining methods is not yet clear, although there is some evidence that Q-banding stains the adenine–thymine rich regions of DNA, and the R-banding thus stains the guanine–cytosine rich regions (Comings, 1978).

The rapid development of these staining procedures meant that fairly precise regions of the chromosomes could be identified and thus a standardised nomenclature was thought desirable. A system was agreed at a Paris conference in 1971, which also incorporated some of the recommendations adopted at earlier meetings. The numbering system has since been adopted for all other species in which the karyotype has been studied. The karyotype of the largest chromosomes of the domestic fowl is illustrated in Fig. 2.8. Chromosomes are numbered in descending size order. A constriction occurs at the region of the centromere dividing the chromosome into two arms, designated '*p*' (petite) for the shorter and '*q*' for the larger arm (Fig. 2.9). The major bands, or 'landmark' bands, are then numbered from the centromere to the distal (telomeric) region of the chromosome. The finer bands are numbered as subsets between two landmark bands.

Progress in defining the karyotype of the domestic fowl has lagged behind work on that of humans. This is partly because there has been a much greater research effort into understanding the human chromosome, but also because the microchromosomes of the domestic fowl are more difficult to study because of their small size. In early studies, the microchromosomes were often viewed as supernumerary chromosomes without typical genetic functions. More recently, genetic functions have begun to be assigned to them, and cytological studies have shown that they undergo the same morphological changes as do the macrochromosomes during mitosis and meiosis (Hutchinson, 1987).

One of the earliest studies using the Q-banding technique was that of Stahl & Vagner-Capodano (1972), who examined chromosomes 1–6 making densitometric traces of the fluorescence staining. The resolution is fairly low and only major bands were discerned. Wang & Shoffner (1974) used the G-banding and C-banding techniques on embryonic material and on feather pulp to examine chromosomes 1–12. They noticed that the W chromosome which is heterochromatic gave a distinct C-band. This can be used as a means of sex identification. Since the W chromosome is so small, without this staining procedure it would be otherwise difficult to discern routinely. Wang & Shoffner (1974) also showed that it is possible to identify translocations using the G-banding method.

A more detailed Q-banding pattern in the domestic fowl is that of Fritschi & Stranzinger (1985) and this is reproduced in Fig. 2.10. In addition to studying the Q-banding pattern they used chromomycin which produces an R-type pattern, and 4,6-Diamidino-2-phenylindol.2HCl (Dapi) which produces a pattern qualitatively similar to that of Q-banding. A region in the chromosomes which can be distinguished using a special staining procedure is the nucleolar organiser region. It is stained using Ag-AS staining (Bloom, 1981). The silver nitrate in this stain is used to differentiate the nucleolar organiser region; the silver interacts specifically with a non-histone protein present in that region. This densely staining area contains multiple copies of the genes, for the synthesis of two types of ribosomal RNA known as 18S and 28S RNA (Muscarella *et al.*, 1985). During mitosis the ribosomal RNA genes are in an inactive state and are contained in the nucleolar organiser regions. The latter appear as secondary constrictions in the metaphase chromosomes. After mitosis the ribosomal RNA genes reactivate, and the ribosome assembly line formed from the nucleolar organiser regions appears in the form of the nucleolus (Sommerville, 1986). The nucleolar organiser region is associated with one of the microchromosomes in the domestic fowl (Bloom & Bacon, 1985), which also contains the genes of the major histocompatibility (B) complex (for details see Chapter 11, section 11.7).

Auer *et al.* (1987), by applying a counterstain-enhanced fluorescence technique to mitogen-stimulated lymphocytes and fibroblasts from domestic fowl, have been able to identify chromosomes 1–18. They have

**Fig. 2.10** Scheme of the Q-banding of a female karyotype of the domestic fowl, showing only the macrochromosomes (Fritschi & Stranzinger, 1985).

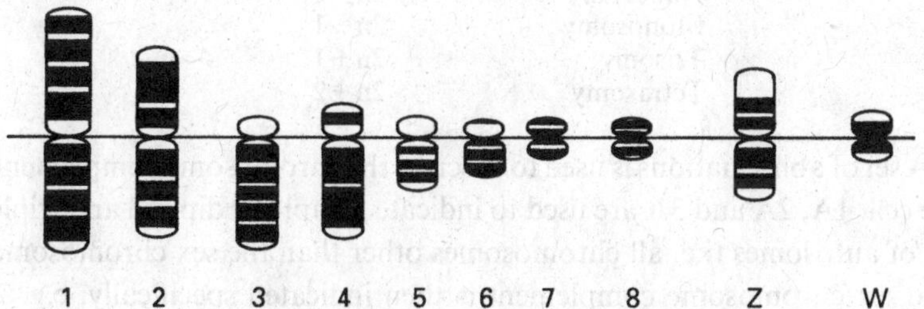

shown that the nucleolar organiser region is located on chromosome 17. This will also enable the precise localisation of several protooncogenes, e.g. c-myc, c-myb and c-src in the future (for review, see Tereba, 1985).

Similar difficulties have arisen in identification of the microchromosomes of other avian species such as the quail, which also has a chromosome number of 78, made up of 20 pairs of macrochromosomes and 19 pairs of microchromosomes. Mayr, Lambrou & Schleger (1989) have used sequential staining with a counterstain-contrasted fluorescent R-banding technique and this has enabled them to identify the largest 19 pairs of autosomes and the sex chromosomes.

## 2.5 Chromosomal abnormalities in the domestic fowl

Chromosomal abnormalities may arise spontaneously or they may be induced by certain chemical or physical treatments. There are a number of different kinds of abnormality, and these can be divided into those which involve a change in the number of chromosomes per cell, and those which involve a change in the structure of a particular chromosome. The latter include **deletions** (where a segment of a chromosome is lost), **additions** (where an additional piece of a chromosome is incorporated), **inversion** (where a piece is removed and reinserted with opposite polarity) and **translocation** (where a piece is removed and reinserted in a different position, either on the same chromosome or moved to a different chromosome). A change in the number of sets of homologous chromosomes is called a change in **ploidy**. **Aneuploidy** is the term used where the total number of chromosomes are not multiples of the haploid number. A list of the different chromosome numbers is given below.

| **Condition** | **Chromosome number** (n = haploid) |
|---|---|
| Haploid | 1n |
| Diploid | 2n |
| Triploid | 3n |
| Tetraploid | 4n |
| Nullosomy | 2n − 2 |
| Monosomy | 2n − 1 |
| Trisomy | 2n + 1 |
| Tetrasomy | 2n + 2 |

A set of abbreviations is used to describe the chromosome complement of the cell. 1A, 2A and 3A are used to indicate a haploid, diploid and triploid set of autosomes (i.e. all chromosomes other than the sex chromosomes). The sex chromosome complement is then indicated specifically, e.g. 2A-

ZZ = a normal diploid male. Where aneuploidy exists the additional or deficient chromosome is indicated by a + or − and its number given, e.g. 2A-ZW + 5 is a trisomic female having an additional chromosome 5. A range of spontaneous chromosome abnormalities has been detected in stocks of poultry; the frequency of the abnormalities varies with the stock or breed, between 0.8% and 14.5% (Bloom, 1981). Bloom suggests that selection may have played a large role in the modulation of these abnormalities. Besides the spontaneous chromosomal abnormalities, a number of abnormalities have been induced by mutagens such as X-rays (Wooster, Fechheimer & Jaap, 1977), ethylmethanesulphonate and triethylenemelamine (Shoffner, 1972). Chromosome rearrangements, especially translocations and inversions, are useful tools for studying a number of genetic problems. For example, if a particular translocation can be identified from the banding pattern, the genes involved will no longer be linked to the original chromosome but to a new chromosome (assuming that it is an interchromosomal translocation). This can be related to the pattern of inheritance of particular characters, which can then be located on a cytologically identifiable chromosome (see Chapter 5, section 5.3). Wooster *et al.* (1977) used X-radiation to induce mutations in sperms which were used to artificially inseminate hens. The surviving progeny were then studied cytologically. From this study they were able to obtain a number of translocations ranging through chromosomes 1–5 and a pericentric inversion in chromosome 2. About 20 different lines of domestic fowl having chromosome rearrangements have been developed (Bitgood & Shoffner, 1990).

Many of the chromosome abnormalities cause serious physical defects and often lead to high mortality rates during incubation. Fechheimer (1981) found that in chicken embryos sampled 16–18 hours after incubation, examination of the macrochromosomes revealed 5.2% having an abnormal karyotype, but in chickens surviving to 3–6 weeks no abnormal karyotypes were found. Haploidy and polyploidy were the most frequent abnormalities. Although some occur in pure form, triploid and diploid–haploid mosaics appear the most frequent type (Bloom, 1972, 1974; Miller, Fechheimer & Jaap, 1976; Fechheimer & Jaap, 1980; Fechheimer, 1981). **Mosaics** are individuals in which not all cells have the same chromosome number. The frequency of triploidy varies from strain to strain, with a maximum of 3.3% (Bloom, 1974). Triploids have a tendency to intersexuality (Abdel Hameed & Shoffner, 1971), having the general appearance of hens but the combs and hackles of a cock and the presence of an ovotestis. Most adults suspected of intersexuality turn out to have the ZZW sex chromosome triplet (DeBoer *et al.*, 1984). Over 80% of triploids originate from faulty disjunction during oogenesis (Bitgood & Shoffner, 1990). Triploids arise mainly from errors in meiosis in females, from suppression of either the first or the second division of meiosis.

Thorne, Collins & Sheldon (1987) have found haploid–diploid, diploid–triploid and haploid–diploid–triploid mosaics that have survived hatching. By intensive selection for six generations, hens which hatch 5–8% triploids have been obtained. The particular subline has a high incidence of meiotic non-disjunction (paired chromosomes not separating and both going to one daughter cell). This appears to be inherited as a single autosomal recessive gene. For an explanation of an autosomal recessive gene see Chapter 3, section 3.2.

A mutation leading to shanklessness has been shown to be due to an inversion in chromosome 2 (Langhorst & Fechheimer, 1985). A number of mosaic abnormalities have also been found. A triploid–diploid mosaic has been reported in which approximately two thirds of the cells had a constitution 3A-ZZZ and one third were diploid having a constitution 2A-ZZ. One of the chromosome translocations described by Wooster *et al.* (1977) is one in which a reciprocal translocation between the long arm of chromosome 1 and a microchromosome occurs, and this has been used to study the effect on viability and fertility. Using cockerels having this translocation Bonaminio & Fechheimer (1988) have shown that the resulting spermatocytes which bear different, unbalanced, genomic contents are not equally viable or fertile.

## References

Abdel Hameed, F. & Shoffner, R. N. (1971). Intersexes and sex determination in chickens. *Science*, **172**, 962–4.

Arthur, R. R. & Straus, N. A. (1983). DNA sequence organisation and transcription of the chicken genome. *Biochimica et Biophysica Acta* **741**, 171–9.

Auer, H., Mayr, B., Lambrou, M. & Schleger, W. (1987). An extended chicken karyotype, including the NOR chromosome. *Cytogenetics and Cell Genetics*, **45**, 218–21.

Bitgood, J. J & Shoffner, R. N. (1990). Cytology and Cytogenetics. In *Poultry Breeding and Genetics*, ed. R. D. Crawford, pp. 401–27. Amsterdam: Elsevier.

Bloom, S. E. (1972). Chromosomal abnormalities in chicken (*Gallus domesticus*) embryos: types, frequencies and phenotypic effects. *Chromosoma (Berlin)*, **37**, 309–26.

Bloom, S. E. (1974). The origins and phenotypic effects of chromosomal abnormalities in avian embryos. In *Proceedings of XV World's Poultry Congress*, pp. 316–20. Washington, DC: McGregor & Warner.

Bloom, S. E. (1981). Detection of normal and aberrant chromosomes in chicken embryos and tumor cells. *Poultry Science*, **60**, 1355–61.

Bloom, S. E. & Bacon, L. D. (1985). Linkage of the major histocompatibility (B) complex and the nucleolar organiser in the chicken. *Journal of Heredity*, **76**, 146–54.

Bonaminio, G. A. & Fechheimer, N. S. (1988). Segregation and transmission of chromosomes from a reciprocal translocation in *Gallus domesticus* cockerels. *Cytogenetics and Cell Genetics*, **48**, 193–7.

Comings, D. E. (1978). Mechanisms of chromosome banding and implications for chromosome structure. *Annual Review of Genetics*, **12**, 25–46.

Crawford, R. J., Krieg, P., Harvey, R. P., Hewish, D. A. & Wells, J. R. E. (1979). Histone genes are clustered with a 15-kilobase repeat in the chicken genome. *Nature*, **279**, 132–7.

Darnell, J., Lodish, H. & Baltimore, D. (1986). *Molecular Cell Biology*. New York: Scientific American Books.

DeBoer, L. E. M., DeGroen, T. A. G., Frankenhuis, M. T., Zonneveld, A. J., Sallevelt, J. & Belterman, R. H. R. (1984). Triploidy in *Gallus domesticus* embryos, hatchlings and adult intersex chickens. *Genetica*, **65**, 83–7.

Fechheimer, N. S. (1981). Origins of heteroploidy in chicken embryos. *Poultry Science*, **60**, 1365–71.

Fechheimer, N. S. & Jaap, R. G. (1980). Origins of euploid chimerism in embryos of *Gallus domesticus*. *Genetica* **52/53**, 69–72.

Fritschi, S. & Stranzinger, G. (1985). Fluorescent chromosome banding in inbred chicken: quinacrine bands, sequential chromomycin and Dapi bands. *Theoretical and Applied Genetics*, **71**, 408–12.

Hutchinson, N. (1987). Lampbrush chromosomes of the chicken, *Gallus domesticus*. *Journal of Cell Biology*, **105**, 1493–500.

Langhorst, L. J. & Fechheimer, N. S. (1985). Shankless, a new mutation on chromosome 2 in the chicken. *Journal of Heredity*, **76**, 182–6.

Lewin, B. (1990). *Genes IV*. Oxford: Oxford University Press.

Mayr, B., Lambrou, M. & Schleger, W. (1989). Further resolution of the quail karyotype and characterisation of microchromosomes by counterstain-enhanced fluorescence. *Journal of Heredity*, **80**, 147–50.

Miller, R. G., Fechheimer, N. S. & Jaap, R. G. (1976). Distribution of karyotype abnormalities in chick embryo sibships. *Biology and Reproduction*, **23**, 526–9.

Muscarella, D. E., Vogt, V. M. & Bloom, S. E. (1985). The ribosomal RNA gene cluster in aneuploid chickens: Evidence for increased gene dosage and regulation of gene expression. *Journal of Cell Biology*, **101**, 1749–56.

Ohno, S. (1961). Sex chromosomes and microchromosomes in *Gallus domesticus*. *Chromosoma (Berlin)*, **11**, 484–98.

Olofsson, B. & Bernardi, G. (1983). Organisation of nucleotide sequences in the chicken genome. *European Journal of Biochemistry*, **130**, 241–5.

Pardue, M. L. & Gall, J. G. (1970). Chromosomal location of mouse satellite DNA. *Science*, **168**, 1356–8.

Pollock, D. L. & Fechheimer, N. S. (1976). The chromosome number of *Gallus domesticus*. *British Poultry Science*, **17**, 39–42.

Shoffner, R. H. (1972). Mutagenic effect of triethylene melamine (TEM) and ethyl methanesulfonate (EMS) in the chicken. *Poultry Science*, **51**, 1865 (Abstract).

Shoffner, R. H., Krishan, A., Haiden, G. J., Bammi, R. K. & Otis, J. S. (1967).

Avian chromosome methodology. *Poultry Science*, **56**, 333–44.
Sommerville, J.(1986). Nucleolar structure and ribosome biogenesis. *Trends in Biochemical Sciences*, **11**, 438–42.
Stahl, A. & Vagner-Capodano, A. M. (1972). Cytogénétique – Etude des chromosomes du poulet (*Gallus domesticus*) par les techniques des fluorescence. *Comptes Rendus Académie des Sciences*, **275**, 2367–70.
Sugarman, B. J., Dodgson, J. B. & Engel, J. D. (1983). Genomic organisation, DNA sequence, and expression of the chicken embryonic histone genes. *Journal of Biological Chemistry*, **258**, 9005–16.
Suzuki, D. T., Griffiths, A. J. F., Miller, J. H. & Lewontin, R.C. (1989). *An Introduction to Genetic Analysis*, 4th edn. San Francisco: W.H. Freeman.
Tereba, A. (1985). Chromosomal localization of protooncogenes. *International Review of Cytology*, **95**, 1–43.
Thorne, M. H., Collins, R. K. & Sheldon, B. L. (1987). Live haploid–diploid and other unusual mosaic chickens (*Gallus domesticus*). *Cytogenetics and Cell Genetics*, **45**, 21–5.
Tjio, J. H. & Levan, A. (1956). The chromosome number of man. *Hereditas*, **42**, 1–6.
Wang, N. & Shoffner, R. N. (1974). Trypsin G- and C-banding for interchange analysis and sex identification in the chicken. *Chromosoma (Berlin)*, **47**, 61–9.
Wooster, W. E., Fechheimer, N. S. & Jaap, R. G. (1977). Structural rearrangements of chromosomes in the domestic chicken: experimental production by X-radiation of spermatozoa. *Canadian Journal of Genetics and Cytology*, **19**, 437–46.

# 3

# *The transmission of inherited characters*

## 3.1 Introduction

Much of the previous chapter dealt with the molecular basis of genetics; ultimately all genetic phenomena should be explicable in molecular terms. Historically, however, it is the branch of genetics now generally referred to as **transmission genetics** which originated the subject, and not until more than half a century after Mendel's work were the beginnings of a molecular explanation possible. In this chapter the phenotypic characters of an organism and their transmission are considered, but not the molecular events that underlie them. The essence of Mendel's findings is encompassed in his principles of inheritance, which are well documented in many biology and genetics textbooks (Strickberger, 1985; Suzuki *et al.*, 1989; Weaver & Hedrick, 1989). The principles are therefore discussed briefly using examples from poultry genetics. This is followed by considering other important aspects of transmission genetics.

## 3.2 Monohybrid crosses

Mendel carried out his experiments using different strains of garden peas. He selected strains having contrasting characters, e.g. green/yellow, round/wrinkled, tall/dwarf, etc. In each case he first tested that the strains bred true. He then proceeded to cross strains having opposing characters, i.e. he made **monohybrid crosses**, and examined the first generation of the progeny (**first filial generation, $F_1$**) which he then selfed to produce the $F_2$ generation. When two strains with opposing characters were crossed, the individuals of the $F_1$ generation did not appear to possess a blend of the two characters, but rather appeared to be entirely that of one character. This gave rise to the idea of **dominant** and **recessive** characters. The $F_2$ generation, obtained from self-crossing the $F_1$, contained the dominant and recessive types in a ratio of 3:1. If the $F_1$ peas were backcrossed with the dominant parental type all the progeny were dominant, but if they were backcrossed with the recessive parental type a 1:1 ratio of dominant:recessive types resulted. From these experiments Mendel postulated (i) that the genetic characters are controlled by factors which are paired in the parents, but which segregate in the gametes, and (ii) when two unlike factors each responsible for a single character are both present in a single individual one factor is dominant and the other recessive. Mendel obtained essentially similar results with all seven pairs of characters from the garden pea, which he tested.

Mendel carried out his experiments between 1856 and 1868 and, as is well known, his findings were not appreciated until their 'rediscovery' at the beginning of the twentieth century. It was at this time that Bateson and Punnett, interested in discovering the principles of inheritance, carried out a similar set of experiments, but in this case using animals, in particular the domestic fowl. At the time several breeds of domestic fowl were available and one of their conspicuous contrasting characters was the shape of the comb. In their first set of experiments they crossed fowl having single combs with those having rose combs (see Fig. 3.1), examined the $F_1$ generation, and then made further crosses, some self-crosses and some backcrosses. Their results were similar to those of Mendel, i.e. the $F_1$ generation contained 100% rose comb, which on selfing gave an $F_2$ having a 3:1 ratio of rose combed to single combed. They also backcrossed the $F_1$ generation with the parental rose comb and obtained 100% rose combed, and when they backcrossed the $F_1$ with the single comb parent they obtained 50% single combed and 50% rose combed. The rose comb is thus seen to be the dominant character and the single comb the recessive character.

The pairs of characters are called **alleles**. In molecular terms the alleles are alternative genes which are located on equivalent positions on the chromosome. The genes are usually given roman letters; the upper case for the dominant alleles and the lower case for the recessive alleles is the system generally used here. An alternative set of symbols which is sometimes preferred is to use $a$ and $a^+$, where the superscript indicates the wild type form, which in this case is dominant, and $A$ and $A^+$ in which case the wild type $A^+$ is recessive. However, for many of the genes studied in the domestic fowl it is not clear which is the wild type form. The wild type is the type from which any mutant form is derived, in this case from the jungle fowl. An example of the use of these latter symbols is given in section 3.6. In the case of rose comb and single comb the symbols $R$ and $r$ are used. The monohybrid crosses are shown below.

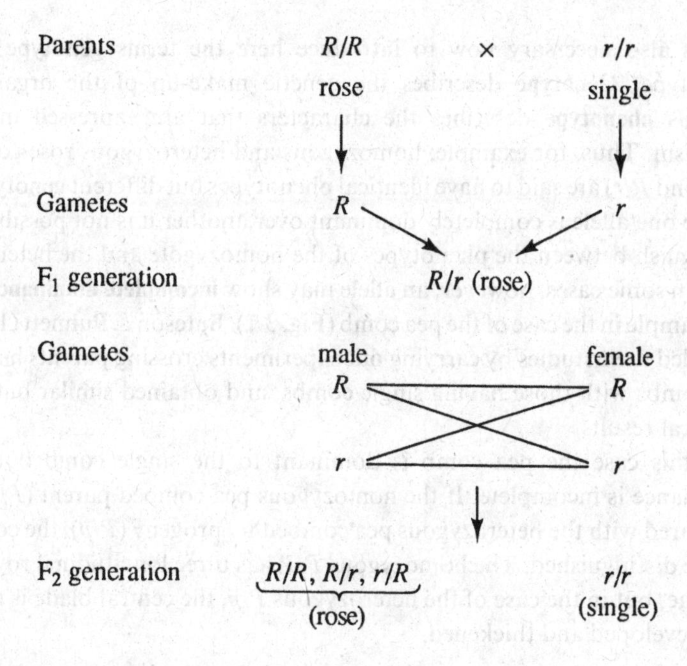


When the $F_1$ generation is self-crossed, both parents produce gametes of the *R* and *r* type, which accounts for the four possible ways in which the genes come together in the $F_2$ generation, three of which produce an individual that is rose combed and one that is single combed.

The two possible backcrosses between the $F_1$ generation and the parents are:

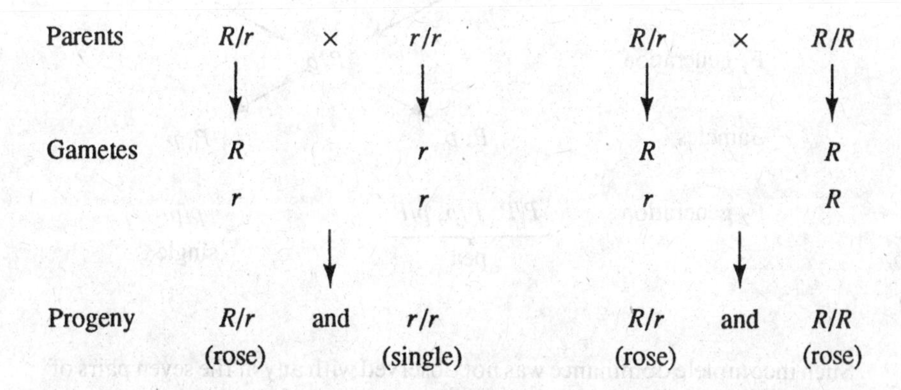


From the above analysis it is clear that a rose comb individual may have either of the two genetic combinations *R/R* or *R/r*; the former is referred to as **homozygous** and the latter **heterozygous**.

It is also necessary now to introduce here the terms genotype and phenotype. **Genotype** describes the genetic make-up of the organism whereas **phenotype** describes the characters that are expressed in the organism. Thus, for example, homozygous and heterozygous rose combs (*R*/*R* and *R*/*r*) are said to have identical phenotypes but different genotypes. Where one allele is completely dominant over another it is not possible to distinguish between the phenotypes of the homozygote and the heterozygote. In some cases, however, an allele may show **incomplete dominance** as, for example in the case of the pea comb (Fig. 3.1). Bateson & Punnett (1908) extended their studies by carrying out experiments crossing parents having pea combs with those having single combs, and obtained similar but not identical results.

In this case the pea comb is dominant to the single comb but the dominance is incomplete. If the homozygous pea combed parent (*P*/*P*) is compared with the heterozygous pea combed $F_1$ progeny (*P*/*p*), the combs can be distinguished. The homozygous *P*/*P* has three longitudinal rows of papillae, but in the case of the heterozygous *P*/*p*, the central blade is more well developed and thickened.

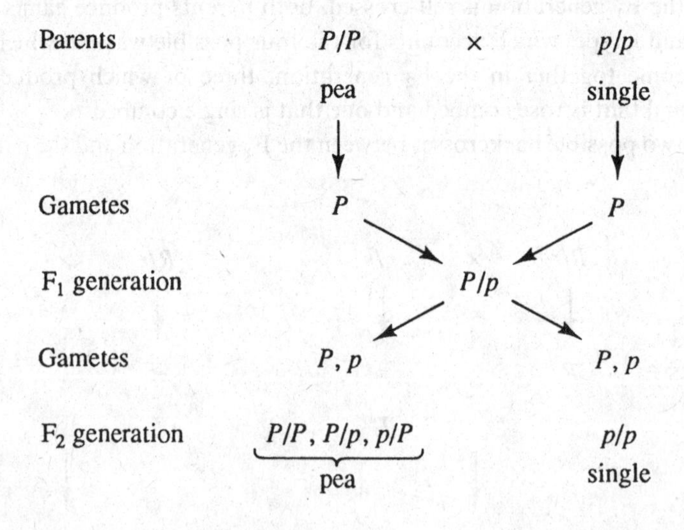


Such incomplete dominance was not observed with any of the seven pairs of alleles of the garden pea which Mendel used. It should also be noted that for the single comb cross with rose comb the symbols used are *r*/*r* whereas for the single comb cross with the pea comb the symbols *p*/*p* are used. This is because two separate pairs of alleles control the rose comb and the pea comb. To possess a single comb requires both pairs of recessive alleles $r/r$ and $p/p$.

Up to this point crosses involving a single pair of genes have been considered; next the more complex situation of dihybrid crosses is described.

## 3.3 Dihybrid crosses

Mendel also carried out dihybrid crosses in which two pairs of contrasting characters were involved, and from this he deduced that the pairs of alleles **assorted independently**. This was another of his important postulates. This is demonstrated below using the symbols $A$, $a$, $B$, $b$ for the two pairs of alleles, where both $A$ and $B$ show complete dominance; the $F_2$ generation is shown using what is described as a Punnett square.

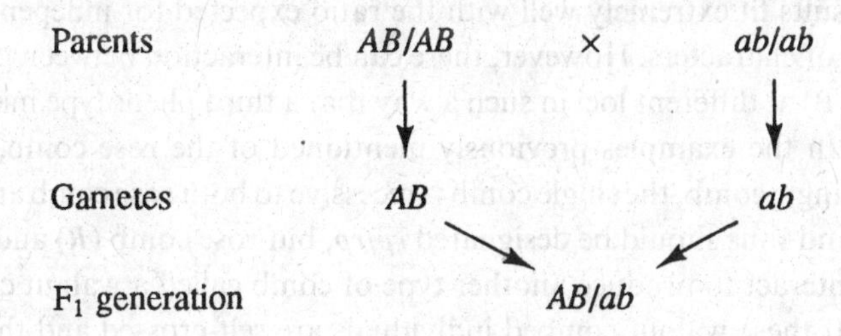


Male gametes

| Female gametes | AB | Ab | aB | ab |
|---|---|---|---|---|
| AB | AB/AB | AB/Ab | AB/aB | AB/ab |
| Ab | Ab/AB | Ab/Ab | Ab/aB | Ab/ab |
| aB | aB/AB | aB/Ab | aB/aB | aB/ab |
| ab | ab/AB | ab/Ab | ab/aB | ab/ab |

$F_2$ generation 9 *AB*/– –, 3 *Ab*/–*b*, 3 *aB*/*a*–, 1 *ab*/*ab*

The '–' means that either allele is possible, e.g. $AB/{-}{-}$ includes $AB/AB$, $AB/Ab$, $AB/aB$, and $AB/ab$.

The $F_2$ generation gives the characteristic 9:3:3:1 ratio consisting of phenotypes showing: both *A* and *B* character (9), *A* and *b* character (3), *a* and *B* character (3) and *a* and *b* character (1).

Bateson carried out the first dihybrid cross with the domestic fowl using White Leghorns and Indian Game. The former have single combs and dominant white plumage; the latter have pea combs and dark plumage. The $F_1$ generation were predominantly white and had pea combs. When self-crossed the results he obtained for the $F_2$ generation were:

| | **Observed number** | **Predicted ratio** | **Predicted numbers** |
|---|---|---|---|
| White, pea comb | 111 | 9 | 106.9 |
| White, single comb | 37 | 3 | 35.6 |
| Dark, pea comb | 34 | 3 | 35.6 |
| Dark, single comb | 8 | 1 | 11.9 |
| Total | 190 | | 190 |

These results fit extremely well with the ratio expected for independent assortment of characters. However, there can be interaction between genes (e.g. A and B) at different loci in such a way that a third phenotype may be generated. In the examples previously mentioned of the rose comb, pea comb and single comb, the single comb is recessive to both rose comb and to pea comb and thus should be designated $rp/rp$, but rose comb ($R$) and pea comb ($P$) interact to produce another type of comb called a walnut comb (Fig. 3.1). If these walnut combed individuals are self-crossed and the $F_2$ generation analysed the ratio is 9 walnut, 3 pea, 3 rose and 1 single. The ratios are still the same as in the previous example but the difference is that a new phenotype (walnut) has been generated. In this case the alleles segregate independently but the products of the genes interact to complement each other in producing a new phenotype: $RP/{-}{-}$ = walnut, $Rp/{-}p$ = rose, $rP/r{-}$ = pea and $rp/rp$ = single.

**Fig. 3.1** Types of comb in domestic fowl.

The 9:3:3:1 ratio is characteristic of the independent assortment of characters, and this was what Mendel found with all seven pairs of alleles from the peas with which he worked. This turned out to be a little fortuitous in that the alleles he happened to choose were either located on separate chromosomes or very distant from one another on the same chromosome. This is very often not the case and independent assortment frequently does not occur. This will be discussed in section 3.5, but before doing so the method of test crossing is considered since it is often necessary to establish the genotype of any particular stock before carrying out breeding experiments.

## 3.4 Test crosses

From the previous section it should be clear that it is not usually possible to distinguish a heterozygote from its dominant homozygote counterpart, since the phenotypes are identical whether one is considering one or more pairs of alleles, e.g. *R*/*r* and *R*/*R* are both phenotypically rose combed. It is often important to be able to make these distinctions in order either to be able to specify the genotypes produced by a particular cross, or to be sure that the initial parents are homozygous for a particular character. Mendel devised a simple test to differentiate the dominant homozygote from the heterozygote. The test is simply to cross the organism concerned with its corresponding recessive counterpart. If more than one locus is being considered, a genotype recessive at each locus is required, e.g. for three loci a triple recessive is required. The example of a phenotypically white feathered, rose comb fowl is illustrated.

(*a*) To differentiate *R*/*r* from *R*/*R* carry out a monohybrid test cross using *r*/*r*. In the case of *R*/*R* all the $F_1$ will possess the rose comb phenotype, whereas in the case of *R*/*r* there will be a ratio of 1:1 rose comb:single comb.

(*b*) To differentiate *RI*/*RI*, *RI*/*Ri*, *RI*/*rI* and *RI*/*ri* where *I* is the gene for dominant white such as occurs in White Leghorns, carry out a dihybrid test cross using the double recessive *ri*/*ri*. The results expected would be as follows:

If *RI*/*RI*–phenotypes all rose comb/white
If *RI*/*Ri*–phenotypes 1:1 rose comb/white:rose comb/dark
If *RI*/*rI*–phenotypes 1:1 rose comb/white:single comb/white
If *RI*/*ri*–phenotypes 1:1:1:1 rose comb/white:rose comb/dark: single comb/white:single comb/dark

These ratios assume the independent assortment of genes.

## 3.5 Linkage

Independent assortment has been mentioned in both sections 3.3 and 3.4. When two pairs of alleles are located on separate chromosomes they assort independently during meiosis. If the $F_1$ progeny from such a cross is selfed the typical 9:3:3:1 ratio is obtained in the $F_2$. When the $F_1$ progeny from a double homozygous cross is testcrossed with the double recessive a ratio of 1:1:1:1 of the four phenotypes is expected. If, however, the pairs of alleles are on the same chromosome assortment is not independent and a different result is produced. This is illustrated by data from Warren & Hutt (1936) concerning two pairs of alleles, one for the presence or absence of a crest (*Cr* and *cr*) and one for white or non-white plumage (*I* and *i*). They testcrossed the double heterozygotes (*Cr, I*/*cr, i*) with the double recessive (*cr, i*/*cr, i*) and the numbers they obtained were:

| Genotype | Observed | Expected* |
|---|---|---|
| *Cr, I*/*cr, i* | 337 | 188.5 |
| *cr, i*/*cr, i* | 337 | 188.5 |
| *Cr, i*/*cr, i* | 34 | 188.5 |
| *cr, I*/*cr, i* | 46 | 188.5 |
| Total | 754 | 754 |

* Expected if independent assortment occurs.

The ratio is far from the 1:1:1:1 expected for independent assortment and it will be noticed that there are far more of the parental genotypes (*Cr, I*/*cr, i* and *cr, i*/*cr, i*) than of the **recombinant** types (*Cr, i*/*cr, i* and *cr, I*/*cr, i*). This is because the genes are **linked** on the same chromosome. The closer the genes are to each other, the less likely it is that they will be separated by crossing over during meiosis. When crossing over occurs it leads to the formation of recombinants. Analysing the frequency with which recombination occurs is used in gene mapping and will be further discussed in Chapter 5.

## 3.6 Test for allelism

When a new character has been discovered and has been established as being inherited in normal Mendelian fashion it is often necessary to know how it is related to other similar characters and in particular whether it is allelic to other known characters, e.g. white skin (*y*) and yellow skin (*Y*). If two or more characters are controlled by genes that are allelic, then the genes would eventually be shown to map at the same locus on the chromosome. This involves a detailed and very time consuming analysis in many cases; there are, however, fairly simple tests available to establish allelism. These are illustrated by an example. Suppose three genes $A$, $A'$, $A''$ are all alleles, i.e. located in the same position on homologous chromosomes. An organism which is homozygous for these genes could be one of $A/A$, $A'/A'$ or $A''/A''$ and one heterozygous could be one of $A/A'$, $A/A''$ or $A'/A''$. If the three possible crosses of homozygous types are made and the $F_1$ generation is self-crossed in each case the $F_2$ will give genotypes like those illustrated below for the $A/A$, $A'/A'$ cross:

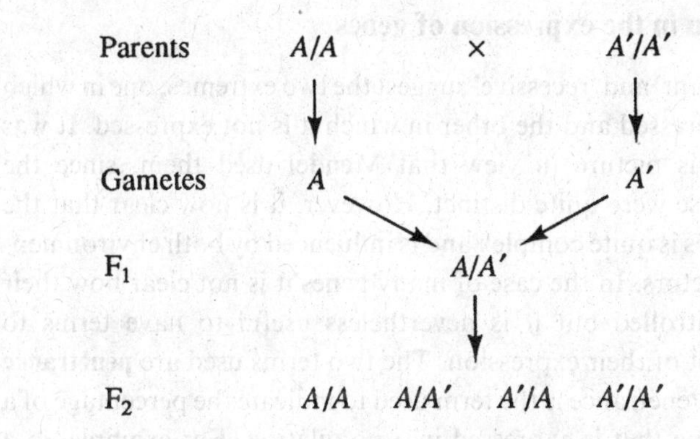


If $A$ is completely dominant to $A'$ then a 3:1 ratio of phenotypes $A$:$A'$ would be expected. If $A$ is incompletely dominant to $A'$ then a 1:2:1 ratio of the phenotypes would be expected with 50% of the progeny having a phenotype intermediate between that of $A$ and $A'$. The 1:2:1 ratio could also arise from complementary gene interaction and therefore does not give a conclusive answer.

Morejohn (1955) used this type of analysis to establish that a number of genes controlling plumage colour were in fact alleles. The genes concerned were designated $e^+$, $e^y$ and $e^b$. The genotype $e^+$ is that of the jungle fowl plumage, genotype $e^b$ has a brown plumage and genotype $e^y$ is yellowish white. Morejohn carried out a large number of crosses including those between the homozygous parents and also backcrosses between parents and the $F_1$ generations. Included below are the three possible crosses between the three heterozygote $F_1$ generations.

| **Crosses** | **Phenotypes** | | | | | **Expected ratio** | **Observed ratio** |
|---|---|---|---|---|---|---|---|
| | $e^+$ | $e^+e^b$ | $e^b$ | $e^be^y$ | $e^y$ | | |
| $e^+e^y \times e^+e^y$ | 66 | | | | 19 | 3:1 | 3.47:1 |
| $e^+e^b \times e^+e^b$ | 12 | 11 | 8 | | | 1:2:1 | 1.5:1.38:1 |
| $e^ye^b \times e^ye^b$ | | | 32 | 50 | 28 | 1:2:1 | 1.14:1.79:1 |

Note that $e^+$ is dominant to $e^y$, $e^+$ is incompletely dominant to $e^b$ and $e^y$ is incompletely dominant to $e^b$.

The ratios obtained for the $e^+e^y$ self-cross and the $e^ye^b$ self-cross are in very good agreement to those expected for alleles. The $e^+e^b$ self-cross is not in such agreement, but this probably stems from the small number of progeny and from the difficulty of differentiating the heterozygous phenotype. If the genes in question are not alleles, but one is epistatic to the other, then different ratios would be obtained. This is described in the next section.

## 3.7 Variation in the expression of genes

The terms 'dominant' and 'recessive' suggest the two extremes, one in which a gene is fully expressed and the other in which it is not expressed. It was probably with this picture in view that Mendel used them, since the characters he chose were quite distinct. However, it is now clear that the expression of genes is quite complex and is influenced by both environmental and genetic factors. In the case of many genes it is not clear how their expression is controlled but it is nevertheless useful to have terms to describe the extent of their expression. The two terms used are **penetrance** and expressivity. Penetrance is the term used to indicate the percentage of a particular genotype that is expressed in a population. For example, in a study of the inheritance of ear tuftedness Somes & Pabilonia (1981) crossed a homozygous male Araucana (*Et*/*Et*) with Leghorn hens (*et*/*et*) and the resulting progeny were 49 tufted and 8 non-tufted. From the genotypes used 57 tufted (i.e. all of the progeny) might have been expected, but the observed result was 14% less. The character tuftedness is said to show reduced penetrance of 14%.

**Expressivity** is used to describe the variation in the extent of expression of a character in a population. For example, in an earlier study Somes (1978) measured the sizes of eartufts (a character for which Araucanas are homozygous) in a population of 663 Araucanas. Expressivity was found to be quite varied, with approximately 25% of the birds having larger tufts on the right side, 25% larger on the left side, 25% with large tufts on both sides and 25% with small to medium sized tufts on both sides.

There are different degrees to which the pairs of alleles in a heterozygote influence the form of the phenotype; these can be thought of as a progression running from overdominance, full dominance, codominance and incomplete dominance to recessive. **Overdominance** is a rare phenomenon in which the dominant allele has a greater effect in the heterozygote than it does in the dominant homozygote. **Full dominance** occurs when the phenotype of the heterozygote and the dominant homozygote are the same. **Codominance** occurs when both pairs of alleles are equally expressed. In **incomplete dominance** the expression of the single dominant allele gives rise to a phenotype for the heterozygote intermediate between the two homozygotes. In contrast to codominance the recessive allele is often a **null allele** that is not expressed. Codominance is shown by the blood group proteins and by the polymorphic egg-white proteins, which are discussed in detail in Chapters 10 and 11. There are a number of antigenic proteins on the surface of the erythrocytes (blood group proteins). If two birds having different allelic erythrocyte antigens are crossed, the progeny will have both types of erythrocyte antigen.

Incomplete dominance occurs in quite a number of cases in the domestic fowl. These include the genes for crest, creeper, frizzling, pea comb, uropygial, hen feathering, muffs and beard, extended black plumage and blue plumage. Several of these are discussed in Chapter 6. One of the earliest examples of incomplete dominance to be discovered was that of the blue Andalusian fowl. In 1906 Bateson and Punnett found that mating blue Andalusian males and females did not result in blue progeny but a 1:2:1 ratio of black:blue:white with a few blue feathers (blue-splashed white). The reason for this plumage pattern is that the blue plumage is that of the heterozygote; the homozygotes are the black and the blue-splashed white. Thus in cases of incomplete dominance such as this, the heterozygote has a phenotype distinct from both homozygotes, and a 1:2:1 ratio occurs instead of the 1:3 ratio. A further example of incomplete dominance is the form of crest. If birds homozygous for crest are crossed with non-crested birds, the $F_1$ generation all have small sized crests. If these are self-crossed then the 1:2:1 ratio occurs in the $F_2$ generation.

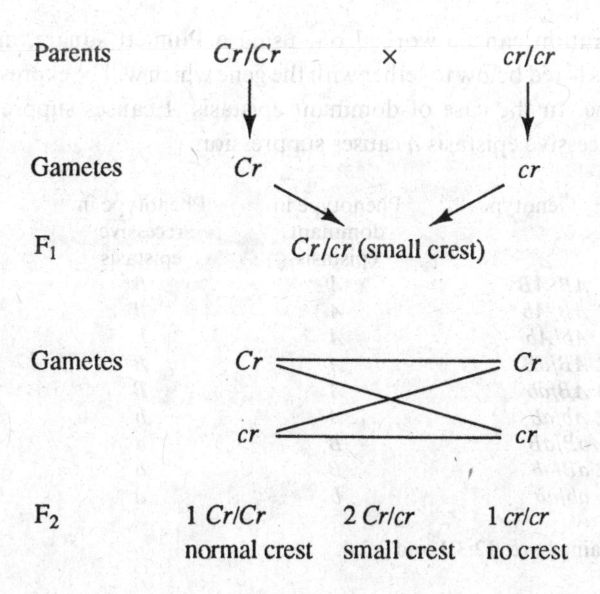

A further phenomenon concerning the expression of a character, distinct from that of the normal dominance relationships, is that of **epistasis**. This phenomenon occurs where one gene is able to suppress the expression of a second, non-allelic, gene. The suppressing gene is known as the epistatic gene (which means 'standing above'), and the gene which is being suppressed is the hypostatic gene (which means 'standing below'). The epistatic and the hypostatic genes have distinct locations on the chromosomes. They normally exist as pairs of alleles, which themselves have dominance relationships. The general principle of epistasis will first be explained and then an example from the genetics of the fowl used to illustrate it.

Suppose there exists an allelic pair of epistatic genes *A* and *a* and an allelic pair of hypostatic genes *B* and *b*. *A* is dominant to *a*, and *B* is dominant to *b*. There are two types of epistasis, **dominant epistasis** and **recessive epistasis**; in the former the epistatic gene is *A*, and in the latter it is *a*. Consider what will happen if a dihybrid cross is made using the homozygous dominant pair and the homozygous recessive pair, and then the $F_1$ progeny self-crossed to generate the $F_2$ generation.

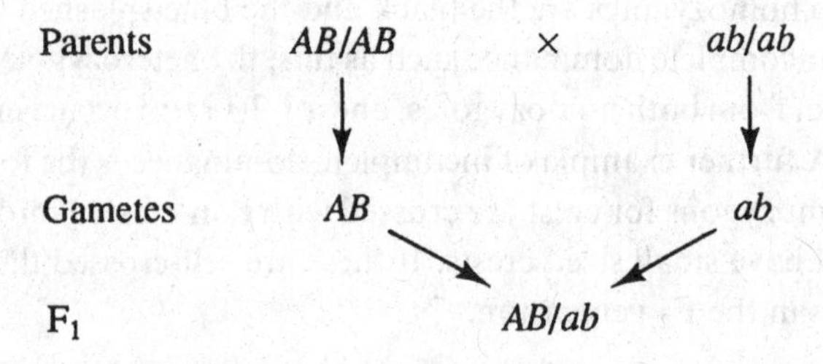


The $F_2$ generation can be worked out using a Punnett square, and its constitution is listed below together with the gene which will be expressed in the phenotype. In the case of dominant epistasis *A* causes suppression, whereas in recessive epistasis *a* causes suppression.

| $F_2$ Genotypes | Phenotype in dominant epistasis | Phenotype in recessive epistasis |
|---|---|---|
| 1 *AB/AB* | *A* | *B* |
| 2 *AB/Ab* | *A* | *B* |
| 1 *Ab/Ab* | *A* | *b* |
| 2 *AB/aB* | *A* | *B* |
| 4 *AB/ab* | *A* | *B* |
| 2 *Ab/ab* | *A* | *b* |
| 1 *aB/aB* | *B* | *a* |
| 2 *aB/ab* | *B* | *a* |
| 1 *ab/ab* | *b* | *a* |

The ratios obtained are: 12:3:1 and 9:3:4.

It can be seen that epistasis gives rise to ratios different from those of a normal dihybrid cross, and also that it is possible to distinguish dominant epistasis from recessive epistasis.

The best known example of epistasis in the domestic fowl is that of the dominant white gene which occurs in the White Leghorn. Its interactions with other genes controlling plumage colour are not fully understood but a simplified picture is given here to illustrate the epistasis; some of the complexities are discussed in Chapter 6. Three sets of genes are concerned with regulating the colour of plumage and its pattern in many of the breeds and these are designated *E*, *C* and *I*. The *E* locus is concerned with the pattern of the feather pigment. For example, if the fowl possesses the *E* gene then it will be self-coloured, that is a uniform colour throughout (E = extended). If it possesses the $e^+$ gene (an allele of *E*) it will have the black-red colour of feather pattern characteristic of the red jungle fowl (see Fig. 6.6). The *C* gene controls the production of the black pigment, eumelanin; thus a fowl having the phenotype *C* will have the ability to produce black pigment, whereas one homozygous for the recessive allele *c* will lack the black pigment and will have white plumage. The third gene, *I*, inhibits the production of black pigment and is epistatic to *C*. A fowl having the genetic constitution *CEi*/*CEi* would have black plumage throughout, whereas a fowl with the genetic constitution $Ce^+i/Ce^+i$ would have the black-red feathering as in the jungle fowl. *CEI*/*CEI* would be white all over as in the White Leghorn because of the epistatic effect of *I*/*I*. Fowl having the genetic constitution *CEI*/*CEi* would be very largely white, but with a few black tips to the feathers. This is due to the incompletely dominant nature of the *I* gene. Fowl with the genetic constitution $Ce^+I/Ce^+I$ would have white and red plumage (known as pile). This is because the *I* gene is effective in suppressing the black pigment, eumelanin, but not the red phaeomelanin (for further details see Chapter 6).

Another example of an epistatic gene is the sex-linked barring gene characteristic of Plymouth Rocks. The sex-linked character of this gene is discussed in Chapter 6. The barring gene *B* is dominant to the non-barring gene *b* and it is epistatic to the *E* gene. Therefore genotype *BE*/*BE* would be barred but genotype *bE*/*bE* would have black plumage throughout.

To conclude this chapter a summary of the characteristics of a range of genes present in the fowl is given (Table 3.1).

Table 3.1. *The inheritance of various genetic traits in the domestic fowl*

| Gene | Symbol | Inheritance | Comment |
|---|---|---|---|
| Rose comb | *R* | A,D | |
| Pea comb | *P* | A,ID | |
| Walnut comb | *PR* | A,DD | |
| Eartuft | *Et* | A,D | Incomplete penetrance, expression variable |
| Uropygial | *U* | A,ID | Bifurcation of oil gland papillae in heterozygote |
| Yellow skin | *w* | A,R | |
| Slow feathering | $K^s$ | S-L,D | At least 4 alleles known |
| Frizzling | *F* | A,ID | |
| Silkiness | *h* | A,R | Characteristic of Silkie breed |
| Crest | *Cr* | A,ID | |
| Muff & Beard | *Mb* | A,ID | |
| Blue | *Bl* | A,ID | Heterozygote has blue feathers |
| Dominant white | *I* | A,ID | Epistatic to *C* |
| Recessive white | *c* | A,R | Hypostatic to *I* |
| Silver | *S* | S-L,D | Recessive allele is gold |
| Extended black | *E* | A,ID | 8 alleles at this locus give different feathering pattern |
| Barring | *B* | S-L,ID | |
| Dwarfism | *dw* | S-L,R | |
| Polydactyly | *Po* | A,ID | 5th toe, expression & penetrance very irregular |
| Blue egg | *O* | A,D | |
| Creeper | *Cp* | A,D | Lethal in homozygote *CpCp* |
| Erythrocyte antigens | *Ea* | A,C | > 10 characterised |
| Serum albumin | $Alb^B$ | A,C | Main plasma protein |
| Immunoglobulin | *IgG* | A,C | Multiple alleles |

Abbreviations: A, autosomal; S-L, sex-linked; D, dominant; ID, incomplete dominant; DD, double dominant; C, codominant; R, recessive.

## References

Bateson, W. & Punnett, R. C. (1906). Experimental studies in the study of physiology. *Poultry Reports to the Evolutionary Committee of the Royal Society*, **3**, 11–30.

Bateson, W. & Punnett, R. C. (1908). Experimental studies in the study of physiology. *Poultry Reports to the Evolutionary Committee of the Royal Society*, **4**, 18–35.

Morejohn, G. V. (1955). Plumage color allelism in the red jungle fowl (*Gallus gallus*) and related domestic forms. *Genetics*, **40**, 519–30.

Somes, R. G. (1978). Ear-tufts: a skin structure mutation of the Araucana fowl. *Journal of Heredity*, **69**, 91–6.

Somes, R. G. & Pabilonia, M. S. (1981). Ear tuftedness: a lethal condition in the Araucana fowl. *Journal of Heredity*, **72**, 121–4.

Suzuki, D. T., Griffiths, A. J. F., Miller, J. H. & Lewontin, R. C. (1989). *An Introduction to Genetic Analysis*, 4th edn. San Francisco: W.H. Freeman.

Strickberger, M. W. (1985). *Genetics*, 3rd edn. New York: Collier Macmillan.

Warren, D. C. & Hutt, F. B. (1936). Linkage relations of crest, dominant white and frizzling in the fowl. *American Naturalist*, **70**, 379–94.

Weaver, R. F. & Hedrick, P. W. (1989). *Genetics*. Iowa: Brown Publishers.

# 4

# *Sex determination and sex-linked inheritance in the domestic fowl*

## 4.1 Introduction

Sexual reproduction ensures the combination of different genotypes, and thus generates progeny having new and varied genotypes. The combination of different genotypes is of greater importance than mutational events in its effect on the rate of evolution (see Chapter 1, section 1.2). In this chapter the current state of understanding of sex-determining mechanisms in the domestic fowl is summarised together with a description of sex-linked inheritance.

The nineteenth century poultry breeder was aware of the pattern of sex-linked inheritance, although the basis of it was not understood. Lancaster (1972) summarised the early reports of sex-linkage which included the inhibitor of dermal melanin in 1850, the silver locus in 1855, cuckoo barring in 1885 and slow feathering in 1885. By 1908 Bateson & Punnett (1908) and also others (see Hutt & Rasmusen, 1982) had shown from cytological studies that the hen is the heterogametic sex, but even by 1949 the nature of the sex chromosomes in the domestic fowl was not clear. Hutt (1949), in his book *Genetics of the Fowl* suggested that the cock has a pair of sex chromosomes (**ZZ**, equivalent to **XX** in mammals) and that the hen has a single sex chromosome (**ZO**, equivalent to **XO**). It was not until the 1960s that Ohno (1961) and Schmidt (1962) detected the smaller W chromosome, thus showing that the male is **ZZ** and the female **ZW**. This is the reverse of the situation in mammals where the female is the homogametic sex (**XX**) and the male the heterogametic sex (**XY**).

## 4.2 Sex determination in the domestic fowl

Although the sex chromosomes in the fowl and other birds are now well characterised, the mechanism of sex determination is still controversial. Sex determination entails a sequence of events. There is the primary genetic event of sex determination, i.e. fertilisation of the egg by the male gamete, which leads to the development and differentiation of the gonads, and eventually to spermatogenesis and ovogenesis. Secondary effects of the development of active gonads are the development of secondary sexual characteristics which are under hormonal control. Many of the genes necessary for sexual development are present on the autosomes, and it is clear that higher organisms have the potential to develop into males or females as a result of the possession of common genes. The sex chromosomes, however, control the trigger mechanism for development which 'switches on' particular blocks of genes necessary for either male or female differentiation.

There are a number of mechanisms for sex determination but the only important ones, as far as the domestic fowl is concerned, are the '**dominance mechanism**' typified by that of humans, and the '**genic balance mechanism**' typified by that of *Drosophila*. In humans the male is the heterogametic sex having XY sex chromosomes, and the female the homogametic sex having 2X sex chromosomes. In the dominance mechanism it is the Y chromosome, and not the presence of more than one X chromosome, which is dominant in sex determination, in this case, maleness. This is evident from the studies of a number of sex chromosome abnormalities; for example, an abnormal human having XXY is found to be male, whereas an abnormal person having XO is female (XO means that there is only a single sex chromosome, X). In the genic balance mechanism, however, it is the ratio of X chromosomes to autosomes that determines the sex, and not the presence or absence of Y. Thus in *Drosophila* an abnormal fly having XXY is female whereas one having XO is male.

The mechanism for sex determination in the domestic fowl has not been conclusively determined, although most evidence favours the genic balance mechanism (White, 1973; McCarrey & Abbott, 1979; Sittmann, 1984). Most of the evidence comes from analyses of individual birds possessing abnormal chromosome patterns. The earliest study is that of Crew (1933). He obtained a single cock which had sex-linked traits indicative of a female. He analysed the progeny of this cock, examining both their cytology and phenotype, and from the results deduced that the original cock was a 2A:ZZW genotype (this abbreviated nomenclature means the following: 2A = 2 homologous sets of autosomes, 2Z and 1W sex chromosomes). This result is consistent with the genic balance mechanism. The W chromosome is not exerting a dominant influence. The weakness of the data is simply that they were obtained from the progeny of a single experimental animal. More recently Abdel-Hameed & Shoffner (1971) carried out karyotyping on 15 fowl from commercial flocks that were sexed as females on hatching but showed malformed gonads which, on examination, were found to comprise testicular tissue. Of the 15, thirteen were 2A:ZZW and two were mosaics. When considering the genic balance it should be noted that for a normal male (2A:ZZ) the ratio of Z chromosomes:autosomes is 1:1 and for a normal female (2A:ZW) the ratio is 1:2. The intersex fowl analysed by Abdel-Hameed & Shoffner (1971) thus had an intermediate ratio of 2:3. However, the genotypes which would be most helpful in discriminating between the two mechanisms would be 2A:ZO and further examples of 2A:ZZW. Bitgood & Shoffner (1990) suggest that the W chromosome has a major influence on determining femaleness, i.e. favouring the dominance mechanism. As evidence they cited that (i) no ZW diploids with male or intersex phenotypes have been reported, and (ii) ZZW triploids and individuals with other sex chromosome mosaicism have different degrees of female phenotypes.

Evidence which also suggests a positive role for the W chromosome comes from studies on the H-W antigen (McCarrey, Abplanalp & Abbott, 1981). Whilst carrying out skin transplantation experiments using inbred strains of mice Eichwald & Silmser (1955) discovered the presence of a minor transplantation antigen which caused the female mice to reject the skin of male mice. Since the rejection occurred only when transplantation was between opposite sexes, from male to female, it was assumed that the antigen was a product of the Y chromosome and it was designated H-Y. In the domestic fowl the rejection was found to occur in the opposite way, i.e. females to males (Gilmour, 1967). Wachtel, Koo & Boyse (1975) developed serological assays for the H-Y antigen and proposed that in mice it was involved in testicular development. An analogous role for the H-W antigen in female chicks is supported by work of McCarrey *et al.* (1981). They carried out similar work to that of Gilmour (1967), except that they used a line of inbred domestic fowl which had been fixed for well-characterised alleles at the histocompatibility locus (B) as well as 10 other blood group loci, thus reducing the possibility of graft rejection through differences in loci other than that for H-W antigen. Nevertheless sex-specific skin graft rejection occurred as found previously. More recent evidence (see Goodfellow, 1986) suggests that the H-Y antigen of mice may not be involved in testis induction but in spermatogenesis. Doubts have been expressed by Muller *et al.* (1979) about the location of the H-W antigen on the W chromosome of the domestic fowl. They found that when embryonic male birds (ZZ) are 'sex-reversed' by application of oestrogens, development of the ovotestis occurs, the presence of the H-W antigen is detected and the birds show sex-linked feathering characteristic of the female although they are still clearly ZZ genotypes lacking the W chromosome.

Another factor which may be involved in the sex determination mechanism is a highly repetitive DNA sequence, Bkm DNA (Mittwoch, 1983). This sequence has been shown by hybridisation studies to be more abundant in the heterogametic sex of both birds and mammals (Jones & Singh, 1982). More research is needed before the mechanism is clarified.

## 4.3 Sex-linked inheritance

The pattern of transmission and expression of certain genetic characters is dependent on the sex of both the parents and the offspring. These are **sex-linked characters**. The pattern depends on the sex because of the differences between the sex chromosomes of the male and female. In the domestic fowl the Z chromosome is much larger than the W chromosome. The W chromosome comprises approximately 1.4% of the total genomic DNA in the female (Tone *et al.*, 1984). The Z chromosome is approximately 2–4 times larger. It is possible to divide the Z and W chromosomes into three regions. First there is a region which is thought to be homologous in the two chromosomes. Although this has not yet been proven in the case of the domestic fowl, it is assumed to be necessary in order for the Z and W chromosomes to pair during meiosis. Also, there is evidence for this in the case of Harris's hawk, where the gene for creatine kinase is located on both Z and W chromosomes (Morozit, Bednarz & Ferrell, 1987). The second region is the large remaining region on the Z chromosome and the third the much smaller remaining region on the W chromosome.

Genes which are present on the homologous region are said to be incompletely or partially sex-linked. This is because they will remain on either the Z or the W chromosome unless crossing over occurs between the Z and W chromosome during meiosis. Thus, for example, if two alleles *A* and *a* were located on the homologous regions of the W and Z chromosomes respectively, they will remain there unless a crossover occurs in that region during meiosis. Genes on the larger remaining region of the Z chromosome will be completely sex-linked since there is no homologous region on the W chromosome with which to cross over. Genes located on the remaining region of the W chromosome are completely W-linked and are known as hologynic genes since they will normally be present only in the female. So far almost all the genes located on the sex chromosomes are fully sex-linked on the Z chromosome. The only gene believed to be located on

Table 4.1. *Examples of sex-linked genes in the domestic fowl*

| Gene | Symbol | Description of the phenotype |
|---|---|---|
| Barring | *B* | Interupts dark pigmentation of feathers causing barring pattern, incomplete dominant |
| Brown Eye | *br* | Iris colour brownish-black |
| Coloboma | *cm* | Erosion of upper beak, variable effects on skeleton resembling teratoma |
| Diplopodia | *dp* | Partial doubling of the foot structure |
| Dwarfism | *dw* | Reduction in body weight of 40% in *dw/dw* males, and 30% in *dw* females |
| Dermal melanin inhibitor | *id* | Inhibits production of melanin in dermis |
| Rate of feathering | $K^n$ | Greatly retards feathering, 4 alleles |
| Light down | *Li* | Lessens brown colour in chick down |
| Naked | *n* | General lack of feathers |
| Paroxysm | *px* | Poor growth, stilted gait, lethal in *px* females |
| Prenatal lethal | *pn* | Lethal at 3½–4 day embryo development |
| Pop-eye | *pop* | Keratoglobus, protrusion of cornea |
| Restricted ovulator | *ro* | Lays few or no eggs |
| Shaker | *sh* | Tremor, particularly in head & neck |
| Silver | *S* | Silver feathering dominant to gold |
| Winglessness | *wl* | Wings & legs rudimentary or absent |

the W chromosome so far is the H-W antigen, but there are doubts about this (see section 4.2). It has been shown that a large proportion of the W chromosome is composed of repetitive sequences (Tone *et al.*, 1984) and it is doubtful whether many genes are located there. Table 4.1 lists the genes that have been located on the Z chromosome.

The transmission of sex-linked characters is next illustrated by the gene for barring, which was one of the earliest to be recognised (Spillman, 1908) and which accounts for the barring of the feathers in Plymouth Rocks, Scots Greys, Scots Dumpies, Barred and Cuckoo Leghorns, Dominiques and Coucou de Malines. The gene for barring (*B*) is sex-linked and dominant to non-barring (*b*). Compare the progenies obtained by crossing a barred cock (*BB*) with a non-barred hen (*b*) with those obtained from crossing a non-barred cock (*bb*) with a barred hen (*B*), and then self-crossing the $F_1$ generations.

| | | |
|---|---|---|
| Parents | $Z^BZ^B \times Z^bW$<br>barred cock, non-barred hen | $Z^bZ^b \times Z^BW$<br>non-barred cock, barred hen |
| Gametes | $Z^B$, $Z^B$ × $Z^b$, $W$ | $Z^b$, $Z^b$ × $Z^B$, $W$ |
| $F_1$ | $Z^BZ^b$ $Z^BW$<br>50% barred cock<br>50% barred hen | $Z^bZ^B$ $Z^bW$<br>50% barred cock<br>50% non-barred hen |
| Gametes | $Z^B$, $Z^b$ × $Z^B$, $W$ | $Z^b$, $Z^B$ × $Z^b$, $W$ |
| $F_2$ | $Z^BZ^B$, $Z^BZ^b$ $Z^BW$ $Z^bW$<br>50% barred cocks<br>25% barred hens<br>25% non-barred hens | $Z^BZ^b$, $Z^bZ^b$ $Z^BW$ $Z^bW$<br>25% barred cocks<br>25% non-barred cocks<br>25% barred hens<br>25% non-barred hens |

An important difference emerges when crossings involve sex-linked genes compared with crossings involving autosomal genes, namely, that in the former the reciprocal crosses give different results. The phenomenon of '**cris-cross**' inheritance is observed in the second example where the father and daughter have the same phenotype, as do the mother and son.

The patterns for sex-linked inheritance in the domestic fowl can be summarised as follows:

(a) *For sex-linked dominant genes*

1. Usually the more frequent phenotype in cocks than hens.
2. Phenotype found in all male offspring when the hen shows the trait.
3. Phenotype fails to be transmitted to any daughter from a hen which lacked the trait.

(b) *For sex-linked recessive genes*

1. Usually the more frequent phenotype in hens than cocks.
2. Phenotype fails to appear in the son unless present in the hen parent.
3. Phenotype appears in hen and daughter only if cock is heterozygous (skip generation inheritance).

## 4.4 Sex-linked genes and gene dosage in the domestic fowl

It should be clear that, for any gene located on the non-homologous region of the Z chromosome, it is possible to have different 'doses' of a gene in the cock and hen. In the cock it is possible to have no copies, one copy or two copies, but in the hen no copies or only one copy is possible (assuming, of course that they have the normal chromosome complement). This phenomenon has been studied more extensively in mammals, where it is apparent that a **dosage compensation mechanism** operates. Dosage compensation ensures that where there are two or more X chromosomes present only one is expressed. Consistent with these observations, densely staining bodies known as **Barr bodies** have been observed in the nucleus. The number of these bodies per nucleus is one less than the number of X chromosomes present. Thus a normal female has a single Barr body and a normal male no Barr bodies. It is generally assumed that the Barr body is the inactive sex chromosome condensed in such a way that at least only part of it can be expressed.

The situation in the domestic fowl and other birds suggests that a similar mechanism does not exist. Cock (1964) reviewed the evidence and concluded that there is no gene dose compensation mechanism in the domestic fowl. The evidence comes from the sex-linked barring gene (*B*) and the slow feathering gene (*K*). The barred hen (*B*) and the heterozygous barred cock (*Bb*) resemble each other in the banding pattern of their feathers, in which the alternate black and white barring patterns are of approximately equal width. The barring gene has the effect of interrupting the deposition of the black melanin pigment, thus a homozygous barred cock (*BB*) has a white band approximately twice the width of the dark band. Thus it is clear that twice the gene dosage has twice the effect on the barring pattern. The importance of the barring pattern is well known to breeders of barred varieties of domestic fowl. The British Poultry Standards stipulate that both cock and hen should have equal dark and light band widths. This is generally achieved by using a hemizygous hen (*B*) and a heterozygous cock (*Bb*). Breeding from these yields 25% *BB* cocks which have too high a proportion of light banding for showing, 25% *Bb* cocks, 25% *B* normal barred hens and 25% *b* non-barred hens. A more detailed account of the practical aspects of barring is given by Carefoot (1985).

More recently Baverstock *et al.* (1982) have obtained evidence that the enzyme aconitase is sex-linked on the Z chromosome and that its concentration in the liver is related to gene dosage. They found that it is possible to distinguish the enzyme aconitase from the domestic fowl ($A_D$) and from the guinea fowl ($A_G$) by their electrophoretic mobilities. They hybridised the domestic fowl and the guinea fowl by artificial insemination and examined the electrophoretic mobilities of aconitase from the surviving hybrids. If the aconitase genes were autosomal then the hybrids of both sexes would be expected to be $A_DA_G$; if, on the other hand, they were located on the Z chromosome the surviving male offspring would be $A_DA_G$, but the female survivors would be either $A_D$ or $A_G$ depending which way the hybridisation was made. The results they obtained supported the second alternative, and they also showed that the amounts of aconitase activity detected in the liver of males were significantly higher than in females. Thus the evidence to date supports the hypothesis that there is no dose compensation in the species of birds so far examined.

A comparison of sex-linked genes in mammals with those in birds suggests that the XY and ZW sex chromosomes have evolved separately (see Bitgood & Shoffner, 1990). Had both X and Z evolved from the same ancestral chromosome, one might expect to find that they contained many of the same sex-linked genes. The genes for the enzymes glucose 6-phosphate dehydrogenase, ornithine transcarbamylase, and one form of phosphoglycerate kinase are sex-linked in mammals, but autosomal in the domestic fowl (Bhatnagar, 1969; Cam & Cooper, 1978). Evidence from a larger number of genes, or from DNA sequence studies, is needed to substantiate this hypothesis.

## 4.5 Sex-limited characters in the domestic fowl

There are certain characters for which the genes are present on the autosomes, but which are expressed only in one sex, in which case the penetrance (see Chapter 3, section 3.7) in the other sex is zero. The lack of expression in one sex is usually brought about as a result of the different concentrations of sex hormones in the two sexes. The different potential for hormone production must ultimately relate back to some genetic mechanism which is triggered by the sex-linked genes. The different hormonal concentrations in the two sexes lead to differences in anatomical development. Two examples of sex-limited genes are those for blue egg and for hen feathering. The blue egg gene is located on autosome 1. Although a cock carrying the blue egg gene is able to transmit this to its offspring, it is never, of course, expressed in the cock or its male progeny. The gene for hen feathering is an autosomal dominant (*Hf*) and its recessive counterpart *hf* leads to cock feathering, but only in the male, as indicated below.

| **Genotype** | **Male phenotype** | **Female phenotype** |
|---|---|---|
| *HfHf* | Hen feathering | Hen feathering |
| *Hfhf* | Hen feathering | Hen feathering |
| *hfhf* | Cock feathering | Hen feathering |

Thus where one would expect cock feathering in the double *hfhf* female it is in fact suppressed since it is a sex-limited gene. The genetics of hen feathering is discussed in more detail in Chapter 6, section 6.3e.

Recently a sex-limited plasma protein related to the complement proteins has been discovered (Wathen *et al.*, 1987). The complement system includes a family of about 20 proteins which are part of the immune system (for details, see Williamson & Turner, 1987; Law & Reid, 1988). Plasma samples from more than 300 inbred domestic fowl were separated by electrophoresis, and then allowed to react to antibody raised against one of the human complement proteins (C4). Two precipitin lines were observed in the samples of plasma from cocks and from immature hens, but in the samples from laying hens a third precipitin line was observed. This suggests that mature hens produce an additional plasma protein structurally similar to the C4 complement protein. This protein is therefore a sex-limited protein; its physiological role is, at present, unknown.

## 4.6 Applications of sex-linkage

The main practical application which makes use of the difference in expression of sex-linked genes is that of sexing one-day-old chicks. Since the advent of sexing by vent examination of one-day chicks, both methods have been widely used. The use of sex-linked markers is less expensive, more convenient and more accurate (Smyth, 1990). The two sex-linked markers most used are those for barring (*B* and $b^+$) and silver (*S* and $s^+$). The barring marker was particularly used for brown egg production during the 1930s–1950s, and is still used in parts of the world today, but the most popular is now the silver marker. It is important for economic production to be rid of excess cocks as soon as possible. The method adopted for many years was to cross a breed having a sex-linked dominant gene with one having the sex-linked recessive gene. In America the most popular cross was that between the Barred Plymouth Rock female (*B*) and the Rhode Island Red male (*bb*) utilising the barring genes. In this the $F_1$ progeny were *Bb* males and *b* females, so that the males were barred, but the females were not. Although the barring pattern is not clear until the downy feathers have been replaced, there is a distinction recognisable on hatching. The *Bb* male chicks have a whitish spot on the occipital region, but this is lacking in the *b* females. In Britain it was more popular to cross the Light Sussex female (*S*) with the Rhode Island Red male (*ss*), this time utilising the silver (*S*) and gold (*s*) sex-linked alleles. The *Ss* male chicks can be distinguished from the *s* female chicks on hatching from the silver and gold plumage, respectively.

The utilisation of the sex-linked genes to sex chicks on hatching had the disadvantage that it meant keeping two separate breeds from which to breed. This led Punnett & Pease (1928) to develop autosexing breeds, that is, breeds in which the male and female chicks of the same breed could be distinguished at hatching. The distinction between males and females at hatching is made easier in chicks of autosexing breeds by having down lighter in colour than in Barred Plymouth Rocks. Punnett and Pease introduced the sex-linked barring gene into a breed having a lighter colour, namely the Golden Campine. They crossed Barred Plymouth Rocks with Golden Campines and then mated the barred progeny with the original Golden Campine. By repeating the procedure for several generations, that is, backcrossing the barred progeny with Golden Campines and then eventually mating the barred male and female progenies, they obtained males having *BB* and *Bb* genes and females having *B* and *b* genes. The homozygous *BB* males and the hemizygous *B* females, if mated, bred true and the chicks could be reproducibly sexed at hatching. In effect what they did was to alter the Golden Campine so that it contained the sex-linked barring gene. Punnett and Pease developed a number of these autosexing breeds, collectively known as Cambridge breeds, but they have become only a curiosity for poultry fanciers. The reasons are: (i) that the development of sexing by vent examination has been much improved, and (ii) that many hybrid strains of domestic fowl have been developed whose egg-laying capacity and growth efficiency far outstrips that of the 'pure breeds' or autosexing breeds.

The sex-linked $S/s^+$ alleles mentioned previously are now most popular for brown egg layers, used together with the Columbian-like restrictor (*Co*) and the dominant white (*I*). In combination with these two autosomal genes (*Co* and *I*) the silver and gold phenotypes are most easily distinguished.

Another sex-linked recessive gene which is being explored for commercial poultry breeding is the dwarfing gene (*dw*), and there is now extensive literature on this genotype (see Merat, 1984, 1990). The aim is to use small hens with which to produce chicks that will grow rapidly and produce large broilers. The dwarf parent hens are more economical on food requirements.

Table 4.2. *Effects of the dwarfing gene* (dw) *in females of laying or broiler type stocks. Data summarised from Merat (1990), and expressed as % deviation from normal females* (Dw+)

| Trait | Light layers | Medium-sized layers | Broiler type |
|---|---|---|---|
| Body weight | −32 | −31 | −31 |
| Egg number | −14 | −7 | +3 |
| Mean egg weight | −8 | −8 | −3 |
| Feed efficiency | +3 | +13 | +37 |

The sex-linked dwarfing gene was first noted by Hutt in 1959. It is a recessive gene (*dw*) whereas the alternative allele *Dw* is non-dwarfing. If a homozygous non-dwarfed cock (*DwDw*) is mated with several dwarf hens (*dw*), the male (*Dwdw*) and female (*Dw*) progeny will be non-dwarfed. The dwarf cocks have body weights reduced by about 40% and the dwarf hens by about 30%; the latter require about 20% less room in the hen house than the corresponding non-dwarf phenotype. Two other alleles, $dw^M$ and $dw^B$, have been reported, but they cause a less pronounced reduction in body size (see Somes, 1990).

The effect of the dwarfing gene on a variety of genetic backgrounds has been studied and a summary of the results is given in Table 4.2 (Merat, 1990). The advantages as an egg layer are better feed conversion, eggs having an increased proportion of polyunsaturated fatty acids, and less space required per laying hen. The disadvantages are lower mean egg weight and lower egg number (Merat, 1990). Measurements of the rate of protein turnover in muscle show that the *dw* gene has a depressing effect on the synthesis of muscle protein, but it does not appear to affect the protein degradation (Jones, Judge & Aberle, 1986; Maeda *et al.*, 1987). The concentrations of triiodothyronine and insulin-like growth factor I circulating in the plasma are lowered and thyroxine slightly raised compared with those in normal birds (Scanes *et al.*, 1984; Decuypere *et al.*, 1986), and these are important hormones regulating growth. Further research in this area is being aimed at optimising the advantages of strains having the *dw* gene.

## References

Abdel-Hameed, M. F. & Shoffner, R. N. (1971). Intersexes and sex determination in chickens. *Science*, **172**, 962–4.

Bateson, W. & Punnett, R. C. (1908). Experimental studies in the physiology of heredity. *Poultry Reports to the Evolutionary Committee of the Royal Society*, **4**, 18–35.

Baverstock, P. R., Adams, M., Polkinghorne, R. W. & Gelder, M. (1982). A sex-linked enzyme in birds – Z chromosome conservation but no dosage compensation. *Nature*, **296**, 763–6.

Bhatnager, M. K. (1969). Autosomal determination of erythrocyte glucose 6-phosphate dehydrogenase in domestic chickens and ring-necked pheasants. *Biochemical Genetics*, **3**, 85–90.

Bitgood, J. J. & Shoffner, R. N. (1990). Cytology and cytogenetics. In *Poultry Breeding and Genetics*, ed. R.N. Crawford, pp. 401–27. Amsterdam: Elsevier.

Cam, A. E. & Cooper, D. W. (1978). Autosomal inheritance of phosphoglycerate kinase in the domestic chicken (*Gallus domesticus*). *Biochemical Genetics*, **16**, 261–70.

Carefoot, W. C. (1985). *Creative Poultry Breeding*. Cliveden, Sandy Bank, Chipping, Preston, UK, published privately.

Cock, A. G. (1964). Dosage compensation and sex-chromatin in non-mammals. *Genetical Research*, **5**, 354–65.

Crew, F. A. E. (1933). A case of non-disjunction in the fowl. *Proceedings of the Royal Society of Edinburgh*, **53**, 89–104.

Decuypere, E., Rudas, P., Huybrechts, L., Mongin, P. & Kuhn, E. R. (1986). Endocrinological effects of the dwarf gene. II. Effect on tissue monodeiodination activity. In *Proceedings of the 7th European Poultry Conference (Paris)*, **2**, 955–9.

Eichwald, E. J. & Silmser, C. R. (1955). (Untitled communication). *Transplantation Bulletin*, **2**, 148–9.

Gilmour, D. G. (1967). Histocompatibility antigen in the heterogametic sex in the chicken. *Transplantation*, **5**, 699–706.

Goodfellow, P. N. (1986). The case of the missing H-Y antigen. *Trends in Genetics*, **2**, 87.

Hutt, F. B. (1949). *Genetics of the Fowl*. New York: McGraw-Hill.

Hutt, F. B. (1959). Sex-linked dwarfism in the fowl. *Journal of Heredity*, **50**, 209–21.

Hutt, F. B. & Rasmusen, B. A. (1982). *Animal Genetics*, 2nd edn, pp. 131–155. New York: John Wiley & Sons.

Jones, S. J., Judge, M. D. & Aberle, E. D. (1986). Muscle protein turnover in sex-linked dwarf and normal broiler chickens. *Poultry Science*, **65**, 2082–9.

Jones, K. W. & Singh, L. (1982). Conserved sex-associated repeated DNA sequences in vertebrates. In *Genome Evolution*, ed. G. A. Dover & R. B. Flavell, pp.135–54. London: Academic Press.

Lancaster, F. M. (1972). Some early records of sex-linked inheritance in fowl. *Journal of Heredity*, **63**, 223–4.

Law, S. K. A. & Reid, K. B. M. (1988). *Complement*. Oxford: IRL Press.

Maeda, Y., Matsuoka, S., Furuichi, N., Hayashi, K. & Hashiguchi, T. (1987). The effect of the *dw* gene on the muscle protein turnover rate in chickens. *Biochemical Genetics*, **25**, 253–8.

McCarrey, J. R. & Abbott, U. K. (1979). Mechanisms of genetic sex determination, gonadal sex differentiation, and germ cell development in animals. *Advances in Genetics*, **20**, 217–90.

McCarrey, J. B., Abplanalp, H. & Abbott, U. K. (1981). Studies in the H-W (H-Y) antigen of chickens. *Journal of Heredity*, **72**, 169–71.

Merat, P. (1984). The sex-linked dwarf gene in the broiler chicken industry. *World's Poultry Science Journal*, **40**, 10–8.

Merat, P. (1990). Pleiotropic and associated effects of major genes. In *Poultry Breeding and Genetics*, ed. R.D. Crawford, pp. 429–67. Amsterdam: Elsevier.

Mittwoch, U. (1983). Heterogametic sex chromosomes and the development of the dominant gonad in vertebrates. *American Naturalist*, **122**, 159–80.

Morozit, D. C., Bednarz, J. C. & Ferrell, R. E. (1987). Sex linkage of muscle creatine kinase in Harris' hawks. *Cytogenetics and Cell Genetics*, **44**, 89–91.

Muller, U., Zenzes, M. T., Wolf, U., Engel, W. & Weniger, J. P. (1979). Appearance of H-W (H-Y) antigen in the gonads of oestradiol sex-reversed male chicken embryos. *Nature*, **280**, 141–3.

Ohno, S. (1961). Sex chromosomes and microchromosomes of *Gallus domesticus. Chromosoma (Berlin)*, **11**, 484–98.

Punnett, R. C. & Pease, M. (1928). Genetic studies in poultry. VI. The gold Barred Rock. *Journal of Genetics*, **19**, 337–50.

Scanes, C. G., Harvey, S., Marsh, J. A. & King, D. B. (1984). Hormones and growth in poultry. *Poultry Science*, **63**, 2062–74.

Schmidt, W. (1962). DNA replication patterns of the heterochromosomes in *Gallus domesticus. Cytogenetics*, **1**, 344–52.

Sittman, K. (1984). Sex determination in birds: progeny of non-disjunction canaries of Durham (1926). *Genetical Research*, **43**, 173–80.

Smyth, J. R. (1990). Genetics of plumage, skin and eye pigmentation in chickens. In *Poultry Breeding and Genetics*, ed. R. D. Crawford, pp. 109–67. Amsterdam: Elsevier.

Somes, R. G. (1990). Mutations and major variants of muscles and skeleton in chickens. In *Poultry Breeding and Genetics*, ed. R. D. Crawford, pp. 209–237. Amsterdam: Elsevier.

Spillman, W. J. (1908). Spurious allelomorphism: results of recent investigations. *American Naturalist*, **42**, 610–5.

Tone, M., Sakaki, Y., Hashiguchi, T. & Mizuno, S. (1984). Genus specificity and extensive methylation of the W chromosome-specific repetitive DNA sequences from the domestic fowl, *Gallus gallus domesticus. Chromosoma (Berlin)*, **89**, 228–37.

Wachtel, S. S., Koo, G. C. & Boyse, E. A. (1975). Evolutionary conservation of H-Y ('male') antigen. *Nature*, **254**, 270–2.

Wathen, L. K., Leblanc, D., Warner, C. M. Lamont, S. J. & Nordskog, A. W. (1987). A chicken sex-limited protein that crossreacts with the fourth component of complement. *Poultry Science*, **66**, 162–5.

White, M. J. D. (1973). *Animal Cytology and Evolution*, 3rd edn. Cambridge: Cambridge University Press.

Williamson, A. R. & Turner, M. W. (1987). *Essential Immunogenetics*, pp. 69–94. Oxford: Blackwell Scientific Publications.

# 5

# *Linkage and chromosome mapping*

## 5.1 The stages in genetic analysis

In the previous chapter sex-linked inheritance was discussed. In sex-linked inheritance the way in which certain characters are transmitted and expressed depends on the sex of the parents and offspring. The mechanisms involved depend on the genes for these characters being carried on one of the sex chromosomes, usually on the Z chromosome in the fowl since it is much larger than the W chromosome. In general, characters will show linked inheritance if the genes responsible are on the same chromosome, whether on the autosomes or sex chromosomes. The closer the genes are on the chromosome the stronger is the linkage. **Linkage** between two genes is a measure of the probability of them being transmitted together to offspring. An important goal of genetic analysis is to determine the positions of all the genes on the chromosomes and how their transmission and expression are controlled. This heightened understanding would enable genetic predictions to be made more accurately.

The process can be divided into a number of stages. (i) The first is to establish that a character is a genetically inherited character, and that it is inherited in Mendelian fashion. Some characters may be environmentally controlled. Others may be polygenetic, i.e. controlled by several genes (these are considered further in Chapter 9) and although each gene follows the normal Mendelian pattern, the overall analysis is rather complex. (ii) The next step is to determine whether the inheritance of a particular character is linked with that of other genetic characters or whether the characters are inherited independently. If they are linked this indicates that the genes are located on the same chromosome, although which chromsome that is will not generally be known at this stage. Analysis for linkage is then extended from two sets of characters to several sets of characters. Gradually a number of linkage groups is built up, each containing the characters whose genes show linked inheritance. Eventually the number of linkage groups found should coincide with the haploid number of chromosomes in that species. For genes that are found to be linked the **recombination frequencies** are determined, and from these it is possible to determine how far they are separated on the chromosome (see section 5.2).

The next problem is then to establish which chromosome corresponds to a particular linkage group; this can be quite difficult in most cases. Linkage groups have now been identified with chromosomes 1, Z, 6, 7, and 17, and in addition certain genes have been assigned to chromosomes 2 and W (Bitgood & Somes, 1990). Identifying the sex chromosomes is usually straightforward since in the heterogametic sex the Z and W (or X and Y in the case of mammals) are very different in size and appearance. It is also straightforward to distinguish a sex-linked character from its mode of transmission. With the autosomes the problem is more difficult and often depends on studying chromosomal abnormalities. Where an abnormal chromosome is found during karyotyping, it is often possible to relate this to the abnormal inheritance of certain characters. Theoretically it should be possible eventually to determine the positions of all the genes in a genome, a goal that has not yet been achieved except in the case of certain viruses. In addition to this there are also the more detailed molecular aspects of the structure and organisation of genes to be resolved. Some of these are considered in the case of the fowl in Chapter 12. In the next sections of this chapter some of the stages in chromosome mapping are considered, with examples from fowl genetics. A chromosome map for the domestic fowl is given in Appendix I.

## 5.2 Recombinant frequencies, two point and three point test crosses

The mechanism by which recombination of chromosomes occurs during meiosis has been described in Chapter 2, section 2.3. The frequency with which recombination occurs between two genes can be used to estimate the distance apart of two genes on a chromosome. An example will now be given to illustrate the method of locating the position of a new gene relative to other already mapped genes on what is described as a **linkage group**. The example is a genetic defect called 'sleepy eye' described by Somes (1968). It was first discovered in 1957. Its characteristic is that the lower eyelid is partially closed at the time of hatching and remains so for the rest of the bird's life. The primary effect of the defect is not lethal, but it affects the bird's field of vision which makes it difficult for it to see its food, and so the mortality rate is higher than in normal chicks. If the lower eyelid is sutured in the open position then the mortality rate is not significantly higher than normal. The first step is to show that this defect is due to a single gene and that it is inherited in normal Mendelian fashion. The sleepy-eyed birds were found to breed true, and when crossed with normal-eyed birds all the $F_1$ generation were normal-eyed, and the $F_2$ generation gave a 3:1 ratio of normal to sleepy-eyed. When the $F_1$ was backcrossed with sleepy-eyed birds then a 1:1 ratio of normal to sleepy-eyed birds was obtained. Taken together this shows that the sleepy-eyed trait is the result of a single autosomal recessive mutation.

The next stage is to test whether the sleepy-eyed gene (*se* is the abbreviation used with 's' in the lower case to indicate recessive) is linked to another well-characterised gene in the fowl. To do this a **two point test cross** is performed. For convenience the other gene will be designated '*A*' and '*a*' for the corresponding dominant and recessive alleles. The sleepy-eyed birds (*se, a*/*se, a*), homozygous at both loci, are first crossed with birds containing the homozygous dominant *A* gene (*Se, A*/*Se, A*). The $F_1$ generation is then backcrossed with the double recessive (*se, a*/*se, a*) and the progeny examined. If gene *se* and gene *A* are on separate chromosomes then they will assort independently and a 1:1 ratio of parental types (*se, a* and *Se, A*) to recombinant types (*se, A* and *Se, a*) would be expected. If, however, the two genes are on the same chromosome then the linkage which was present in the original parents will remain intact unless crossing over occurs during meiosis, and there will thus be more of the parental types than recombinants. The closer the two genes are together, the less likely that a chiasma will have formed between them during meiosis, and thus the fewer recombinants there will be. Somes (1968) made a number of two point test crosses using different marker genes; an extract from the data is given in Table 5.1.

The data for the crosses with the dominant white, creeper, duplex comb, extended black and rose comb are sufficiently close to 1:1 that they suggest independent assortment of the genes. Since the segregation is a random event, the values will not necessarily be exactly 1:1. The larger the sample, the closer to unity the ratio would be expected to be. In order to test that the deviation is not significantly different from 1:1 a Chi squared test has to be performed (details are given in Appendix III). The data from the pea comb and the white skin crosses are, however, significantly different from 1:1, thus indicating linkage. To ascertain how close the linkage is, the percentage of recombinants is calculated: for the pea comb cross it equals $(872 \times 100)/(872+1063)=45.06\%$, and for the white skin it equals $(240 \times 100)/(240+314)=43.3\%$. The percentage of recombinants gives a quantitative measure of the distance separating the genes concerned. For convenience in drawing a genetic map it is useful to have a scale; that chosen is to define **one map unit (mu)** as the distance of separation which would give 1% recombination frequency.

Although this has demonstrated, in essence, how the distance separating two genes may be determined, there are complications, and it is not possible precisely to relate recombinant frequencies directly to the physical distance separating two genes. Complications arise because crossing over may occur at more than one location between the two genes in question, and these will not be recognised if only two pairs of genes are being studied. It can be seen from Fig. 5.1 that a double crossing over between two loci *a*/*A* and *b*/*B* appears the same as if no crossing had occurred, and that a triple crossing over appears the same as a single crossing over, and so on for odd and even numbers of crossovers. The frequency of multiple crossovers increases the further apart the two loci are. As a rough guide it is reasonable to assume that a double crossover is unlikely if the two genes in question are less than 5 mu apart. The observed recombinant frequency can never be more than 50%. Thus if two genes are located at the extreme ends of a chromosome the recombinant frequency would be approaching 50%. If, on the other hand, a series of loci is mapped throughout the length of a chromosome, the sum of

**Fig. 5.1** Single, double and triple crossing over. (N.B. The single and triple crossovers appear as recombinants between A/a and B/b, whereas the double does not.)

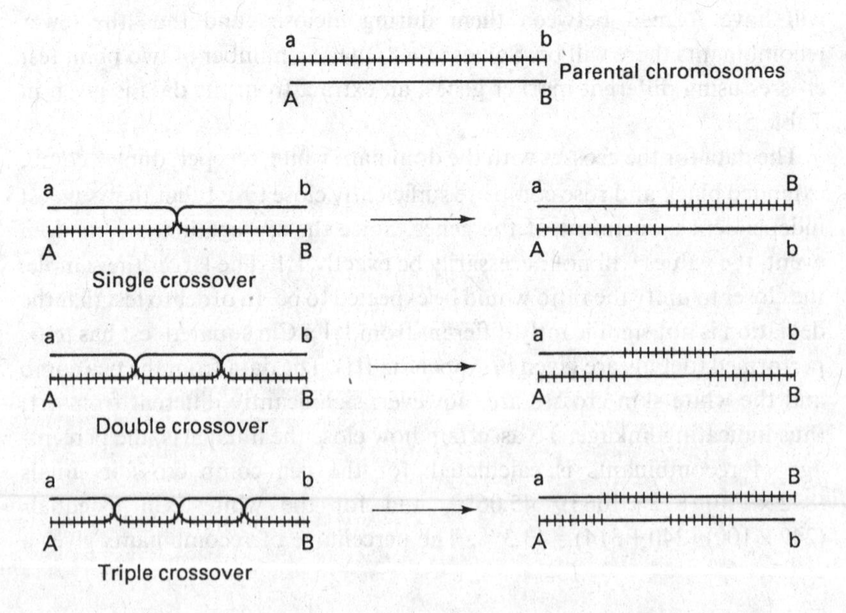

Table 5.1. *Tests for the linkage of sleepy-eye with other traits*

| Marker gene | Number of parental phenotypes | Number of recombinant phenotypes |
|---|---|---|
| Dominant white (*I*) | 614 | 616 |
| Creeper (*Cr*) | 187 | 198 |
| Duplex comb (*D*) | 194 | 211 |
| Extended black (*E*) | 354 | 352 |
| Rose comb (*R*) | 89 | 89 |
| Pea comb (*P*) | 1063 | 872 |
| White skin (*w*) | 314 | 240 |

From Somes, 1968.

these distances could be much greater than 50 mu, and would depend on the physical size of the chromosome.

This relationship is illustrated in Fig. 5.2. Thus it can be seen that a linear relationship holds well over short distances but becomes increasingly inaccurate at larger distances. A second factor affecting the relationship is that crossing over events may not occur with equal frequency in all parts of the chromosome.

From analysis of the data in Table 5.1 we can say that the sleepy-eyed gene is located in the same linkage group (and hence chromosome) as the genes for pea comb and white skin and the distances of separation are 45

**Fig. 5.2** Relationship between recombination (%) and linkage map distance (%). Linkage map distance is obtained by summation of recombination frequencies between a series of linked loci.

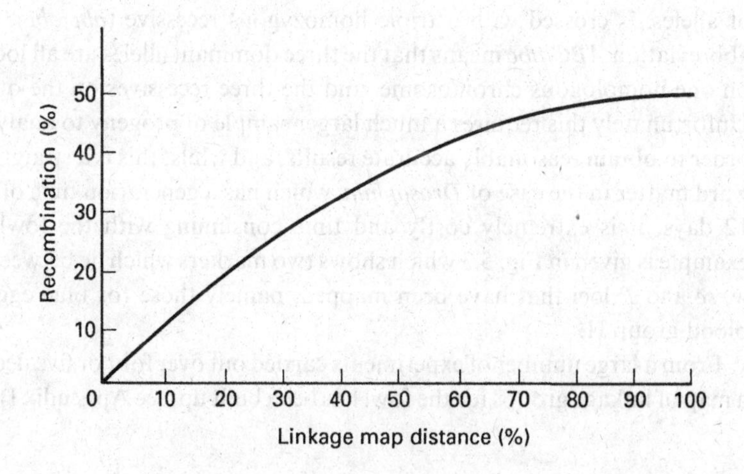

and 43 mu respectively. Both these distances may be underestimates if double crossing over events occur between the gene loci. We can also say that these three genes are on different linkage groups from all the other genes given in the table.

Another question arises: what is the distance separating the P gene and the W gene ? The arrangement could be either of those shown below.

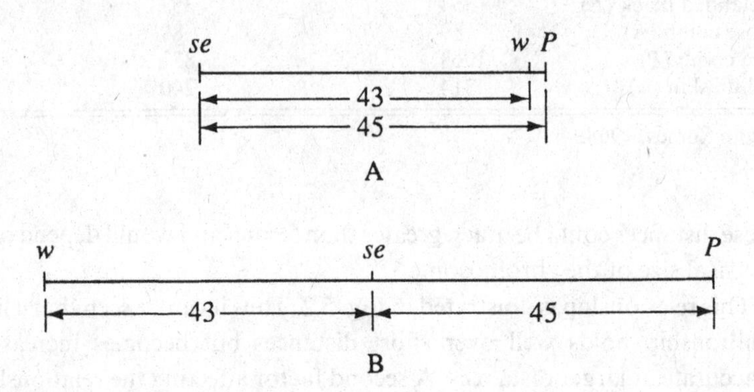


In (A) the two genes would be very closely linked and with the appropriate test cross one would expect only 2% recombinants (45–43%), whereas in (B) the two genes are so far apart that they would segregate almost independently and it would be difficult to demonstrate any linkage. The latter situation in fact occurs.

When two genes are widely separated on a chromosome the mapping distance can be more accurately determined by choosing a third marker located between the two genes and carrying out a **three point test cross**. In this, a trihybrid of the type *ABC*/*abc* where *A*,*a*; *B*,*b* and *C*,*c* are three pairs of alleles, is crossed with a triple homozygous recessive (*abc*/*abc*). (The abbreviation *ABC*/*abc* means that the three dominant alleles are all located on one homologous chromosome, and the three recessives on the other.) Unfortunately this requires a much larger sample of progeny to analyse in order to obtain reasonably accurate results, and whilst this is a straightforward matter in the case of *Drosophila*, which has a generation time of only 12 days, it is extremely costly and time consuming with the fowl. An example is given in Fig. 5.3 which shows two markers which lie between the *w*, *se* and *P* loci that have been mapped, namely those for blue egg and blood group H.

From a large number of experiments carried out over four or five decades a map of linkage groups for the fowl has been built up (see Appendix I). The process has been painfully slow, so much so that Hutt (1949) reckoned a decade per research worker per locus was the going rate, the rate limiting step being the generation time of the fowl. Not only does it require one year per generation, but for some genes the phenotype is not clear until the progeny matures, e.g. blue egg. There are new methods that are now being applied which will dramatically speed up genetic analysis; these are discussed in sections 5.4 and 5.5 of this chapter.

## 5.3 Relating linkage groups to individual chromosomes

The next stage in analysis is to try to relate the different linkage groups to chromosomes which can be cytologically identified and to locate genes at precise positions on the arms of particular chromosomes. Most detail is available for chromosome 1 and the Z chromosome, although a mutation termed shankless has been assigned to chromosome 2 (Langhorst & Fechheimer, 1985) and some genes have been assigned to microchromosomes, e.g. the 18S and 28S RNA and the major histocompatibility (B) complex have been located on an acrocentric microchromosome (Bloom & Bacon, 1985). The method which has been applied in the case of chromosome 1 has been to use chromosomal inversions and translocations (Bitgood *et al.*, 1980) as described below.

The method is analogous to determining the recombinant frequency of two phenotypic characters, but differs in that it involves determining the recombinant frequency between a phenotypic character and a chromosomal inversion or translocation. If the character in question is on a different chromosome from the inversion or translocation then there will be independent assortment giving a recombinant frequency of 50%, but if both are on the same chromosome then linkage is expected and a recombinant frequency of less than 50% will result. This means that the chromosome pattern or karyotype, usually the G-banding pattern (Wang & Shoffner, 1974), has to be examined in the progeny. This can easily be done by making a microscope preparation of the feather pulp, embryos or domestic fowl cells grown in culture (MacGregor & Varley, 1988; Bitgood & Shoffner,

**Fig. 5.3** Part of the chromosome 1 Map (Linkage Group III); *w*, white skin; *EaH*, erythrocyte alloantigen H; *se*, sleepy eye; *O*, blue egg; *P*, pea comb. The complete chromosome map is given in Appendix 1.

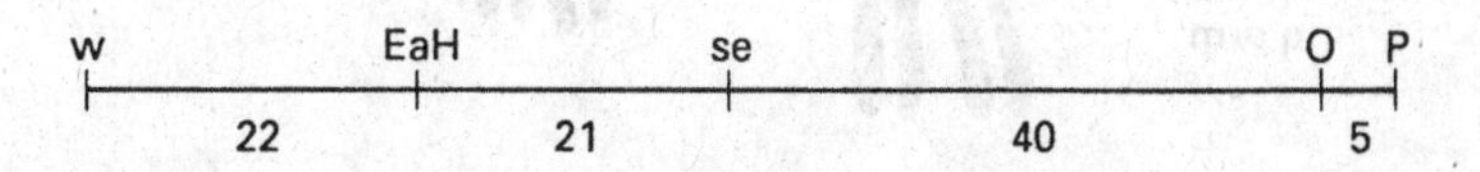

1990). Bitgood *et al.* (1980) used five strains of fowl, each having rearrangements in chromosome 1, to establish that the linkage group containing the genes for pea comb and blue egg was chromosome 1. One of the strains they used contained a translocation between chromosome 1 and the Z chromosome, t(Z;1)(p+;p−) as shown in Fig. 5.4. (The following conventions are used to describe chromosome rearrangements: (i) the type of rearrangement is first indicated; t = translocation, inv = inversion, etc., (ii) this is followed by an indication of which chromosome(s) are involved, and these are given in parentheses, e.g. Z and 1, and (iii) the arms of the chromosomes involved are indicated, e.g. p+ = short arm increase, q− = long arm decrease.)

The strain containing the translocation t(Z;1)($p^+$;$p^-$)was crossed with a strain having a normal chromosome arrangement, and having the gene to be mapped (see the example below). The $F_1$ cross was thus heterozygous both for the chromosomal rearrangement and for the alleles about to be mapped. This $F_1$ generation was then backcrossed to stock having the recessive phenotype and normal chromosomal arrangement. This gives rise to two parental types and two recombinant types in the $F_2$ generation. The recombinant frequency can then be used to map the distance between the point of translocation and the gene responsible for the phenotype.

The example now given illustrates this type of mapping in the case of the gene for pea comb (*P*). The symbols 'N' and 'T' are used for normal and translocated chromosome arrangement, and the data are those of Bitgood *et al.* (1980). N.B. The italicised symbol '$p^+$' refers to the recessive allele for normal comb (single) which is the wild type, and should not be confused with the bold '**p**$^+$' which indicates a chromosome transfer, such as a translocation on to the p arm of a chromosome.

**Fig. 5.4** Diagram showing translocation between chromosome 1 and the Z chromosome, t(Z;1)($p^+p^-$). In each case a pair of normal chromatids and a pair of translocated chromatids are shown for chromosomes 1 and Z.

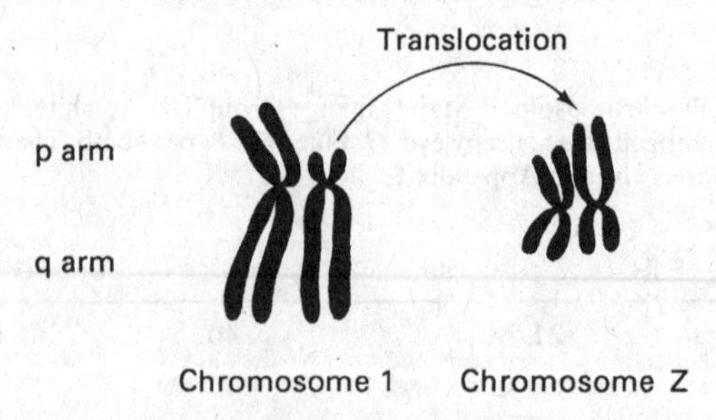

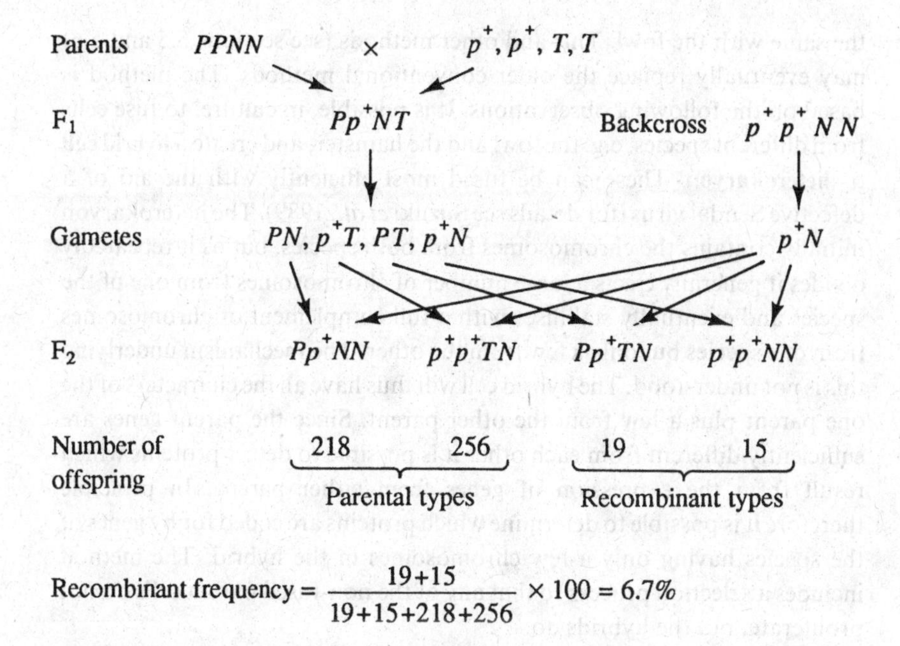


Therefore the translocation is 6.7 mu from the pea comb locus on chromosome 1. Similar experiments showed that the blue egg gene (*O*) is very close to the *P* locus, approximately 0.6 mu away. Taken together with the morphology of the chromosomes, Bitgood *et al.* (1980) concluded that both genes are on the proximal one third of the short arm of chromosome 1.

Using a further set of translocations on the Z chromosome Bitgood (1985) was able to show that the sex-linked genes, barring (*B*), recessive white skin (*y*) and dermal melanin inhibitor ($id^+$), are located on the long arm of the Z chromosome, whereas the silver gene (*S*) is located on the short arm. Langhorst & Fechheimer (1985) have shown that a mutation characterised by lack of shanks (shankless, *shl*) is invariably associated with a pericentric inversion on chromosome 2, thus it is possible to assign this gene to chromosome 2. (A pericentric inversion is one that includes the centromere in the inverted fragment.) Chromosomal abnormalities have been found for all the macrochromosomes except 6 (Wooster, Fechheimer & Jaap, 1977), so it should be possible in principle to determine on which chromosome the other linkage groups belong.

## 5.4 Chromosomal mapping by using somatic cell hybridization

Somatic cell hybridisation is a newer method of genetic analysis which has made an enormous impact on human genetic analysis and promises to do the same with the fowl. This and other methods (see sections 5.5 and 5.6) may eventually replace the older conventional methods. The method is based on the following observations. It is possible, in culture, to fuse cells from different species, e.g. the fowl and the hamster, and create a **hybrid cell** or **heterokaryon**. These can be fused most efficiently with the aid of a defective Sendai virus (for details see Suzuki *et al.*, 1989). The heterokaryon initially contains the chromosomes from both species, but as it repeatedly divides it generally ejects a large number of chromosomes from one of the species and eventually stabilises with a full complement of chromosomes from one species but only a few from the other. The mechanism underlying this is not understood. The hybrid cell will thus have all the characters of the one parent plus a few from the other parent. Since the parent genes are sufficiently different from each other it is possible to detect proteins which result from the expression of genes from either parent. In principle therefore it is possible to determine which proteins are coded for by genes of the species having only a few chromosomes in the hybrid. The method includes a selection process so that any of the non-fused parent cells do not proliferate, but the hybrids do.

This method has been applied using chicken erythrocytes fused to mouse cell mutants by Leung *et al.* (1975) and chicken cells fused to Chinese hamster cells by Kao (1973) and by Palmer & Jones (1986). Leung *et al.* (1975) used the technique to determine the location of the gene for the enzyme thymidine kinase. They fused chicken erythrocytes with a thymidine kinase deficient mutant of a mouse cell line, and then selected for hybrids using a medium that would only allow cells having thymidine kinase present to grow and divide. The hybrid cells were shown to contain thymidine kinase which had the characteristics of the chicken. Examination of the karyotypes of the hybrids showed that they possessed the full complement of mouse cell chromosomes plus two or three microchromosomes from the chicken. The thymidine kinase gene is thus located on one of the microchromosomes. To confirm and further identify the microchromosome involved, a counter selection procedure was adopted. The hybrid cells were grown in a medium containing 5-bromodeoxyuridine. This is an analogue of thymidine and lethal to cells which *contain* thymidine kinase since the 5-bromodeoxyuridine becomes incorporated into DNA. The hybrid cells surviving this procedure were found to have lost a microchromosome and have no detectable chicken thymidine kinase. This confirms the location of thymidine kinase on one of the microchromosomes. The procedure can clearly be extended to any other detectable gene products that are located on the same microchromosome as thymidine kinase.

In an analogous set of experiments, Kao (1973) fused Chinese hamster mutant cells with chicken erythrocytes and followed similar selection procedures. It was then possible to show that genes concerned with the biosynthesis of adenine were located on chromosomes 7 and 8. In a more recent study Palmer & Jones (1986) have shown that phosphoglucomutase, serum albumin, vitamin D binding protein and phosphoribosyl pyrophosphate amidotransferase genes are located on chromosome 6. They were able to obtain somatic cell hybrids in which chromosome 6 was the only chromosome from the domestic fowl present in the hybrid. The presence of chicken phosphoglucomutase was demonstrated by cellulose acetate electrophoresis, since it had a different electrophoretic mobility from the hamster enzyme. The presence of the serum albumin gene was detected using radioactively labelled chicken serum albumin complementary DNA (cDNA) as a probe (see section 5.5). The vitamin D binding protein had previously been demonstrated to be linked to that for serum albumin using electrophoretic methods (Juneja *et al.*, 1982).

## 5.5 Chromosome mapping by the use of cDNA

A method which at present has not been widely applied to the domestic fowl, but which may have wider application in the future, is that of identifying a gene directly by use of a radioactively labelled gene copy. This method differs from those described in the previous section, in that it detects the gene sequence rather than the product of gene expression. Its application is limited to those genes where a complementary DNA (cDNA) probe is available. In the fowl it has been applied in the case of haemoglobin, certain egg proteins, β-actin, ribosomal RNA, avian leukosis viral genes and cellular oncogenes. Where a specialised cell produces a single protein or a few proteins in large amounts it is usually feasible to isolate the mRNA for those proteins, since there will generally be a much higher concentration of those particular mRNAs present. Once the mRNA has been isolated it can be copied into a radioactively labelled cDNA using an enzyme called reverse transcriptase. This group of enzymes is normally associated with RNA viruses since the latter become transcribed into DNA during infection. The radioactively labelled cDNA can then be used as a probe to identify the gene to which it is complementary. The process is known as **hybridisation**.

Two methods have been used to locate the genes on the appropriate chromosome. The first (Hughes *et al.*, 1979) is to fractionate the chromosomes by centrifugation on a sucrose density gradient. In this process the larger macrochromosomes sediment more rapidly than the smaller ones. Fractions of the gradient can then be sampled to see whether they hybridise with the labelled cDNA probe. By using this method of chromosome fractionation (Hughes *et al.*, 1979) it has been shown that $\alpha^A$ and $\alpha^D$ globin genes are on microchromosomes 10–15 whereas $\beta$, $\rho$ and $\epsilon$ globins are located on macrochromosome 1 or 2. Messenger RNAs for the egg-white proteins, ovalbumin, ovomucoid and transferrin are made in relative abundance after oestrogen stimulation of the oviduct and the appropriate cDNA probes thus can be generated. By using the chromosome fractionation method these genes were found to be located as follows: ovalbumin, chromosome 2 or 3; transferrin, chromosomes 9–12, and ovomucoid, chromosomes 10–15. A disadvantage of this method at present is that it does not permit complete separation of all of the chromosomes.

The three different ribosomal RNAs (5S, 18S and 28S) are transcribed on tandem copies of the genes. With these maximum hybridisation occurred in the region of the sucrose gradient containing chromosomes 10–15.

Alternatively, the cDNA may be used to hybridise with chromosomes *in situ*. The preparation is then stained to show up the chromosomes and by autoradiography the position of the radioactively labelled cDNA located. This has been done in the case of avian leukosis viral genes and cellular oncogenes (Tereba & Astrin, 1982), and more recently with $\beta$-actin (Shaw, Guise & Shoffner, 1989). The genes for actin constitute a multigene family of 6 to 11 genes. Actins serve a number of functions besides that of muscle contraction, and these include maintenance of cytoskeletal structure, cell motility and chromosome movement. Using a cDNA probe for $\beta$-actin Shaw *et al.* (1989) located the gene sequences on the long arm of chromosome 2 (2q) and on one of chromosomes 9–12.

## 5.6 DNA fingerprinting in the domestic fowl

A method of gene mapping which has evolved in the past few years depends on detecting changes in the molecular structure of the gene. This method, often referred to as DNA fingerprinting, was devised by Jeffreys, Wilson & Thein (1985) and was initially applied to the human genome, but has since been used on vertebrates, including the domestic fowl. It depends on making specific cuts in DNA using restriction endonucleases, separating the fragments by agarose gel electrophoresis, and then detecting specific fragments using radioactively labelled probes. The result of these procedures on a particular DNA is to give an almost unique pattern of radioactive labelling for the DNA from one individual, hence the term 'fingerprint'.

**Restriction endonucleases** each recognise a specific sequence of generally between 4 and 6 bases, and hence cut the DNA on average into about 250

($\simeq 4^4$) to 4000 ($\simeq 4^6$) base pairs in length. These fragments are separated on agarose gel electrophoresis on a size basis. The haploid DNA content of the human genome is about $2.4 \times 10^9$, and therefore if it were cut every 1000 base pairs would yield $2.4 \times 10^6$ fragments, which would be far too many to resolve on a gel. If, however, a probe consisting of an oligonucleotide containing radioactive label is used, then only fragments containing sequences complementary to the probe will be detected. The radioactive label ensures that sequences will be detected even if present only in minute amounts, using autoradiography. In this way the number of fragments detected is manageable.

The particular probes that Jeffreys (1987) used were tandem repeats of short DNA sequences, or minisatellites. These sequences were found to hybridise with a large number of different loci throughout the human genome. These minisatellites do not constitute a true family of sequences; specifically, they are not derived from one another. They are, however, related in the sense that they are based on very similar core sequences which are in the region of 15 bases long. The probes Jeffreys *et al.* (1985) initially used were isolated from an intron on the human myoglobin gene, and are referred to as 33.6 and 33.15. They comprise four repeats of a 33 base pair sequence. Each 33 base pair sequence contains the common core of 11–18 bases having the following sequence **GGAGGTGGGCAGGAG**, or minor variations on this sequence. An alternative probe (Vassart *et al.*, 1987) which has also been found useful is the repeat sequence found present in the gene coding for protein III in the bacteriophage M13. The consensus sequence, of which there many repeats in M13, is **GAGGGTGGXGGXTCT**. It will hybridise with many loci in vertebrate DNA but gives a pattern distinct from the 33.6 and 33.15 probes.

The number of repeats of these sequences is quite variable between alleles, and because certain alleles have more copies than others they can be distinguished by their size difference. The function of these repeat sequences is unknown, but it is possible they may be recombinant signals, that promote unequal exchanges between chromatids, and which have converged through evolution. Since there are a large number of loci which hybridise to these minisatellites, and because of their size variation, the pattern from one particular DNA after subjection to this protocol is almost unique. Jeffreys (1987) has calculated that the probability of two unrelated human DNAs coinciding is $5 \times 10^{-19}$. The method has been used in humans to determine whether same sex twins are identical or dizygotic, in establishing parentage and other family relationships, and also to locate defective genes in inherited diseases.

It has been found that sequences that hybridise with these human probes

(33.6 and 33.15) also occur widely in many vertebrates. This means that DNA fingerprinting can be carried out in other vertebrates using the human probes.

Two recent studies have been made on the domestic fowl (Hillel *et al.*, 1989; Kuhnlein *et al.*, 1989). Hillel *et al.* used the three probes, 33.6, 33.15 and M13 phage, and found the 33.6 most useful. They found that with the domestic fowl (broilers and layers), the Muscovy duck and turkeys they could resolve about 20 bands quite distinctly, and the fraction of these bands shared by genetically unrelated individuals ranged between 0.21 and 0.27 in the domestic fowl and turkey, but were somewhat higher in Muscovy ducks (0.64). From these data they deduced that the probability of obtaining two identical patterns of these bands was $1.15 \times 10^{-9}$ in ducks, and $8.7 \times 10^{-17}$ in broilers. Many of the minisatellites were larger than those obtained in humans. There was a low incidence of linkage, i.e. the bands to which the probes became attached showed little evidence of linkage. This suggests that the minisatellites in the domestic fowl are not tightly clustered, but are spread throughout the genome. This could be useful since it is possible that the minisatellites are present on most chromosomes and if found linked to other genes could be used as markers. They could also be used as markers in selection programmes aimed at selection either for or against a particular gene or character.

**Fig. 5.5** Genetic relationships among four strains of domestic fowl. Strains S and K (White Leghorns) were derived from a common base in 1936, S being selected for susceptibility to Marek's disease, K being selected for resistance to Marek's disease, high egg production and egg weight. Strain 7 (White Leghorn) has a broad genetic base and has been maintained since 1960 without selection. Strain NH (New Hampshire) is a different breed from the others and has been maintained as a closed population for more than 20 generations. The numbers indicate the genetic distance (Kuhnlein *et al.* 1989).

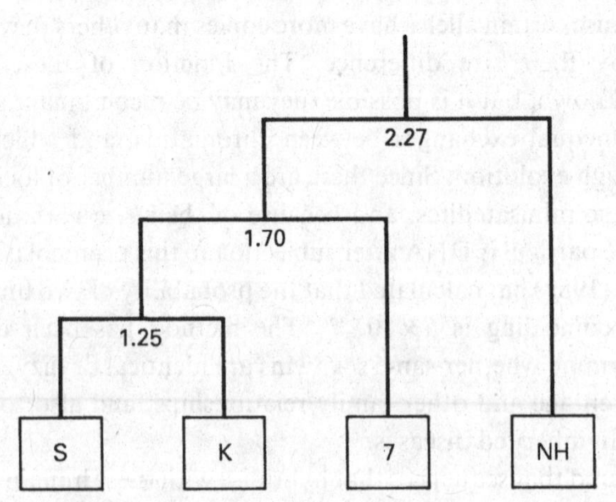

The work of Kuhnlein *et al.* (1989) has focused on inbred strains of the domestic fowl. They were able to show that the higher the degree of inbreeding, the fewer polymorphisms were detected by the DNA fingerprinting. They devised a method of computing an index of genetic distance from the frequencies of bands on the fingerprint pattern. The genetic distance is a measure of how closely related genetically two different groups are. They were able show that the known genetic relationships among four strains agreed well with the genetic distances computed from the fingerprints (Fig. 5.5). It is also possible to produce a calibration curve relating the degree of inbreeding to the average band frequency and allelic frequency (Kuhnlein *et al.*, 1990). This could then be used to assess the extent of inbreeding from analysis of the DNA fingerprints.

## References

Bitgood, J. J. (1985). Additional linkage relationships within the Z chromosome of the chicken. *Poultry Science*, **64**, 2234–8.

Bitgood, J. J. & Shoffner, R. N. (1990). Cytology and cytogenetics. In *Poultry Breeding and Genetics*, ed. R. D. Crawford, pp. 401–27. Amsterdam: Elsevier.

Bitgood, J. J., Shoffner, R. N., Otis, J. S. & Briles, W. E. (1980). Mapping of the genes for pea comb, blue egg, barring, silver, and blood groups A, E, H, and P in the domestic fowl. *Poultry Science*, **59**, 1686–93.

Bitgood, J. J. & Somes, R. G. (1990). Linkage relationships and gene mapping. In *Poultry Breeding and Genetics*, ed. R. D. Crawford, pp. 469–95. Amsterdam: Elsevier.

Bloom, S. E. & Bacon, L. D. (1985). Linkage of the major histocompatibility (B) complex and the nucleolar organizer in the chicken: Assignment to a microchromosome. *Journal of Heredity*, **76**, 146–54.

Hillel, J., Plotzy, Y., Haberfield, A., Lavi, U., Cahaner, A. & Jeffreys, A. J. (1989). DNA fingerprints of poultry. *Animal Genetics*, **20**, 145–53.

Hughes, S. H., Stubblefield, E., Payvar, F., Engel, J. D., Dodgson, J. B., Spector, D., Cordell, B., Schimke, R. T. & Varmus, H. E. (1979). Gene localization by chromosome fractionation: Globin genes are on at least two chromosomes and three estrogen-inducible genes are on three chromosomes. *Proceedings of the National Academy of Sciences of USA*, **76**, 1348–52.

Hutt, F. B. (1949). *Genetics of the Fowl*. New York: McGraw-Hill.

Jeffreys, A. J. (1987). Highly variable minisatellites and DNA fingerprints. *Biochemical Society Transactions*, **15**, 309–17.

Jeffreys, A. J., Wilson, V. & Thein, S. L. (1985). Hypervariable 'minisatellite' regions in human DNA. *Nature*, **314**, 67–73.

Juneja, R. K., Sandberg, K., Anderson, L. & Gahne, B. (1982). Close linkage between albumin and vitamin D binding protein (*Gc*) loci in chicken. *Genetical Research*, **40**, 95–8.

Kao, F. T. (1973). Identification of chick chromosomes in cell hybrids formed between chick erythrocytes and adenine-requiring mutants of Chinese hamster cells. *Proceedings of the National Academy of Sciences of USA*, **70**, 2893–98.

Kuhnlein, U., Dawe, Y., Zadworny, D. & Gavora, J. S. (1989). DNA fingerprinting: a tool for determining genetic distances between strains of poultry. *Theoretical and Applied Genetics*, **77**, 669–72.

Kuhnlein, U., Zadworny, D., Dawe, Y., Fairfull, R. W. & Gavora, J. S. (1990). Assessment of inbreeding by DNA fingerprinting: Development of a calibration curve using defined strains of chickens. *Genetics*, **125**, 161–5.

Langhorst, L. J. & Fechheimer, N. S. (1985). Shankless, a new mutation on chromosome 2 in the chicken. *Journal of Heredity*, **76**, 182–6.

Leung, W. C., Chen, T. R., Dubbs, D. R. & Kit, S. (1975). Identification of chick thymidine kinase determinant in somatic cell hybrids of chick erythrocytes and thymidine kinase-deficient mouse cells. *Experimental Cell Research*, **95**, 320–6.

MacGregor, H. & Varley, J. (1988). *Working with Animal Chromosomes*, 2nd edn. Chichester: John Wiley.

Palmer, D. K. & Jones, C. (1986). Gene mapping in chicken–Chinese hamster somatic cell hybrids: Serum albumin and phosphoglucomutase-2 structural genes on chicken chromosome 6. *Journal of Heredity*, **77**, 106–8.

Shaw, E. M., Guise, K. S. & Shoffner, R. N. (1989). Chromosomal localization of chicken sequences homologous to the $\beta$-actin gene by *in situ* hybridization. *Journal of Heredity*, **80**, 475–8.

Somes, R. G. (1968). Sleepy-eye, an eyelid mutant of the fowl. *Journal of Heredity*, **58**, 375–8.

Suzuki, D. T., Griffiths, A. J. F., Miller, J. H. & Lewontin, R. C. (1989). *An Introduction to Genetic Analysis*, 4th edn. San Francisco: W.H. Freeman.

Tereba, A. & Astrin, S. M. (1982). Chromosome clustering of five defined endogenous retroviral loci in White Leghorn chickens. *Journal of Virology*, **43**, 737–40.

Vassart, G., Georges, M., Monsieur, R., Brocas, H., Lequarre, A. S. & Christophe, D. (1987). A sequence in M13 phage detects hypervariable minisatellites in human and animal DNA. *Science*, **235**, 683–4.

Wang, N. & Shoffner, R. N. (1974). Trypsin G- and C-banding for interchange analysis and sex identification in the chicken. *Chromosoma (Berlin)*, **47**, 61–9.

Wooster, W. E., Fechheimer, N. S. & Jaap, R. G. (1977). Structural rearrangements of chromosomes in the domestic chicken: experimental production by X-irradiation of spermatozoa. *Canadian Journal of Genetics and Cytology*, **19**, 437–46.

# 6

# *Genes controlling feathering and plumage colour*

## 6.1 Introduction

Feathers are probably the most complex derivatives of the integument to be found in any vertebrate and they are certainly one of the most striking anatomical features of birds. They are of great importance both to the poultry fancier and to the commercial poultryman. For the former much of the emphasis in breeding is to obtain a plumage pattern agreed upon by the 'experts' as the standard. For the commercial poultryman the feathers are important for two reasons: (i) since they are approximately 90% protein they represent a dietary protein input which will not be recovered at the end of the day as edible protein, (ii) they may also be a useful indicator of the growth rate and the sex of the bird. In this chapter both the structure and distribution of feathers are considered, and also the colours of the plumage. Although there has been quite a lot of research into the genes controlling feathering and plumage colour they are still, by general genetic standards of the 1990s, not well understood. There are are some instances where the genes controlling certain characters are well established, their alleles known, their dominance relations understood, and the genes in question have been mapped, but there are many other examples where it is not yet clear how many genes control a particular character and how they interact.

## 6.2 The structure and growth of feathers

On a structural basis feathers can be divided into three main categories, which on detailed examination can be further subdivided (Stettenheim, 1972). The main categories are (i) **Contour** feathers, which comprise the main body feathers including the remiges of the wings and the rectrices of the tail, (ii) **Down** feathers, and (iii) **Filoplume** feathers. The down feathers cover the whole of the body in the newly hatched chick and they form the undercoat in the adult lying beneath the contour feathers, particularly on the abdomen. The filoplumes are the smallest feathers, rather hair-like in structure. Their function is not known for certain but it seems most likely they have a sensory function. The filoplumes which lie beneath the contour feathers have several nerve endings in their follicle wall which probably sense the movement of the contour feathers enabling the latter to adjust position for flight, insulation or bathing.

The largest and most conspicuous feathers are the remiges and rectrices and it is these that have been most studied genetically. A diagram of a typical contour feather is given in Fig. 6.1. It has a central vane with branches called **barbs**. Typically on a flight feather the barbs are joined together by a system of interlocking **barbules**. Most of the barbules of flight feathers carry little hook-like structures called **barbicels**. The barbicels may become unhooked and then reformed during the bird's preening activity. In actively flying birds the barbicels extend throughout the length of the flight feathers, but in the fowl they are present only in the central region of the feathers. In the down feathers the central rachis is either very short, in which case it is shorter than the longest barbs, or it is absent altogether. Down feathers also lack barbicels.

The feathers develop in the feather follicles, which are embedded in the epidermis and dermis. The follicles have a rich blood supply to ensure adequate nourishment, and also have a complex innervation. At the base of each follicle there is a collar of epidermal cells that divide and differentiate

**Fig. 6.1** Diagram of a typical contour feather.

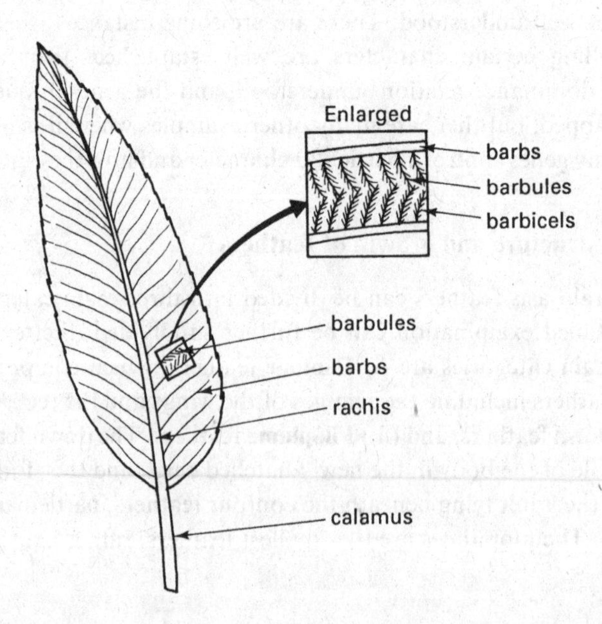

to give rise to the feathers. The growing barbs and associated cells migrate around the collar in a dorsal direction and then up the developing rachis (Fig. 6.2). Those barbs that are initially at the tip of the rudimentary rachis (Fig. 6.2*A*) eventually become the barbs at the tip of the fully developed feather (Fig. 6.2*B*). It can be seen that the tips of the barbs mature ventrally on the collar, whereas the base of the barbs arises dorsally closest to the developing rachis. This is important in relation to feather pigmentation, since pigment introduced at different points on the collar will appear in different regions of the fully developed feather. This is discussed further in section 6.4a. Feather growth shows diurnal rhythm. The development of a new feather generally pushes out the old feather, as occurs during moulting. After a feather has fully developed the germinal activity of the follicle ceases for a variable period, i.e. until the next moult. For a fuller account of the structure of feathers see Stettenheim (1972).

## 6.3 Genes controlling feathering

The structure of feathers, their pattern, growth rate and ultimate size are under genetic control, but each of these may also be influenced to some degree by environmental conditions. Most of the genes that have been studied to date have either been those which are present in certain breeds e.g. Silkie, Frizzle, or genes present in mutants which have arisen sponta-

**Fig. 6.2** A feather follicle showing the growth of a feather. *A*, The epidermal collar showing the rudimentary rachis; *B*, a developing feather. Compare the positions of the barbs 1, 2, and 3, and also the distal (a), central (b) and proximal (c) parts of the barbs in the rudimentary feather and developing feather, in order to understand the sequence of growth.

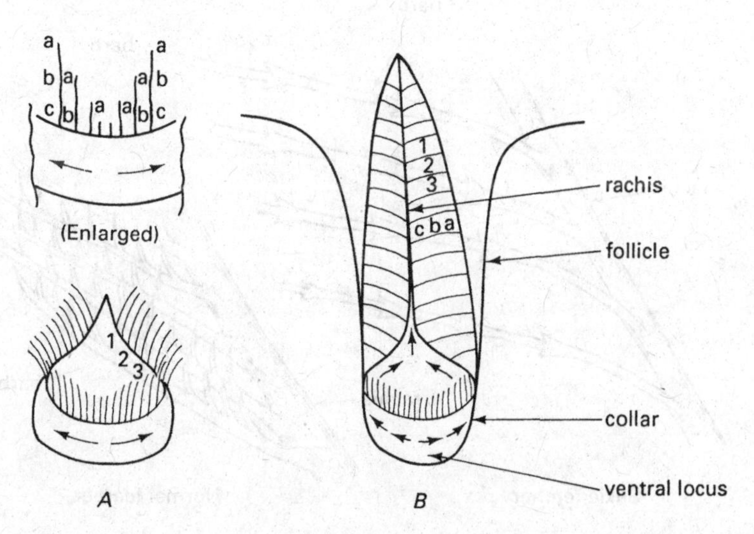

neously in a population. Some of the most studied ones are described in this section.

### 6.3a *Silkiness, h*

The Silkie breed has been known for several hundred years; it has the characteristic very fluffy plumage and a ragged appearance to the ends of its wings and tail feathers. The fluffiness is due to a lack of barbicels on the contour feathers. As a result the barbs are not attached to one another but often appear to bifurcate and not lie in a single plane as do typical flight feathers (see Fig. 6.3). The recessive character of silkiness was noticed by Darwin and Tegetmeier. It is controlled by a single autosomal recessive gene '*h*' which has been mapped on chromosome 1 (see Appendix I).

### 6.3b *Frizzling, F*

This is also largely a characteristic of a single breed, known as the Frizzle fowl. The shafts of the contour feathers, instead of being straight, are curved. The feathers are more delicate and the barbs eventually wear off the remiges of the outer primaries, leaving the birds unable to fly. This character is controlled by an incompletely dominant autosomal gene which has been mapped on linkage Group II (see Appendix I), which also contains the genes for Crest (*Cr*) and Dominant White (*I*). There are three

**Fig. 6.3** A comparison between a Silkie feather and normal feather. A section of a contour feather is shown in each case. Note the more open structure of the Silkie feather, together with the longer barbs that are lacking in barbules.

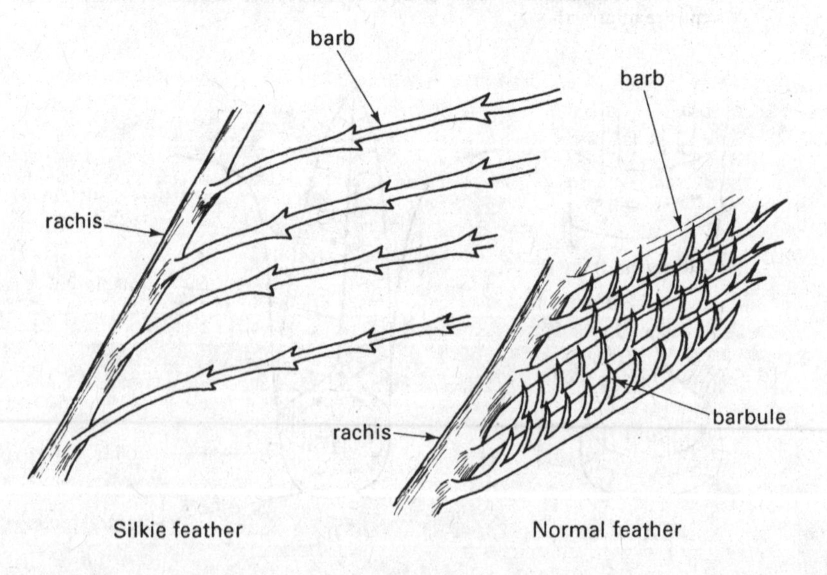

phenotypes corresponding to *F/F*, *F/f* and *f/f*. The extent of the frizzling is so pronounced in the homozygous *F/F* that the feathers break very easily and the birds very soon lose most of their feathers. Unless they are kept in a very protected environment then the *F/F* is lethal. The poultry fancier usually exhibits the heterozygote *F/f* which has less pronounced curling of the feathers but is rather more sturdy. The effects of the Frizzle gene are modified by an autosomal recessive modifier (*mf*) as described by Hutt (1949). The *mf* gene modifies heterozygotes (*F,mf/f,mf*) so that they are almost indistinguishable from the wild type.

### 6.3c *Genes causing varying degrees of nakedness*

Although there are probably a few genes which cause nakedness to varying degrees in the fowl, only two pairs of alleles have been studied to any extent by the geneticist. These are the sex-linked recessive gene, *n*, causing a general nakedness, and the autosomal incompletely dominant gene, *Na*, causing nakedness primarily around the head and neck. Chicks having the gene *n* hatch with varying degrees of nakedness ranging from a few scattered feathers to up to 75% normal down. Adults have more feathers than the chicks although some are almost completely naked. Pterylosis, i.e. the distribution of feather tracts throughout the skin, is normal, but the feather follicles are probably in some way defective, hence the lack of feathers. The female, although hemizygous and having only one *n* gene, shows a greater deficiency of feathers than the homozygous male.

The gene for naked neck (*Na*) is an autosomal dominant and is mapped on Chromosome 1, although it appears to show no significant linkage with two well-characterised genes on chromosome 1, namely those for pea comb (*P*) and blue egg (*O*) (Bitgood, Otis & Shoffner, 1983). The naked neck occurs in a number of breeds including Transylvanian Naked Neck, Malay Game, Cou Nu du Forez (France) and Shingangadi (Zaïre) and also in a number of local populations throughout the world (Merat, 1986). Unlike the gene for nakedness (*n*), that for naked neck (*Na*) is characterised by an almost complete lack of feather follicles in the head and neck region (Hutt, 1949). Recently there has been interest in introducing the *Na* gene into broilers and layers for rearing in warm climates (Merat, 1986). In a homozygous fowl, *Na/Na*, there is an approximately 40% reduction in feathering and in the heterozygote, *Na/na*, approximately 30%. When normal broilers are reared in a hot environment there is generally a reduced growth rate and reduced body size. Hanzl & Somes (1983a, b) introduced the *Na* gene into a broiler strain of White Plymouth Rocks and found that overall in a hot environment the *Na/Na* birds achieved significantly greater body weights than the *na/na* birds while containing significantly less lipid but more moisture and ash. Presumably in a hot environment less feathering is required to maintain body temperature and use can be made of this by rearing birds with the *Na* gene so that a smaller proportion of the dietary intake is required for feather production.

### 6.3d Genes affecting the rate of feather growth

The best characterised alleles which affect the rate of feather growth are the *K* locus for rate of feathering, and the tardy locus *T*. Serebrovsky (1922) discovered a difference in the rate of feathering between Mediterranean breeds such as Leghorns, Anconas and Minorcas and the Asiatic breeds and demonstrated that this was due to a single pair of sex-linked genes. The genes concerned, *K* (slow feathering) and $k^+$ (rapid feathering) have been mapped on the Z chromosome close to the genes for sex-linked silver and gold. The rate of feathering locus is an example of multiple alleles, since a third and fourth allele, $K^n$ and $K^s$ have been discovered more recently (Somes, 1969; McGibbon, 1977). The order of dominance of the four alleles is: $K^n$ (extremely slow feathering), $K^s$ (slow feathering), *K* (late feathering) and $k^+$ (rapid feathering).

The rate of feathering locus is interesting for two main reasons: (i) it is a clear case where a particular growth rate is controlled at a single locus, whereas growth parameters are usually controlled at a number of loci, and (ii) it can be used to sex chicks at an early age. The latter is useful in the case of White Leghorns, which cannot be sexed by the colour of their plumage because of their dominant white gene. Both homozygous male chicks ($k/k$) and hemizygous $k^+$ females show little tails and wing feathers extending to the end of the body by the time they are 8–10 days old. By contrast $K/K$, $K/k^+$ and *K* chicks have no tails and very short wings at the same age. If a rapid feathering male $k^+/k^+$ is crossed with a slow feathering female *K* the $F_1$ generation yields slow males ($K/k^+$) and rapid feathering females ($k^+$) which can be sorted with 95% accuracy at 8 days.

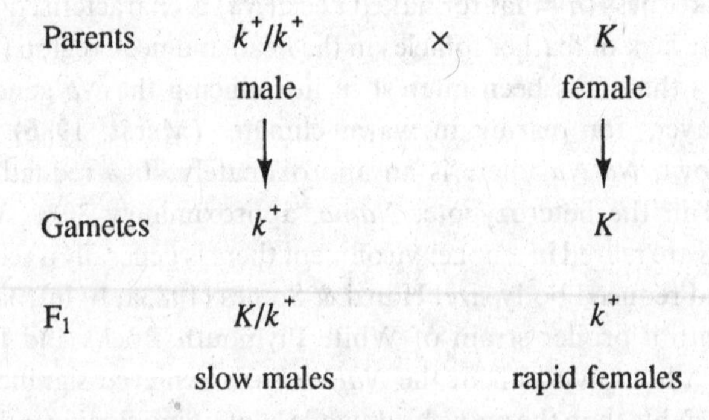

The contrast between $k^+$ and $K^n$ is even greater so that the phenotypes can be distinguished with a higher degree of certainty earlier. This greater accuracy enables a more accurate mapping of the *K* locus and the nearby Silver *S* locus which are only 1.1 mu apart (Somes, 1969). With the extremely slow feathering gene, feathering is so retarded that females may remain naked well into adult life. The trait is also characterised by hypertrophied uropygial glands and reduced comb development. Dunnington & Siegel (1986) have examined the effect of introducing differences in rate of feathering genes into broiler stock to see whether these genes had any effect on growth, feed efficiency or disease resistance. They concluded that the feathering genotype did not appear to interact with these characters. It has been observed (Bacon *et al.*, 1988) that in many White Leghorn strains, the female progeny of slow feathering hens (*K*) produce fewer eggs and have higher mortality than progeny of rapid feathering hens ($k^+$). This has been attributed to a higher incidence of viral infection, which is attributed in turn to tight linkage between the *K* locus and the *ev* locus (endogenous viral infection).

Three alleles have been described for the tardy locus (*T*) (Jones & Hutt, 1946): *T*, *t* and $t^s$. These are autosomal genes and the wild type *T* is dominant to the other two, $t^s$ being dominant to *t*. The tardy gene (*t*) and the retarded gene ($t^s$) become expressed only in the presence of the rapid feathering $k^+k^+$ and $k^+$ genotypes. They show up as slow feathering chicks with ragged ends to their wings in a flock of otherwise rapid feathering birds. The superscript in $t^s$ stands for 'secondary', indicating that it is the secondary feathers that are most affected. Recently the tardy locus has been mapped. Bitgood, Klorpes & Arias (1987) have used a chromosome translocation, *t(Z;1)* to determine the position of the tardy locus on chromosome 1. For details on the use of chromosome translocations, see Chapter 5, section 5.3. The *t* locus is found to lie at least 40 mu from the genes for melanotic (*Ml*), pea comb (*P*) and blue egg shell colour (*O*). The *t* locus is on the long arm (q arm), whereas *Ml*, *P*, and *O* lie on the short arm (p arm), the order being *t-centromere–Ml–P–O*.

### 6.3e *Hen feathering, Hf*

Most breeds of domestic fowl show sexual dimorphism in feathering. The main differences between normal cock and hen feathering are that in the male the tail and neck feathers are long and curved, whereas in the female the corresponding feathers are short and stand erect. However, in the Sebrights and Golden Campines the cocks have the characteristic hen feathering. This also occurs occasionally in Hamburgs and Silver-Laced

Wyandottes and can be introduced into other breeds by suitable crossings. Both genetic and hormonal factors control this sexual dimorphism. The early work on hen feathering has been well described by Hutt (1949). The first indications as to how the feathering was controlled were obtained by Morgan and by Punnett and Bailey in the 1920s. Within the last decade more extensive studies have been made and much is now known about the underlying controls for hen feathering (George, Noble & Wilson, 1981; Leshin *et al.*, 1981; George & Wilson, 1982; Somes *et al.*, 1984).

The inheritance of hen feathering is autosomal dominant and is sex-limited. (See Chapter 4, section 4.5). Although the genes are not carried on the sex chromosome (as are sex-linked characters) the character is expressed only in one sex. In this case the recessive genes (*hf*) which allow cock feathering, whether present in the male or female, are expressed only in the male. Any particular breed will be expected to be homozygous for either *Hf*/*Hf* or *hf*/*hf* as is the case for any character that has been selected for. Thus for breeds other than Sebright or the Golden Campine both hens and cocks will be homozygous for cock feathering (*hf*/*hf*). The different possible combinations are given in Chapter 4, section 4.5.

The question arises, why is the cock feathering gene (*hf*) sex-limited in this way? Feathering in the fowl is controlled hormonally, as are typical secondary sexual characteristics. Thus in a breed showing normal sexual dimorphism in feathering, if the cock is castrated, its feathering remains the typical cock feathering, whereas if a hen is castrated its feathering becomes the typical cock feathering after it has moulted. The ovary produces a number of chemically related hormones collectively known as **oestrogens**, and the testis produces **androgens** including **testosterone**. It therefore seems that lack of circulating testosterone does not affect cock feathering, whereas lack of circulating oestrogens causes hen feathering to revert to cock feathering. The male and female hormones are synthesised in the gonads and thence are secreted and circulate in the blood in very low concentrations.

Both male and female hormones are synthesised from the same precursor, cholesterol. The simplified biosynthetic sequence is as follows:

$$\text{Cholesterol} \longrightarrow \text{Testosterone} \longrightarrow 17\beta\text{-Oestradiol}$$

There are in fact a number of steps in the conversion of cholesterol to testosterone. It can be seen that the male hormone is an intermediate in the formation of the female hormone, and very low concentrations of testosterone can be detected in the ovaries of normal hens. George *et al.* (1981) have found that if a castrated male Sebright showing cock feathering is administered testosterone it resumes the hen feathering pattern. The reason for this is that the Sebright has the ability, unusual in the domestic fowl, of being able to convert testosterone to 17$\beta$-oestradiol in the skin. The enzyme responsible for this conversion, aromatase, has been measured in fibroblasts from the skin of the domestic fowl. Leshin *et al.* (1981) have found that aromatase activity is several hundred times higher in fibroblasts from Sebright and Golden Campine than in normal domestic fowl. George *et al.* (1990) have shown that the same allele causes hen feathering in both the Sebright and the Golden Campine. The process in the male Sebright is summarised below.

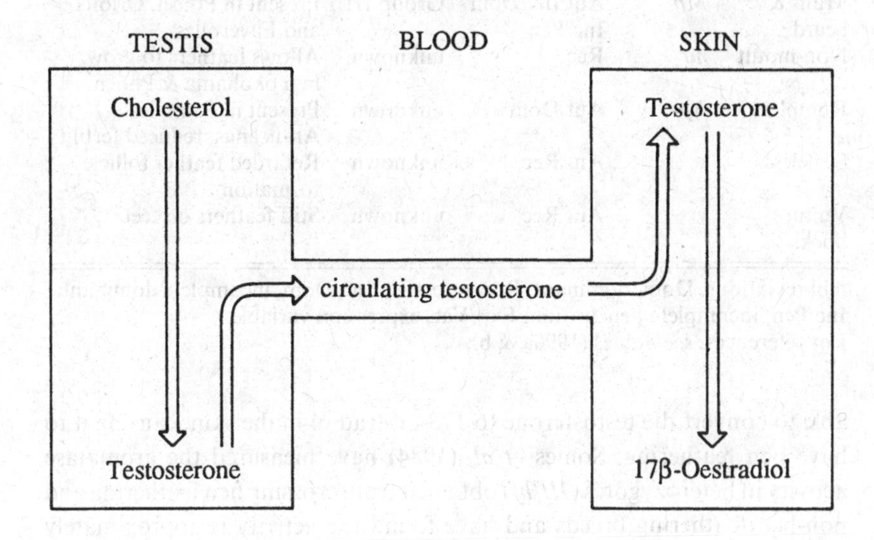


A Sebright cock has hen feathering because it is able to produce 17$\beta$-oestradiol in the skin from testosterone. If castrated, it changes to cock feathering, since the skin does not have a supply of testosterone to convert to 17$\beta$-oestradiol. It has also been shown that if skin is transplanted from a Sebright cock to another breed of cock, the latter showing cock feathering, then the transplanted skin still maintains the hen feathering, since the transplant still has high aromatase activity.

A 'normal hen' (i.e. not Sebright or Golden Campine) having the genes *hf/hf* will show hen feathering because of the oestrogens made in the ovary which circulate to the skin. A 'normal' cock having the genes *hf/hf* produces testosterone which circulates in the blood but this cock lacks the ability to convert it to 17$\beta$-oestradiol in the skin. The Sebright cock, on the other hand, although producing testosterone, which causes it to have the normal comb and wattles of a cock, and be able to produce fertile sperms, is also

Table 6.1. *Genes affecting feather structure and distribution*

| Gene | Symbol | Inheritance | Linkage | Comments |
|---|---|---|---|---|
| Congenital baldness | *ba* | Rec Exp/Var Inc/Pen | Sex-link | Bald spots, lacking feather follicles in top head |
| Crest | *Cr* | Aut Inc/Dom | Group II | Present in Silkies, Houdan & Polish |
| Displastic remiges | *dr* | Aut Rec | unknown | Necrosis & eventual loss of remiges |
| Long tail | *Gt* | Dom | unknown | Feathers continue to grow, in Yokohama & Phoenix |
| Muffs & beard | *Mb* | Aut Inc/Dom Inc/Pen | Group III | Present in Polish, Orloffs and Faverolles |
| Non-moult | *mt* | Rec | unknown | Allows feathers to grow, in Yokohama & Phoenix |
| Rumpless | *Rp* | Aut Dom | unknown | Present in Rumpless Araucanas, reduced fertility |
| Scaleless | *sc* | Aut Rec | unknown | Retarded feather follicle formation |
| Vulture hocks | *v* | Aut Rec | unknown | Stiff feathers on feet |

Abbreviations: Dom, dominant; Rec, recessive; Inc/Dom, incomplete dominant; Inc/Pen, incomplete penetration; Exp/Var, expression variable.
For references, see Somes (1990a & b).

able to convert the testosterone to 17β-oestradiol in the skin, causing it to have hen feathering. Somes *et al.* (1984) have measured the aromatase activity in heterozygotes (*Hf/hf*) obtained from crossing hen feathering and non-hen feathering breeds and have found the activity is approximately half that of the homozygous *Hf/Hf*. The 50% activity is sufficient for full expression of the hen feathering phenotype. They have been unable to demonstrate linkage with any known loci, but were able to conclude that hen feathering is an autosomal codominant trait.

### 6.3f Other genes affecting feather structure and distribution

There are several other genes which are known to affect feather structure and distribution; some of these are included in Table 6.1.

The crest, which is present in breeds such as Silkies, Houdans, Crevecoeurs and Polish, is controlled by an autosomal incompletely dominant gene, thus the heterozygotes (*Cr/cr*) have a smaller crest. However, there is considerable variation in the size of the crests among the different breeds, suggesting that there may be additional modifying genes (Somes, 1990b). There seems to be a linkage between the presence of a crest and the tendency to cerebral hernia but the precise relationship is not clear. The muffs and beards characteristic of Houdans, Faverolles, MilleFleur, Orloffs and Araucanas are controlled by an incompletely dominant autosomal gene.

Apart from the hen feathering mentioned above, two other conspicuous tail features are the extraordinarily long tails which have been developed in breeds such as the Yokohama, and the lack of a tail in the Rumpless varieties. The long tails develop partly as a result of having the correct genotype but also require a special rather artificial environment in which to mature (for details see Carefoot, 1985). The two genes are a recessive non-moulting gene, *mt*, and a dominant non-limited growth gene, *Gt*. Together these ensure that the tail feathers continue to grow from year to year. The rumpless condition is thought to be controlled by a dominant gene, *Rp*. Being rumpless, it not only lacks normal tail feathers but it also lacks the uropygeal gland or preening gland (Somes, 1990b). It is said to confer the advantage of being able to escape foxes, but also to have difficulty in mating (Carefoot, 1985), clearly a mixed blessing!

There are many other aspects to feathering which have not been investigated in depth. For example exhibition fowl are usually classified into hard feather and soft feather; the former include the game fowl which have shorter and more brittle feathers emerging from the body at a narrower angle (Carefoot, 1985). Although hard is known to be dominant to soft, for full understanding further genetic research is needed.

## 6.4 Plumage colour

Before considering plumage colour in detail it is first important to consider what causes the appearance of particular colours. When white light falls on an object the different colour components may be absorbed, reflected, refracted or scattered in different directions by the object. How the colour appears then depends on which wavelengths of light eventually strike the retina. The colours of feathers are divided into two main groups: structural colours and pigmentary colours (Veevers, 1982). Structural colours are often produced by very small particles or air pockets in the feathers having dimensions similar to the wavelengths of light, and these cause the scattering of light. Structural colours may also arise from thin films having different refractive qualities. These produce the iridescent blues and greens that change in appearance depending on the angle of vision. The white appearance of some feathers is caused by the reflectivity of some of their surfaces. However, most of the important feather colours in fowl result from the presence of pigments. The carotenoids contribute to a limited extent to the yellow and orange colour of some feathers but by far and away the most important are the melanins which are the principal black and red pigments in feathers.

### *6.4a The production of melanins*

The melanins are synthesised by a group of specialised cells, present in the dermis and epidermis, known as melanocytes. It is believed that the pigment-producing cells in vertebrates have a common origin developing from the neural crest (Bagnara *et al.*, 1979). The partially differentiated cells migrate to their ultimate location in the skin and there complete differentiation. There are two principal types of melanocytes responsible for the formation of feather pigment: those producing the black melanin, eumelanin, and those producing the red melanin, phaeomelanin. The melanins are found in intracellular granules called melanosomes; those containing eumelanin are generally rod-shaped, about 5 nm in diameter, and those containing phaeomelanin ovoid or spherical, about 3 nm in diameter. The development of the mature melanosome has been studied by a combination of biochemical and genetic techniques and by electron microscopy. The chemical structures of neither eumelanins nor phaeomelanins have yet been fully resolved largely due to their insolubilities which prevent adequate purification and subsequent structural studies, although nuclear magnetic resonance spectroscopy offers the prospect of solving the structure of phaeomelanin (Chedekel *et al.*, 1987). However, the early steps in their biosynthetic pathway are well understood.

For the production of both eumelanin and phaeomelanin two separate components have to be brought together, namely the melanin pigment and the matrix to which it is attached. It has been suggested that the difference between eumelanin and phaeomelanin lies in the nature of the matrix to which it is attached (Brumbaugh, Wilkins & Moore, 1979). Phaeomelanin biosynthesis also requires sulphydryl-containing compounds, which interact with DOPA to form cysteinyl-DOPA. Melanogenesis (the production of melanin pigments) involves activities at two different subcellular locations; melanin biosynthesis from tyrosine occurs in the subcellular organelle called the Golgi system, and the matrix is formed in the smooth endoplasmic reticulum (Brumbaugh *et al.*, 1979). The formation of melanin from tyrosine is shown on Fig. 6.4; it can be seen that the key enzyme for the first step is tyrosinase (also known as DOPA oxidase). A second enzyme, dopachrome oxidoreductase is also required for the pathway (Barber *et al.*, 1985), but the remaining steps appear to be spontaneous, i.e. are not enzyme catalysed. The matrix or premelanosome forms the substructure to which the melanin is attached. It then fuses with Golgi associated vesicles which contain tyrosinase, and forms the melanosomes. An albino mutant of the quail has been shown to contain tyrosinase in the Golgi apparatus, but appears defective in the transport of tyrosinase to the melanosomes (Yamamoto *et al.*, 1987). A hypothetical scheme (Brumbaugh *et al.*, 1979) is given in Fig. 6.5.

Wilkins, Brumbaugh & Moore (1982) have studied the steps of melanin synthesis in melanocytes by detailed genetic analysis. They used somatic cell hybridisation (see Chapter 5, section 5.4) to hybridise different genotypes having different deficiencies in melanin production. They then examined the heterokaryons (hybrid cells having nuclei from different origins) to see whether complementation had occurred. If the two geno-

**Fig. 6.4** Formation of melanin from tyrosine.

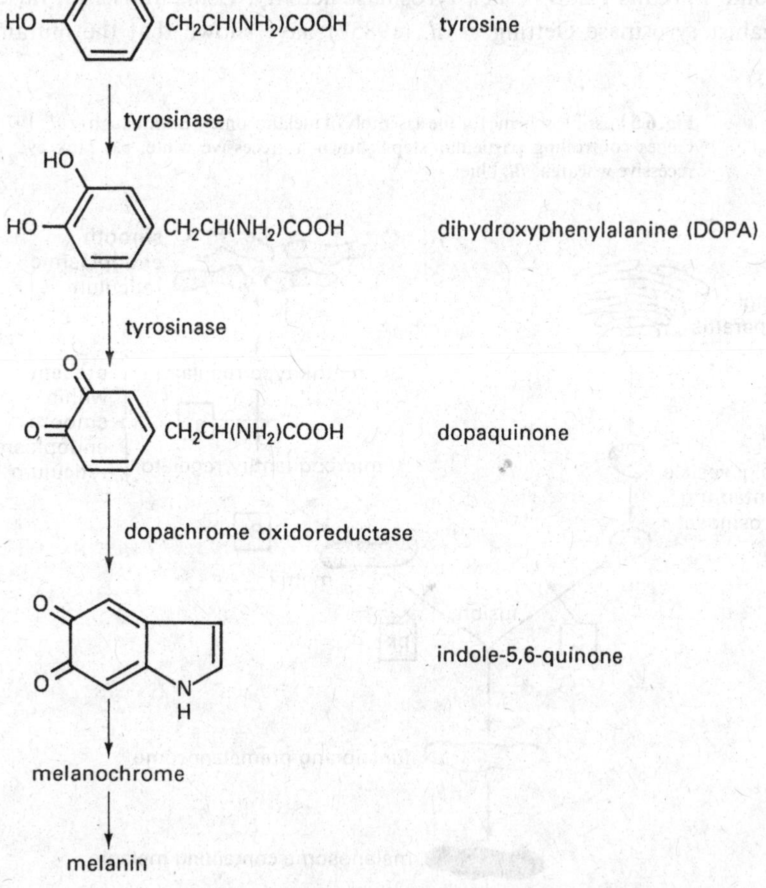

types used in the fusion experiment are deficient in different stages of the process then they might be expected to complement one another. As a result of one series of experiments Brumbaugh *et al.* (1979) deduced that there are two complementation groups between the genes *c* (recessive white), $e^y$ (recessive wheaten), *Bl* (blue) and *pk* (pink eye), the *c* gene complementing each of the three others. It was shown in a separate experiment that these four characters are at unlinked loci. From their results they were able to propose a hypothesis for the relationship between these genes (Fig. 6.5). This should be regarded as a working hypothesis and the process may well be more complex, but it helps to focus on the direction of future experimentation. By using a phorbol ester that inhibits melanogenesis Oetting, Langner & Brumbaugh (1985a) compared the proteins synthesised by inhibited and uninhibited melanocytes and have found at least nine proteins besides tyrosinase that are putative melanogenic proteins. The *C* locus is most probably the locus for the structural gene for tyrosinase. Two albino mutants *c* and $c^a$ lack tyrosinase activity. Using antiserum raised against tyrosinase Oetting *et al.* (1985b) have shown that the mutants

**Fig. 6.5** Possible scheme for the assembly of melanosomes (Brumbaugh *et al.*, 1979). Genes controlling particular steps shown: *c*, recessive white; *pk*, pink eye; $e^y$, recessive wheaten; *Bl*, blue.

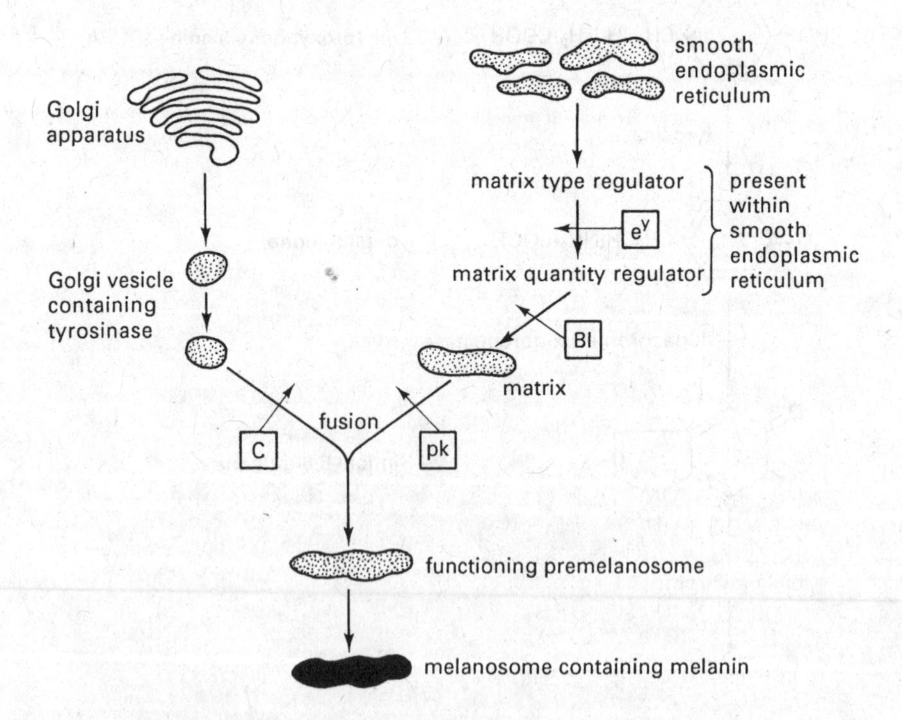

contain cross-reacting material, i.e. proteins recognised by antibodies raised against the enzyme tyrosinase. The mutants are therefore assumed to produce inactive tyrosinases, possibly due to amino acid substitution at the active site. For a review on the use of melanocytes in culture to study the genetics of pigment biosynthesis, see Brumbaugh & Oetting (1986).

Apart from the production of the eumelanin and phaeomelanin there is the question of how the melanins are transferred to the growing feathers. The melanocytes congregate beneath the feather germs which lie at the base of the feather follicles. Adjacent to these melanocytes is another type of cell, the keratinocyte. The keratinocytes are responsible for the synthesis of keratin, the principal structural protein of the feathers. Both of these are located in the collar (Fig. 6.2). The melanocytes have a large number of fine projections or dendrites that are responsible for the transfer of melanins to the keratinocytes and hence to the developing feather. There are clearly genes controlling the structure of the dendrites and also the release of melanins from the melanosomes. Mutations at these control points will also affect the colour of the plumage. This is discussed in connection with lavender plumage in section 6.5a.

## 6.5 Genes affecting plumage colour

When considering the different alleles which affect plumage colour, the question arises as to which allele is the wild type. Once this has been established or agreed, it is then possible to standardise the symbols used for each allele, indicating dominant/recessive relationships by a capital or small letter and the wild type with the appropriate superscript, '$^{+}$' as discussed in Chapter 3, section 3.2. The question therefore is, what is a wild type domestic fowl? For this, both Jaap & Hollander (1954), and Morejohn (1955) have taken the red jungle fowl (*Gallus gallus*) as the wild type, and this has generally been adopted with respect to alleles for plumage colour.

Plumage colour is considered under three separate headings: (i) the overall colour, (ii) the pattern of colour distribution throughout the contour feathers, and (iii) the pattern of colour distribution within the individual feathers. There is a quite extensive literature on plumage colour, and only the most studied examples are considered here, but for a comprehensive review, see Smyth (1990).

### *6.5a The overall plumage colour*

The two principal pigments which account for the plumage colour are eumelanin and phaeomelanin. These give rise primarily to black and red;

however, both can be modified by decreasing or interrupting the flow of pigment into the keratinocytes, causing the production of brown, blue, buff and yellow. Complete absence of melanins will lead to white plumage. It is clear that both the colour and its distribution throughout the plumage are the result of the interactions involving a number of loci. Using the model in Fig. 6.5 as a base, the way in which some of the genes for plumage colour may exert their effect is discussed. Assuming all the steps in the sequence (Fig. 6.5) are operational, then either phaeomelanin or eumelanin is produced. Brumbaugh *et al.* (1979) have suggested that the switch between eumelanin and phaeomelanin is controlled by the *E* locus, and this is discussed in more detail in section 6.5b since it is primarily concerned with colour distribution. There are several alleles at the *E* locus giving rise to a range of black and red coloured birds. The dominant allele '*E*' is a self-coloured black whereas the most recessive allele '$e^y$' (wheaten) gives rise to a mainly reddish coloured bird.

As might be expected from the scheme (Fig. 6.5) there are a number of mutants which can give rise to white plumage, or almost white plumage. The recessive white '*c*' accounts for the plumage of the white varieties of Plymouth Rocks, Wyandottes, Sussex, Orpingtons, Silkies, Frizzles, Cochins, Pekins, Minorcas, Jersey Giants, Dorkings and Japanese bantams. It appears to exert an epistatic effect on the *E*, *B* and *S* loci. The melanosomes lack tyrosinase and so melanin is not produced. There are four known alleles: $C^+$ (the wild type which permits colour), *c* (recessive white), $c^a$ (albino) and $c^{re}$ (red-eyed). The '$c^a$' was shown to be a third allele by Brumbaugh, Barger & Oetting (1983); it differs from the recessive white in being a more extreme form. The recessive white affects only the feathers whereas the albino ($c^a$) affects also the skin and iris. The fourth allele, $c^{re}$ resembles the recessive white, *c* except that the former has red eyes whereas the latter has dark pigmented eyes (Smyth, Ring & Brumbaugh, 1986). The order of dominance is $C^+ > c > c^{re} > c^a$.

The second white locus is the dominant white that is characteristic of the White Leghorns. It is also an important contributor to several patterned breeds, e.g. Old English and Modern Red Pile Game. It is given the symbol *I* for inhibitor of colour. It is able to suppress the production of black pigment (eumelanin) but has little effect on red (phaeomelanin), and appears epistatic to the *E*, *B* and *S* loci. Its mode of action appears to be that it suppresses the formation of melanosomes in certain genotypes. Brumbaugh & Lee (1975) have shown that there is reduced $^3$H-leucine incorporation into melanosomes of genotype *I*/*I* when compared with the *i*/*i* genotypes, indicative of a lower rate of biogenesis. The third white locus is the so called pink-eye mutant '*pk*' which also causes reduction in pigmentation. This is thought to be a premelanosome mutant in which the melanosomes do not fully mature. The mutants contain tyrosinase but $^{3}$H-DOPA is poorly incorporated into the melanosomes. It is thought to be a structural mutant preventing the proper formation of the matrix.

The fourth gene locus on Fig. 6.5 is the blue locus characteristic of the Blue Andalusian breed. This is an incompletely dominant autosomal gene. The heterozygote *Bl/bl* gives rise to blue feathers with a black edging. The homozygote *Bl/Bl* is what is described as splashed white, i.e. mainly white with an occasional black edged feather, and the homozygous *bl/bl* is normally black. A histological examination of blue feathers shows that they contain the eumelanin granules but that they are more sparsely distributed than in the case of black feathers; also they are oval rather than rod-shaped. Being heterozygous, the blue feathered birds do not breed true but produce 25% black, 50% blue and 25% splashed white.

One of the earliest sex-linked genes to be studied (Sturtevant, 1912) was the gene for silver '*S*' and its allele, gold, '*s*'. Brumbaugh (1971) has shown that it has no effect on eumelanin synthesis, but affects phaeomelanin biosynthesis. This is illustrated by comparing the effects of introducing an '*S*' gene and introducing an '*I*' separately into cocks having black-red plumage; the former becomes what is described as silver duckwing and the latter a pile (Fig. 6.6). Understanding the mode of action of the *S* gene is complicated because it appears to be strongly influenced by modifying genes (Smyth, 1990). The transmission of the *S* gene differs, being sex-linked (see Chapter 4, section 4.3).

Finally the lavender gene (*lav*), an autosomal recessive, is a pigmenting gene which causes dilution of both black and red pigmentation. Black feathers become grey and red feathers become buff. Histological examination of the developing feathers reveals that '*lav*' causes a reduction in the transfer of melanins to the surrounding keratinocytes, although it does not seem to affect pigment granule synthesis (Brumbaugh, Chatterjee & Hollander, 1972). In the wild type, melanosomes are produced in the perinuclear region of the melanocytes and are then translocated to the tips of the dendrites. The tips are pinched off and phagocytosed by the adjacent keratinocytes, that eventually migrate up the developing rachis (Fig. 6.2). In the *lav/lav* genotype translocation from the perinuclear region to the dendrite tip is defective, and melanosomes build up in the perinuclear region (Mayerson & Brumbaugh, 1981).

### *6.5b Colour distribution throughout the plumage*

It is generally more difficult to envisage the genetic mechanism controlling a colour pattern than it is to understand how a particular shade of colour arises. A scheme analogous to that in Fig. 6.5 for the overall colour has not

**Fig. 6.6** The effect of introducing the *S* gene and the *I* gene into a cock having black-red plumage. The figure shows the main groups of feathers of the fowl, and the table below, the colours of particular regions. The precise colours may vary between breeds because of differences in expressivity attributable to their different genetic backgrounds. The introduction of the S gene to black-red plumage gives rise to silver duckwing, and the I gene to pile plumage.

| Feathers | Black-red | Silver duckwing | Pile |
|---|---|---|---|
| Neck hackle | orange | silver-white, black striped | orange-yellow |
| Breast & legs | black | black | white |
| Wing bow | red | silver-white | red |
| Wing bar | green-black | metallic blue | white |
| Wing feathers | orange | silver-white | orange |
| Saddle hackle | red | silver-white, black striped | red |
| Tail | black | black | white |

Table 6.2. *Plumage patterns for different alleles at the* E *locus*

| Description | Symbol | Male | Female |
|---|---|---|---|
| Extended black | $E$ | Black breast, legs and tail; coloured neck and saddle hackles | Black all over except possibly neck hackle |
| Wild | $e^+$ | Black breast, legs, wing bar and tail; red neck, saddle, wing bows and feathers | Salmon breast, brown darkly stippled body |
| Brown | $e^b$ | Black breast, legs and tail; golden striped with black neck and saddle hackles; red wing bow and feathers | Brown body with black stippling most prominent on neck; dark brown tail |
| Buttercup | $e^{bc}$ | Similar to $e^b$ | Similar to $e^b$, but with coarser stippling and less black |
| Recessive wheaten | $e^y$ | Black breast, legs and tail; mahogany neck and saddle hackles; blue wing bar; brown wing bows and feathers | Brown body with ligher hackles heavily stippled with black |

Refer to Fig. 6.6. for the main groups of feathers in the fowl. The feather patterns described above may become modified by interactions between the *E* locus and other loci.

yet been proposed for the pattern of colour distribution. Presumably the mechanism must entail having a non-uniform distribution of the eumelanosomes and phaeomelanosomes throughout the skin. There are a number of known loci that affect colour distribution and undoubtedly others remain to be discovered. The most studied is the '*E*' locus at which there have been reported to be at least eight alleles (Smyth, 1990). There are also a number of eumelanin restrictors, i.e. loci that restrict the distribution of eumelanin, and therefore interact with the *E* locus; those most studied are the Columbian (*Co*), Mahogany (*Mh*) and Dark Brown (*Db*).

The *E* locus or 'Extender' is so described since the dominant allele *E* extends the plumage colour throughout the body. Thus, for example, an *E,Bl*/*E,bl* genotype would be expected to have a uniform blue plumage throughout (Somes, 1988). The existence of five alleles was first demonstrated by Morejohn (1955) and subsequently confirmed (Smyth, 1965). These were: *E* (extended black), $e^+$ (wild), $e^b$ (brown), $e^s$ (speckled) and $e^y$ (recessive wheaten) in order of their dominance relationships. In the absence of other major modifying genes they give the plumage patterns given in Table 6.2. Three more have been added, i.e. dominant wheaten

($e^{Wh}$), buttercup ($e^{bc}$) and birchen ($E^R$) (Smyth *et al.*, 1980). Four additional ones have been proposed but their relationship to the *E* locus has still to be established (see Smyth, 1990). The approximate order of dominance is $E > E^R > e^+ > e^b > e^s > e^{bc} > e^y$, with $e^{Wh}$ probably lying below $E^R$ (Smyth, 1990).

The wild '$e^+$' plumage is that found in the red jungle fowl and similar to the '$e^b$', or brown allele that is typically found in the brown Leghorn. To add to the complexity of having at least eight alleles, because each allele does not show complete dominance, the heterozygotes may show intermediate patterns. For example, $e^{Wh}$ interacts with $e^+$ and $e^b$ to give different phenotypes. The degree of dominance also appears to be affected by modifying genes (Carefoot, 1981). The genes of the *E* locus affect not only the adult plumage, but also that of the chicks.

The second most studied locus concerned with colour distribution is a restricting rather than an extending locus, the Columbian restriction locus. The mutant *Co* is an autosomal dominant to the wild type $co^+$. It causes restriction of pigmentation of the plumage to the hackles, wing tips, foot and tail (Fig. 6.7). Smyth (1990) points out that the name originates from the Columbian Exposition at the Chicago World's Fair in 1893, where Wyandottes with that phenotype were first exhibited.

**Fig. 6.7** Plumage pattern for the Columbian restriction locus.

The *Co* gene is more restricting in the males than in females. It is seen in breeds such as the White Sussex, Columbian Plymouth Rock, and Light and Buff Brahmas. In earlier studies it was believed that the Columbian restriction locus was an allele of the *E* locus, but it is now clearly established as a distinct locus (Smyth & Somes, 1965), which is able to modify expression of some of the alleles at the *E* locus (Smyth, 1970), although neither of the loci has yet been mapped. The *Co* gene enhances the difference between the silver (*S*) and gold ($s^+$) genes and this helps in sexing chicks of the genotypes *S*/*s* and *s* (Malone & Smyth, 1979). Because of its ability to enhance sexing in conjunction with the *S* and *s* genes, it is useful to know whether the *Co* genotype is in any way linked to the genes controlling body weight, since this could affect the application of the results in commercial breeding. It has, however, been found to have no significant effect on body weight (Fox & Smyth, 1984).

Another restriction locus is the dark brown Columbian (*Db*). It has the effect of restricting the distribution of black in the black-red male entirely to its flight and tail feathers. It is an incompletely dominant autosomal gene (Moore & Smyth, 1972). Its expression is modified by the presence of particular genes at the *E* locus. This has been studied by Campo & Alvarez (1988) in a Spanish breed known as Villafranquina. When Villafranquina males were crossed with tester strains having either $e^{bc}$ or $e^{b}$ at the *E* locus the progeny had plumage similar to the Villafranquina breed. Evidence suggests that the Villafranquina breed contains the *Db* allele and either $e^{b}$ or $e^{bc}$ allele, but it is difficult to distinguish between the two.

The third restriction locus is Mahogany (*Mh*). It restricts the amount of eumelanin in the plumage, particularly on the breast, back and wing bows and fronts of females, and the breast of the male. It is incompletely dominant.

A list of possible genotypes involving the *E* locus and restriction loci in certain breeds is summarised below; for further details see Smyth (1990).

Buff or Light Brahma: $e^{b}/e^{b}$ $Co/Co$ $db^+/db^+$ $mh^+/mh^+$
New Hampshire: $e^{Wh}/e^{Wh}$ $Co/Co$ $db^+/db^+$ $Mh/?$
Rhode Island Red: $e^{y}/e^{y}$ $Co/?$ $db^+/db^+$ $Mh/Mh$
Prat: $e^{Wh}/e^{Wh}$ $Co/Co$ $db^+/db^+$ $mh^+/mh^+$
Vasca: $e^{Wh}/e^{Wh}$ $Co/Co$ $db^+/db^+$ $Mh/?$
Villafranquina: $e^{b}/e^{b}$ $co^+/co^+$ $Db/Db$ $mh^+/?$
Buff Minorca: $e^{y}$ *or* $e^{bc}/?$ $Co/Co$ $db^+/db^+$ $Mh/Mh$

The only other extending locus to be established to date is the melanotic locus (*Ml*) which extends the black into normally red areas of pile zoned fowl. A linkage distance of approximately 10 map units between the *Ml* and the *Db* loci has been proposed (Carefoot, 1987a). It is incompletely dominant so that while *Ml*/*Ml* gives rise to black plumage, the heterozygote *Ml*/*ml* appears as dark brown (Moore & Smyth, 1971).

### 6.5c Genes affecting the pattern within the feathers

Patterns within the feathers include barring, mottling, pied, spangling and lacing. The sex-linked gene for barring was one of the earliest to be investigated, by Spillman in 1908. It is the pattern characteristic of Barred Plymouth Rocks, Scots Greys, Scots Dumpies, Barred and Cuckoo Leghorns and Dominiques. The gene is incompletely dominant and sex-linked. Its action is to inhibit deposition of melanin, giving rise to a series of white bands usually on a black background. The inhibition is caused by autophagic degeneration of the melanocytes at the proximal end of the black band. Few if any melanocytes exist in the white bands (Bowers, 1988). The barring gene also reduces the number and melanin content of the choroidal melanophores (the choroid is the middle layer of the eye, external to the retina). This may be either by preventing the initial migration from the neural crest, or by premature degeneration of the melanocytes, though at present it is not clear which (Schreck & Bowers, 1989).

It is not surprising, since barring is incompletely dominant, that the homozygous male (*B*/*B*) has wider white bands than the hemizygous female (*B*). The heterozygous male (*B*/*b*) will have a pattern closer to that of the hemizygous female. The barring pattern depends on the *B* gene and also on the growth rate. Sharper barring patterns appear with slower growth rates. Attaining the barring patterns stipulated by the Poultry Standards requires attention to both genetics and the environment to attain the optimal growth rate. The details of how this is attained are well described by Carefoot (1985). In addition to the differences in barring between homozygous adult males and hemizygous adult females there are differences in the down feathers of the chick. The male chick has a larger patch of light coloured down on the head than the female chick. The sex-linked barring gene was utilised by Punnett and Pease in developing autosexing breeds in the late 1920s as described previously in Chapter 4, section 4.6.

A second but distinct barring gene is the autosomal barring gene (*Ab*) which is found in Pencilled Hamburgs and Gold, Silver and Chamois Campines, and all partridge varieties. It is an autosomal dominant.

Another pattern is mottling in which the melanin deposition is interrupted in such a way as to leave the tips of feathers in all regions deficient in pigment, and thus white. It is controlled by an autosomal recessive gene *mo*. It is present in such breeds as the Ancona and Mottled Houdan and accounts for the tricolour patterns including speckled, mille fleur and apical spangling (Somes, 1980). Another gene which was believed to account for the black and white colour distribution in the Exchequer Leghorn was the pied gene (*pi*) (see Hutt, 1949). Carefoot (1987b) obtained evidence that the *pi* and *mo* are identical. He crossed an Exchequer Leghorn male with an Ancona female. If the two loci are distinct, then the genotypes of the birds would be $Mo^+/Mo^+$ *pi*/*pi* and *mo*/*mo* $Pi^+/Pi^+$ respectively. The $F_1$ progeny were 5 males and 3 females, all of which had black and white plumage intermediate between that of pied and mottled. This result would suggest that the *Mo* and *Pi* loci are not distinct, since no progeny having completely pigmented plumage occurred, as would be predicted for $Mo^+/$ *mo* $Pi^+/pi$. An alternative explanation suggested by Smyth (1990) is that pied and mottled are different incompletely dominant genes.

The inheritance of the lacing pattern in Wyandottes has been studied by Moore & Smyth (1972) and more recently by Carefoot (1986). Other patterns such as spangling and pencilling are described by Smyth (1990).

Many of the alleles that determine plumage pattern clearly interact with other genes to determine the phenotype. To date there is very little information on the location of these genes on particular chromosomes. Also, with few exceptions (see section 6.4a) the mechanisms that underlie these patterns are unclear in most cases. Brumbaugh *et al.* (1979) have suggested a mechanism to account for the action of the *c*, $e^y$, *Bl*, and *pk* genes (see Fig. 6.5) on the basis of their somatic cell hybridisation experiments. This, or similar cellular/molecular approaches, will ultimately have to be adopted in order to clarify the interactions involved at other gene loci.

## References

Bacon, L. D., Smith, E., Crittenden, L. B. & Havenstein, G. B. (1988). Association of the Slow feathering (*K*) and an endogenous viral (*ev21*) gene on the Z chromosome of chickens. *Poultry Science*, **67**, 191–7.

Bagnara, J., Matsumoto, J., Ferris, W., Frost, S., Turner, W., Tchen, T. & Taylor, J. (1979). Common origin of pigment cells. *Science*, **203**, 410–15.

Barber, J. L., Townsend, D., Olds, D. P. & King, R. A. (1985). Decreased dopachrome oxidoreductase activity in yellow mice. *Journal of Heredity*, **76**, 59–60.

Bitgood, J. J., Klorpes, C. A. & Arias, J. A. (1987). Tardy feathering locus (*t*) located in chromosome 1 in the chicken. *Journal of Heredity*, **78**, 329–30.

Bitgood, J. J., Otis, J. S. & Shoffner, R. N. (1983). Refined linkage value for pea comb and blue egg: Lack of effect of pea comb, blue egg, and naked neck on age at first egg in the domestic fowl. *Poultry Science*, **62**, 235–8.

Bowers, R. R. (1988). The melanocyte of the chicken. In *Advances in Pigment Cell Research*, ed. J. Bagnara, pp. 49–63. New York: Alan R. Liss.

Brumbaugh, J. A. (1971). The ultrastructural effects of the *I* and *S* loci upon black-red melanin differentiation in the fowl. *Developmental Biology*, **24**, 392–412.
Brumbaugh, J. A., Barger, T. W. & Oetting, W. S. (1983). A 'new' allele at the *C* pigment locus in the fowl. *Journal of Heredity*, **74**, 331–6.
Brumbaugh, J. A., Chatterjee, G. & Hollander, W. F. (1972). Adendritic melanocytes: a mutation in linkage group II of the fowl. *Journal of Heredity*, **63**, 19–25.
Brumbaugh, J. A. & Lee, K. (1975). The gene action and function of two dopa oxidase positive melanocyte mutants of the fowl. *Genetics*, **81**, 333–47.
Brumbaugh, J. A. & Oetting, W. S. (1986). What can we learn from chick embryo melanocytes? *BioScience*, **36**, 381–7.
Brumbaugh, J. A., Wilkins, L. M. & Moore, J. W. (1979). Genetic dissection of eumelanogenesis. *Pigment Cell*, **4**, 150–8.
Campo, J. L. & Alvarez, C. (1988). Genetics of the black-tailed red plumage pattern in Villafranquina chickens. *Poultry Science*, **67**, 351–6.
Carefoot, W. C. (1981). Notes on the 'wheaten' plumage phenotype of the domestic fowl. *British Poultry Science*, **22**, 499–502.
Carefoot, W. C. (1985). *Creative Poultry Breeding*. Published privately.
Carefoot, W. C. (1986). Pencilled and double-laced plumage pattern phenotypes in the domestic fowl. *British Poultry Science*, **27**, 431–3.
Carefoot, W. C. (1987a). Test for linkage between the eumelanin restrictor (*Db*) and the eumelanin extension (*Ml*) genes in the domestic fowl. *British Poultry Science*, **28**, 69–73.
Carefoot, W. C. (1987b). Evidence that the mottled (*mo*) and pied (*pi*) plumage genes of the domestic fowl are identical. *British Poultry Science*, **28**, 753–4.
Chedekel, M. R., Subbarao, K. V., Bhan, P. & Schultz, T. M. (1987). Biosynthetic and structural studies on pheomelanin. *Biochimica et Biophysica Acta*, **912**, 239–43.
Dunnington, E. A & Siegel, P. B. (1986). Sex-linked feathering alleles ($K$, $k^+$) in chicks of diverse genetic backgrounds. 1. Body temperatures and body weights. *Poultry Science*, **65**, 209–13.
Fox, T. W. & Smyth, J. R. (1984). The Columbian restriction gene, *Co*, and early growth rate in the domestic fowl. *Poultry Science*, **63**, 586–8.
George, F. W., Matsumine, H., Mcphaul, M. J., Somes, R. G. & Wilson, J. D. (1990). Inheritance of henny feathering trait in the Golden Campine chicken – Evidence for allelism with the gene that causes henny feathering in the Sebright bantam. *Journal of Heredity*, **81**, 107–10.
George, F. W., Noble, J. F. & Wilson, J. D. (1981). Female feathering in Sebright cocks is due to conversion of testosterone to estradiol in skin. *Science*, **213**, 557–9.
George, F. W. & Wilson, J. D. (1982). Developmental pattern of increased aromatase activity in the Sebright bantam chicken. *Endocrinology*, **110**, 1203–7.
Hanzl, C. J. & Somes, R. G. (1983a). The effect of the Naked Neck gene, *Na*, on growth and carcass composition of broilers raised in two temperatures. *Poultry Science*, **62**, 934–41.
Hanzl, C. J. & Somes, R. G. (1983b). Organoleptic and cooked meat characteristics of Naked Neck broilers raised in two temperatures. *Poultry Science*, **62**, 942–6.
Hutt, F. B. (1949). *Genetics of the Fowl*. New York: McGraw Hill.
Jaap, R. G. & Hollander, W. F. (1954). Wild type as standard in poultry genetics. *Poultry Science*, **33**, 94–100.

Jones, D. G. & Hutt, F. B. (1946). Multiple alleles affecting feathering in the fowl. *Journal of Heredity*, **37**, 197–205.

Leshin, M., Baron, J., George, F. W. & Wilson, J. D. (1981). Increased estrogen formation and aromatase activity in fibroblasts cultured from the skin of chickens with the Henny feathering trait. *Journal of Biological Chemistry*, **256**, 4341–4.

Malone, G. W. & Smyth, J. R. (1979). The influence of the *E*, *Co*, and *I* loci on the expression of the silver (*S*) and gold ($s^+$) alleles in the fowl. *Poultry Science*, **58**, 489–97.

Mayerson, P. L. & Brumbaugh, J. A. (1981). Lavender, a chick melanocyte mutant with defective melanosome translocation: a possible role for 10 nm filaments and microfilaments but not microtubules. *Journal of Cell Science*, **51**, 25–51.

McGibbon, W. H. (1977). A sexed-linked mutation affecting rate of feathering in chickens. *Poultry Science*, **56**, 872–5.

Merat, P. (1986). Potential usefulness of the *Na* (naked neck) gene in poultry production. *World's Poultry Science Journal*, **42**, 124–42.

Moore, J. W. & Smyth, J. R. (1971). Melanotic, key to a phenotypic enigma in the fowl. *Journal of Heredity*, **62**, 214–9.

Moore, J. W. & Smyth, J. R. (1972). Inheritance of Silver-laced Wyandotte plumage pattern. *Journal of Heredity*, **63**, 179–84.

Morejohn, C. V. (1955). Plumage color allelism in the Red Jungle fowl. *Genetics*, **40**, 519–30.

Oetting, W. S., Churilla, A. M., Yamamoto, H. & Brumbaugh, J. A. (1985b). Pigment locus mutants of the fowl produce enzymatically inactive tyrosinase-like molecules. *Journal of Experimental Zoology*, **235**, 237–45.

Oetting, W., Langner, K. & Brumbaugh, J. A. (1985a). Detection of melanogenic proteins in cultured chick-embryo melanocytes. *Differentiation*, **30**, 40–6.

Schreck, R. E. & Bowers, R. R. (1989). Effect of the Barring gene in eye pigmentation in the fowl. *Pigment Cell Research*, **2**, 191–201.

Serebrovsky, A. S. (1922). Crossing over involving three sex-linked genes in chickens. *American Naturalist*, **56**, 571–2.

Smyth, J. R. (1965). Allelic relationship of genes determining extended black, wild type and brown plumage pattern in the fowl. *Poultry Science*, **44**, 88–98.

Smyth, J. R. (1970). Genetic basis for plumage color patterns in the New Hampshire fowl. *Journal of Heredity*, **61**, 280–3.

Smyth, J. R. (1990). Genetics of plumage, skin and eye pigmentation in chickens. In *Poultry Breeding and Genetics*, ed. R. D. Crawford, pp. 109–67. Amsterdam: Elsevier.

Smyth, J. R., Classen, H., Malone, G. W. & Moore, J. W. (1980). Genetics of the Buttercup plumage pattern. *Poultry Science*, **59**, 2373–8.

Smyth, J. R., Ring, N. M. & Brumbaugh, J. A. (1986). A fourth allele at the C-locus of the chicken. *Poultry Science*, **65** (suppl. 1), 129.

Smyth, J. R. & Somes, R. G. (1965). A new gene determining the columbian feather pattern in the fowl. *Journal of Heredity*, **56**, 151–6.

Somes, R. G. (1969). Delayed feathering, a third allele at the *K* locus of the domestic fowl. *Journal of Heredity*, **60**, 281–6.

Somes, R. G. (1980). The mottling gene, the basis of six plumage color patterns in the domestic fowl. *Poultry Science*, **59**, 1370–4.

Somes, R. G. (1988). *International Registry of Poultry Genetic Stocks. Storrs Agricultural Experimental Station Bulletin*, 476.
Somes, R. G. (1990a). Mutations and major variants of plumage and skin in chickens. In *Poultry Breeding and Genetics*, ed. R. D. Crawford, pp. 169–208. Amsterdam: Elsevier.
Somes, R. G. (1990b). Mutations and major variants of muscles and skeleton in chickens. In *Poultry Breeding and Genetics*, ed. R. D. Crawford, pp. 209–37. Amsterdam: Elsevier.
Somes, R. G., George, F. W., Baron, J., Noble, J. F. & Wilson, J. D. (1984). Inheritance of the henny-feathering trait of the Sebright bantam chicken. *Journal of Heredity*, **75**, 99–102.
Spillman, W. J. (1908). Spurious allelomorphism. *American Naturalist*, **42**, 610–15.
Stettenheim, P. (1972). The integument of birds. In *Avian Biology*, volume II, ed. D. S. Farnes, J. R. King & K. C. Parkes, pp. 2–64. New York: Academic Press.
Sturtevant, A. H. (1912). An experiment dealing with sex linkage in birds. *Journal of Experimental Zoology*, **12**, 499–518.
Veevers, G. (1982). *The Colours of Animals*. London: Arnold.
Wilkins, L. M., Brumbaugh, J. A. & Moore, J. W. (1982). Heterokaryon analysis of the genetic control of pigment synthesis in chick embryo melanocytes. *Genetics*, **102**, 557–69.
Yamamoto, H., Ito, K., Ishiguro, S. & Takeuchi, T. (1987). Gene controlling a differentiation step in the quail melanocyte. *Developmental Genetics*, **8**, 179–85.

# 7

# *Muscle, nerve and skeleton*

This chapter focuses on the genes affecting three important tissues of the body, namely muscle, the nervous system and the skeleton. Each will be considered in turn.

## 7.1 Muscle

The muscles usually make up a large proportion of the body weight (*c.* 40%) and also about 40% of the body protein. Muscle development is very important in the production of broilers, since meat is often the main protein component of the average human diet in many countries. There are three basic types of muscle: skeletal or voluntary muscle, cardiac muscle, and smooth or involuntary muscle. Skeletal and cardiac muscle are striated in appearance, whereas smooth muscle is non-striated, reflecting its less regular structure. Smooth muscle is found particularly in the gut and the lining of blood vessels, whereas skeletal muscle is found in association with the skeleton.

Within skeletal muscle the constituent fibres may be subdivided into the two principal types, red fibres and white fibres. The difference in colour is dependent on the larger amount of myoglobin (see Chapter 10, section 10.2) and cytochromes in the red fibres. The white fibres have a poor supply of mitochondria and are often referred to as fast twitch muscles. They are able to undergo short, but not sustained bursts of activity. During such periods they metabolise anaerobically and degrade glucose to lactic acid. The red fibres function aerobically and contain a large number of mitochondria. They are referred to as slow twitch muscles, and are able to sustain activity for much longer periods. Each individual muscle may have a mixture of both types of fibre, e.g. pectoral muscle is predominantly (>90%) white fibres, whereas the gastrocnemius is predominantly red fibres. The proportion of the different types of fibre in a given muscle is thought to be under genetic control, since the differences in composition exist before hatching. There is also believed to be an element of environmental control since the proportion of red to white may be influenced by exercise.

There are two aspects of muscle genetics which have been particularly studied in the domestic fowl; they are the genetics of muscle development and of muscular dystrophy. As a tissue, muscle has a number of advantages as source material for developmental genetics. Firstly, it can undergo complete development from undifferentiated myoblasts in culture medium, which simplifies its study. Secondly, it is a fairly homogeneous tissue comprising mainly a single cell type, and thirdly the principal proteins present in muscle have been thoroughly characterised from a biochemical standpoint.

Mononucleated myoblasts, the progenitors of myofibrils, can be obtained in reasonable quantity by dissociating chick embryo muscle into suspensions of single cells. These suspensions can then undergo the terminal stages of myogenic differentiation in culture by forming first myotubes and then spontaneously contracting muscle fibres. The developmental stages are summarised in Fig. 7.1. If cocultured with explants of neural tissue, active neuromuscular junctions may also be formed. It is thus a good tissue with which to study developmental genetics, that is, how the genes are programmed to be 'switched on' and 'switched off' during differentiation. When myoblasts are grown in culture they tend to differentiate, and cease proliferation after 50–80 doublings. If they are required to stay in continuous culture without differentiating, then transformed myoblasts (myoblasts that have been transformed into cancerous myoblasts and will grow continuously in a suitable medium) have to be used.

The interest in muscular dystrophy of the domestic fowl is that its study may provide useful information for understanding human muscular dystrophy.

### 7.1a *Muscle proteins*

The principal proteins present in skeletal muscle are those which form an integral part of the muscle fibre, and those enzymes which are either directly or indirectly involved in the generation of ATP, the primary source of energy used in contraction. In a study of mRNAs from domestic fowl muscle Hastings & Emerson (1982) distinguished 18 types, of which six were identified as the mRNAs for myosin(2), actin, tropomyosin and troponin C and troponin I. However, if the developing myoblasts were studied rather than the differentiated muscle, as many as 17000 different RNA species could be detected, but after differentiation into myofibrils only about 2500 sequences could be detected (Paterson & Bishop, 1977). Of these mRNAs, six were present in very high concentration, corresponding to the principal structural proteins of muscle. An outline of the structure of a skeletal muscle fibre is shown in Fig. 7.2, and the location of the proteins within it is given on Table 7.1.

Actin (25%) and myosin (55%) make up the principal proteins of the thin and thick filaments, respectively, and are the main contractile elements. Associated with the actin on the thin filaments are the regulatory proteins tropomyosin and troponin. α-Actinin is associated with the Z-disc. β-Actinin acts as a capping protein for the thin filaments, thereby stabilising their length and preventing the addition of further actin monomers that would lengthen the filaments (Funatsu, Asami & Ishiwata, 1988). Two

**Fig. 7.1** Stages in the development of muscle fibres. Stem cells differentiate to form myoblasts. The myoblasts are capable of fusing with other myoblasts eventually to form myotubes. Once differentiated myotubes are formed they no longer undergo nuclear division. Foetal myotubes produce the foetal isoforms of muscle proteins, whereas the adult myotubes produce the adult isoforms and have fully functioning myofibrils.

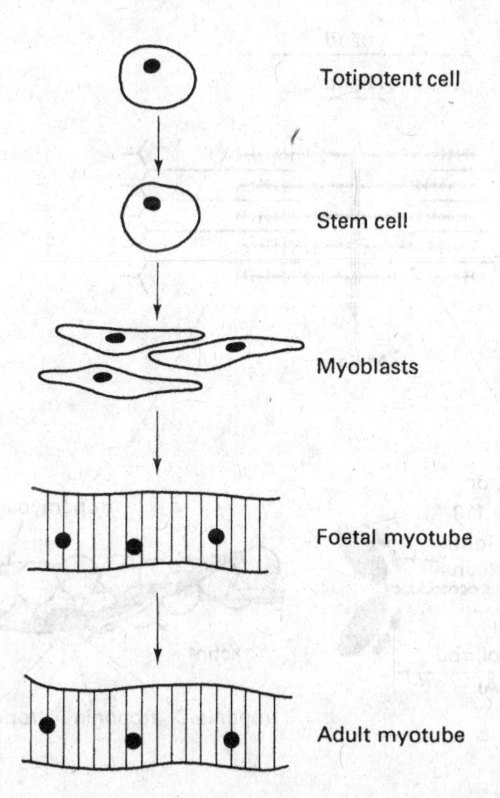

further proteins that have been studied in skeletal muscle of the domestic fowl are the M-line proteins. The M-line is a dark line found in the middle of the H-zone. Only two have been clearly identified; these are the MM isozymes of creatine kinase and myomesin. It is estimated that the M-line creatine kinase makes up only 5% of the total creatine kinase present in the skeletal muscle of the domestic fowl (Wallimann, Turner & Eppenberger, 1977). Myomesin makes up less than 0.3% of the protein in skeletal muscle. It may have a role in maintaining the thick filaments in register. In cardiac muscle, although there is no visible M-line, myomesin is found strongly bound to the myofibrils (Eppenberger, 1981). Myomesin is lacking in smooth muscle. Outside muscle tissues, myomesin is only found in the thymus, which contains myogenic cells.

Although the muscle proteins are well characterised, the genetics of their

**Fig. 7.2** Schematic diagram of the organisation of skeletal muscle. *A* shows the overall arrangement of thin and thick filaments; *B*, arrangement of the myosin light and heavy chains that make up the thick filaments; *C*, the polymerised actin molecules together with tropomyosin and troponins C, I, and T that make up the thin filament.

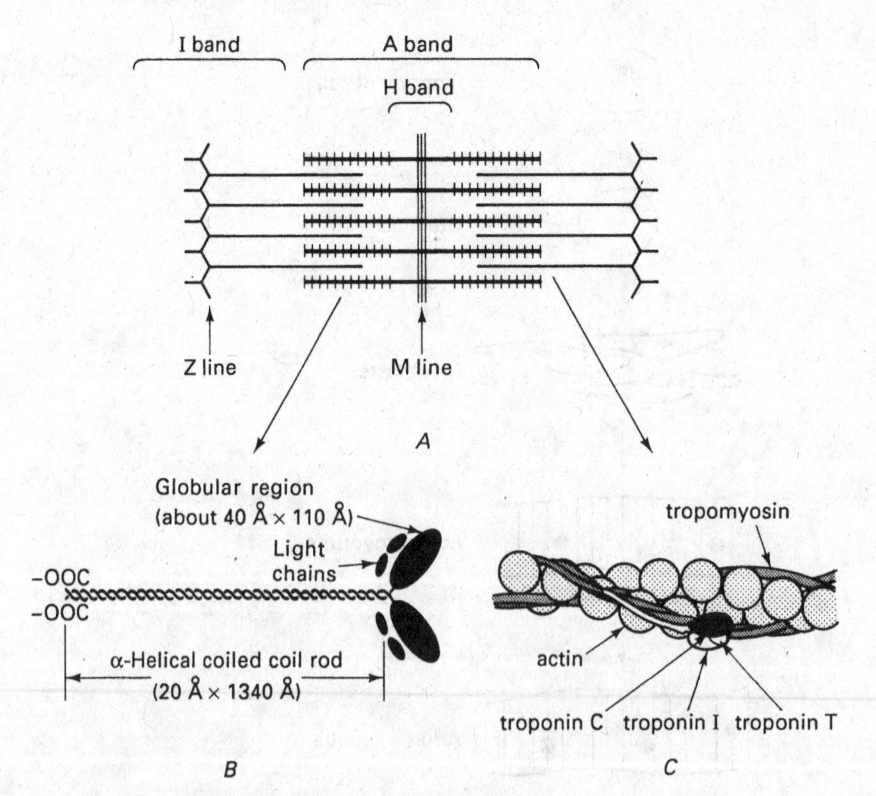

Table 7.1. *Protein components of skeletal muscle myofibril*

| Protein | Location | Function |
|---|---|---|
| Myosin | Thick filament | Contraction |
| Actin | Thin filament | Contraction |
| Tropomyosin | Thin filament | Regulation of contraction |
| Troponin | Thin filament | Regulation of contraction |
| α-Actinin | Z disc | |
| β-Actinin | Thin filament | Capping protein at pointed end of filament |
| Myomesin | M-line | Filament assembly |
| MM-creatine kinase | M-line | Enzyme |

development is poorly understood. The ideal approach is to obtain mutants that are arrested at a particular stage in the developmental programme. This approach has been very successful in understanding the developmental genetics in lower organisms. For example, bacterial sporulation has been studied by obtaining a series of mutants that are blocked in different stages of development, and it has been possible to establish the sequence in which genes are 'switched on' during this process (see Peberdy, 1980). This approach has not yet been very fruitful in the case of avian or mammalian systems, since it has not been possible so far to isolate a wide spectrum of mutants with developmental defects. This may be because such mutants are usually lethal. Muscular dystrophy, although a genetic defect, causes degeneration of normal muscle rather than prevention of formation of fully differentiated muscle.

An interesting feature of the development of muscle fibres is the existence of isoforms of many of the proteins involved, e.g. isoforms of actin, myosin, tropomyosin, α- and β-actinin, and creatine kinase. Isoforms of proteins are proteins that are very closely related in structure to one another and serve the same role in the system. There are often different isoforms present in different tissues, and also there is often a developmental sequence in which one isoform is replaced by another during embryogenesis. Some of these developmental changes will now be considered and also the relationships between the genes for the different isoforms.

### *7.1b The expression of isoforms of myosin and actin during differentiation*

The principal protein of muscle, myosin, comprises six polypeptide chains; two are the large subunits and four are the small subunits. The two large subunits, which are identical, are referred to as myosin heavy chains (MHC). The two pairs of smaller chains can be released from the intact myosin by different chemical treatments, either alkali-released or DTNB (dithio-bis[nitrobenzoate]) released (see Fig. 7.2). There are two isoforms of the alkali-released light chains referred to as $LC_1$ and $LC_3$. Both forms are found in skeletal muscle, whereas cardiac muscle has a light chain homologous to $LC_1$, and gizzard has one homologous to $LC_3$. The DTNB-released light chain found in skeletal and gizzard muscle is $LC_2$. Cardiac muscle has two isoforms, $LC_{2A}$ and $LC_{2B}$. These each comprise 163 amino acid residues, and differ from one another in 18 positions (Matsuda *et al.*, 1981). All forms of the light chains are believed to have arisen from a single ancestral form. The numbering $LC_1$, $LC_2$ and $LC_3$ is in the order of their anionic mobility at pH 8.3. The alkali-light chains are concerned with ATP hydrolysis during muscle contraction, and the DTNB-light chains have a regulatory role in contraction.

There are a number of ways in which the translation of these isoforms can be programmed. It may involve the selection of one gene from within a multigene family. This appears to occur with myosin heavy chain (MHC). Other possibilities are either a rearrangement of the DNA itself, as happens with immunoglobulins (see Chapter 11, section 11.4) or that there are alternative mechanisms for splicing RNA. This last mechanism is particularly common in muscle. At present about 50 genes in different tissues and organisms are known to generate isoforms through alternative splicing (see Fig. 7.3). The alternative splicing mechanism is known to occur with MHC (muscle heavy chain) and alkaline MLC (muscle light chain), and also in tropomyosin and skeletal and cardiac troponin T (Breitbart, Andreadis & Nadal-Ginard, 1987).

The differentiation of muscle fibres occurs in a number of distinct stages. The initial undifferentiated mesenchyme transforms into unicellular myoblasts. The myoblasts fuse to form multinucleate myotubes which eventually form the fully differentiated contractile apparatus (Fig. 7.1). During these processes the predominant isoform of myosin changes. The embryonic form of myosin heavy chain is present until hatching, when a neonatal form appears, and by 8 weeks after hatching a third form, that is indistinguishable from the adult, predominates (Bandman, Matsuda & Strohman, 1982). Cerny & Bandman (1987) have shown that the muscle–nerve interaction during development plays a role in determining which isoform is expressed. The nerve is involved in the repression of the neonatal form. The significance of these changes in isoform is not clear, but they represent part of the developmental programme. A major change that occurs at about the time of hatching is that the muscles become load bearing for the first time.

Nabeshima *et al.* (1984) have studied in detail the switch between the $LC_1$ and $LC_3$ that occurs during embryonic muscle development. $LC_1$ predominates in early embryonic muscle, but $LC_3$ appears in significant amounts after 15 days and is then present in comparable amounts to that in the adult. Nabeshima *et al.* (1984) have sequenced cDNA clones for $LC_1$ and $LC_3$ (see Chapter 12, section 12.3b). They found that the 3′ ends of the nucleotide sequences in both are identical, whereas there is a difference in the 5′ end. (Since polynucleotides, e.g. mRNA and DNA, are asymmetrical the two ends of each strand can be distinguished. These are referred to as the 3′ and 5′ ends according to whether the 3′ or 5′ hydroxyl group on the ribose or deoxyribose sugar at each end is not involved in the internucleotide link.) Both proteins are coded for by a single gene which is 18 kb (kilobases) in size. This gene has two transcriptional initiation sites from which 17.5 and 8 kb precursors are transcribed. These RNAs are then processed differently to form the two mRNAs encoding the two distinct light chains $LC_1$ and $LC_3$ (Fig. 7.3). It has been shown that the expression of cardiac $LC_2$ depends on a DNA sequence of 150 bp (base pairs) which is situated upstream from the initiation site for mRNA synthesis on the structural gene for $LC_2$ (Arnold *et al.*, 1988).

The myosin heavy chain is encoded in a multigene family, but the precise

**Fig. 7.3** Alternative splicing of myosin alkali light chain gene (Nabeshima *et al.*, 1984). There are two alternative modes of splicing the RNA to give the mRNAs. The mechanism initiating this alternative splicing is at present unclear; it may be a consequence of the alternative initiation sites for transcription.

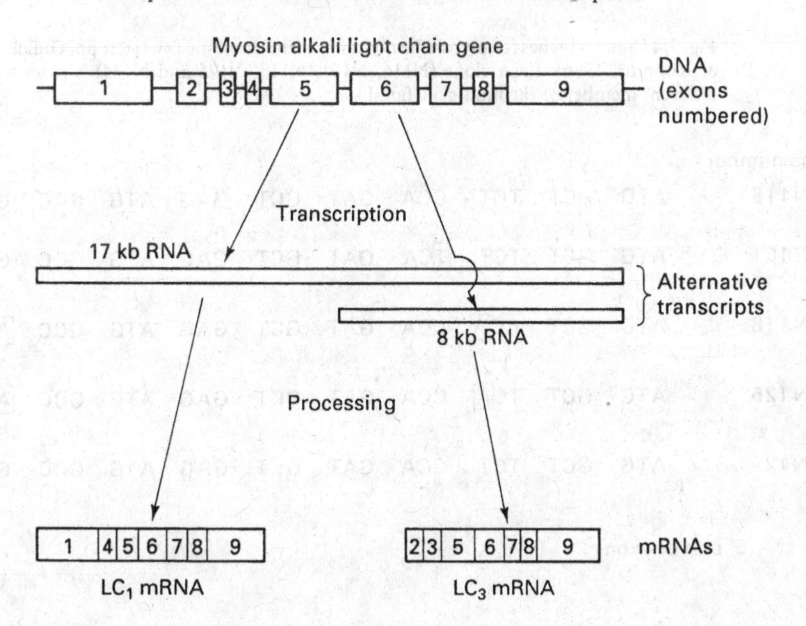

number of genes is at present uncertain. It is clear that there is a high degree of homology at the 5′ end of the gene sequences. This is apparent from the organisation of the 5′ ends of three myosin heavy chains from skeletal muscle, one cardiac muscle and one embryonic muscle, as seen in Fig. 7.4. Robbins *et al.* (1986) used a cloned fragment of the 5′ end of the adult gene to screen two genome libraries (see Chapter 12, section 12.2) for homologies. They found 31 unique non-overlapping clones, and seven of these fall within a well defined subgroup. It is thus clear that myosin heavy chain genes form a multigene family. It is not yet clear how many different genes there are, especially as some of the clones may be pseudogenes.

There are between 6 and 11 isoforms of actin, the second most abundant protein in muscle. Actin not only occurs in myofibrils, but also forms part of the cytoskeleton, in both muscle and non-muscle cells. Actin from skeletal, cardiac and smooth muscle is referred to as $\alpha$-actin, and cytoskeletal actin as $\beta$- and $\gamma$-actin. The actin multigene family has 8–10 different loci. $\beta$-Actin genes have been located by *in situ* hybridisation (Shaw, Guise & Shoffner, 1989) on the long arm of chromosome 2 and one of the smaller chromosomes (9–12). The genes for $\beta$- and $\gamma$-actin are expressed in the myoblasts, but the $\alpha$-skeletal and $\alpha$-cardiac actin genes are not expressed until fusion to myotubes has occurred. This sequence has been demonstrated both during normal embryonic development, and in proliferating myoblasts in culture (Hayward & Schwartz, 1986). Saborio *et al.* (1979) have shown that in 8-day embryos almost all the actin is the $\beta$-type, but after 8 days there is a continuous increase in the amount of $\gamma$-actin. The

**Fig. 7.4** 5′-end of genes for myosin heavy chain in the domestic fowl (Kropp, Gulick & Robbins, 1986). Each clone (N116, N101, N118, N125 and N124) is from a different member of the multigene family.

| Clone number | | | | | | | | | | |
|---|---|---|---|---|---|---|---|---|---|---|
| N116 | ATG | GCT | TCT | CCA | GAT | GCT | GAG | ATG | GCC | GCC |
| N101 | ATG | GCT | TCT | [T]CA | GAT | GCT | GAG | ATG | GCC | G[T]C |
| N118 | ATG | GCT | [A]C[A] | CCA | GAT | GCT | GAG | ATG | GCC | [A][T][A] |
| N125 | ATG | GCT | TC[A] | CCA | GAT | GCT | GAG | ATG | GCC | [A][T][A] |
| N124 | ATG | GCT | TCT | CCA | GAT | GCT | GAG | ATG | GCC | GCC |

↑ 5′ end of exon

myoblasts contain mainly $\beta$- and $\gamma$-actin prior to cell fusion, but after cell fusion the amounts of $\beta$- and $\gamma$-actin decline and there is a large increase in $\alpha$-actin. Presumptive $\alpha$-actin is detectable in the 17–20 day old chick embryo (Saborio *et al.*, 1979).

### *7.1c Expression of other proteins in muscle during differentiation*

The formation of the myofibrils requires the formation of two structural elements, namely the Z-disc and the M-band (Fig. 7.2). The proteins in the M-band have been studied during myofibrillogenesis in the domestic fowl. Perriard *et al.* (1986) studied the expression of three muscle specific genes for the M-band proteins: myomesin, M-protein and the M isoenzyme of creatine kinase. They showed clearly that myomesin ($M_r = 185\ 000$) is immunologically distinct from the M-protein ($M_r = 165\ 000$). They used antibodies to detect the expression of the appropriate proteins. In embryonic heart muscle the myomesin is expressed 6–7 days before the M-protein. In embryonic pectoral muscle myomesin also accumulates before the M-protein. With creatine kinase, in contrast to myomesin and the M-protein, the gene structure is known and, as with many other muscle proteins, isoforms exist. The B form of creatine kinase is expressed in many adult non-skeletal tissues and also in undifferentiated myogenic cells, but once differentiation occurs, the B form is repressed and the activation of expression of the M-form triggered. The switch in isoforms occurs at the transcriptional level. This has been demonstrated by detecting which mRNAs are expressed at various stages of myogenesis using cultured cells (Kwiatkowski *et al.*, 1985). The mRNAs were detected by *in situ* hybridisation using cDNA. Only the M form of creatine kinase is found in asssociation with the M-band. Creatine kinase is a dimer and exists in three isoenzyme forms, MM, MB, and BB. In the early stages of myogenesis of skeletal muscle only the BB isoenzyme is expressed, then all three forms, and by the time of hatching only the MM isoenzyme is expressed.

There are also multigene families of troponin and tropomyosin, in which the different isoforms are expressed at different stages of development. $\beta$-Actinin also exists in three isoforms, $\beta_{I}$, $\beta_{II}$ and $\beta_{III}$. $\beta_{I}$ is present in breast muscle from the earliest embryonic stages through to the adult, $\beta_{II}$-actinin becomes detectable in the 10-day embryo, increases during further development and is retained in the adult, and the $\beta_{III}$ form steadily decreases during development and is only detectable up to 10 days after hatching (Asami, Funatsu & Ishiwata, 1988).

One protein which has been studied, that is neither a structural component of the myofibrils nor necessary for energy generation, is thymidine kinase. This enzyme is a useful index of cell replication, since it is responsible for catalysing the synthesis of a DNA precursor. As terminal differentiation of the myoblasts occurs the gene products for muscle functioning increase, whereas those for cell division decrease. Thymidine kinase expression is attenuated and there is a rapid decline in the production of mRNA for thymidine kinase (Merrill, Hauschka & McKnight, 1984). The changes in muscle proteins during development are summarised in Fig. 7.5.

### *7.1d Muscular dystrophy*

Muscular dystrophy was first reported in the domestic fowl by Asmundson & Julian (1956). It is controlled by a single autosomal recessive gene, *am*. As mentioned previously, it is not of value in the study of muscle development, since it represents a defect in muscle maintenance. However, it has been studied as an important animal model of a genetically inherited myopathy. The disease generally affects the white fast twitch muscles. Affected birds show symptoms at about 2–3 weeks after hatching, when they show an inability to right themselves after being placed on their backs. Experiments have been performed to test whether it is of myogenic or neurogenic origin. Linkhart *et al.* (1976) carried out experiments in which wing limb buds were

**Fig. 7.5** Expression of some of the muscle proteins during myogenesis.

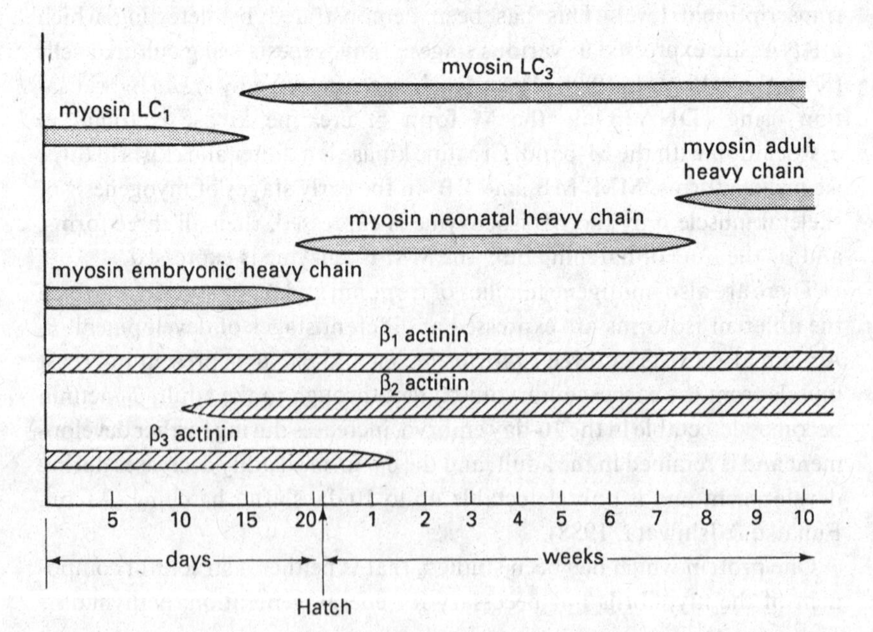

transplanted between normal and dystrophic 3.5-day-old embryos. The results showed that when wing buds from dystrophic embryos were transferred to normal ones, dystrophic muscles developed, but when normal wing buds were transferred to dystrophic embryos, the transferred wing buds developed normally, suggesting a myogenic origin. However, experiments by Rathbone *et al.* (1975), in which neural tubes from normal and muscular dystrophic chick embryos were transplanted into normal chick embryos at a much earlier stage of development (2–4 days), the dystrophic neural tissue induced a high thymidine kinase activity in the hosts, characteristic of muscular dystrophy. These experiments thus suggest the reverse conclusion, namely that the neural tubes are responsible for the course of development.

That the nervous tissue has a controlling influence has also been suggested by measurements of the acetylcholinesterase activity in fast twitch muscles. There are three isoenzymes of acetylcholinesterase in chick embryo and two of these almost disappear by 2 weeks in the normal chick, but their concentration remains high in muscular dystrophic chicks (Wilson *et al.*, 1970). However, denervation of normal chick fast twitch muscle at 2 days results in the acetylcholine esterase activity remaining high. Although the primary biochemical defect in muscular dystrophy is still unclear, it has been shown that the viscosity of the cell membrane is significantly higher in membranes from muscular dystrophic chicks (Sha'afi *et al.*, 1975). The increased viscosity has been shown to affect the influx of potassium ions into red blood cells, and it may also affect potassium ion influxes into other tissues.

The early hypertrophy or atrophy of the muscle fibre associated with muscular dystrophy, either of which may occur depending on the genetic background, is the result of secondary gene interactions (Wilson *et al.*, 1988). Genomic fragments of DNA that contain the sequence coding for myosin heavy chain have been isolated from normal and muscular dystrophic domestic fowl. Those from normal fowl have an adjacent sequence coding for tcRNA 102 (translation control RNA) that was found lacking in muscular dystrophic fowl. This lack of tcRNA prevented *in vitro* transcription of the myosin heavy chain genes (Zezza & Heywood, 1986).

There have been suggestions that muscular dystrophy is associated with abnormally low levels of circulating immunoglobulins. A number of different strains of dystrophic birds have been kept in different research institutes, each strain originally derived from the *am* spontaneous mutation reported by Asmunden & Julian (1956). The strains so derived do not now possess the same immunoglobulin patterns. The Storrs strain of muscular dystrophic fowl has been found to have lower circulating immunoglobulin levels (Sanders, Kline & Morton, 1980), small thymus lobes (Cosmos *et al.*, 1977), reduced size of bursa (Befus *et al.*, 1981) and their immature macrophages or monocytes are less readily activated than in normal strains (Chu & Dietert, 1989). How these relate to the prime cause of the disease is at present not known.

## 7.2 The nervous system

The genetic aspects of the nervous system that have been studied in the domestic fowl are somewhat different from those of muscle. A number of mutants involving the nervous system have been investigated. These are mutants which have generally arisen spontaneously in flocks and in some cases have been propagated in order to understand the pathology and genetics of the lesion. Generally they have not been of value in understanding the developmental programme for the nervous system, and the molecular bases of the lesions are not understood. The expression of the gene for nerve growth factor has, however, been studied in early chick embryo. The messenger RNA for nerve growth factor can be detected as early as 3.5 days' incubation. The nerve growth factor has a role in the development of both the peripheral and central nervous system (Ebendal & Persson, 1988). The mode of inheritance has been determined in most cases, but linkage relations only in a few. A total of twenty mutants has been described; some are summarised in Table 7.2, but full details are given by Crawford (1990). Of the twenty, ten of the mutants cause some kind of tremor and seven cause convulsive seizures. Most of those causing tremors involve the cerebellum. The most probable reason for cerebellar lesions being common is that a malfunctioning cerebellum is usually not lethal. The cerebellum is situated near the top of the brain stem behind the cerebral hemispheres. The role of the cerebellum is in the fine control of muscular coordination. For example, when reaching out for a distant object, the cerebellum ensures that the limb moves smoothly to the precise position, not overreaching or moving jerkily. The removal of the cerebellum results in jerky and poorly coordinated movement. A particularly important type of cell within the cerebellum which is involved in this coordination is the Purkinje cell. Several of the mutants involve defects or loss of Purkinje cells. Many of these cause the birds to adopt abnormal postures; some of these are illustrated in Fig. 7.6.

Of the seven mutations described that cause convulsive seizures, only two have been studied in any detail, namely paroxysm (*px*) and epileptiform seizures (*epi*). Interest in these stems from their possible use as models for human epilepsies. Most of the studies on these have focused on the

Table 7.2. *Neurological traits in the domestic fowl*

| Trait name | Symbol | Inheritance | Lesion |
|---|---|---|---|
| Pirouette | *pir* | A/R | Tremor, whirling about |
| Paroxysm | *px* | S-L/R | Tetanic spasm |
| Shaker | *sh* | S-L/R | Rapid movement of head, loss of Purkinje cells |
| Jittery | *j* | S-L/R | Head retraction & rapid shaking. Purkinje cell degeneration |
| Epileptiform seizure | *epi* | A/R | Seizures triggered by nervous fatigue |
| Congenital quiver | *cq* | A/R | Continuous tremor, degeneration of neurons, loss of Purkinje cells |
| Congenital tremor | *ct* | A/R | Tremor, myelination defect |
| Faded shaker | *fs* | A/R | Tremor, myelination defect |
| Arched neck | | several A | No gross lesions of CNS |
| Cerebellar degeneration | | S-L/R | Atrophy of cerebellum |
| Crazy | *cy* | A/R | Uncontrolled body movements |

Abbreviations: A, autosomal; S-L, sex-linked; R, recessive.
For references, see Crawford (1990).

physiological and behavioural aspects, rather than the genetics, and in neither case is the primary lesion known. In paroxysm there is evidence of higher levels of the inhibitory neurotransmitter GABA ($\gamma$-aminobutyric acid) (Firman & Beck, 1984), and in epileptiform seizures lower than normal levels have been detected (see Crawford, 1990).

Several of these mutants show defects other than those of the nervous system. Congenital quiver, which is apparent at hatching, although an

**Fig. 7.6** Neurological defects in the domestic fowl.

autosomal recessive mutation, appears to have more pleiotropic effects in the male than in the female. The homozygous males are sterile, in contrast to the homozygous females. The sterility is due to abnormal spermatogenesis. Histological examination of the neurons in congenital quiver shows degeneration of both the fibre tracts and the neurons. There is also a loss of Purkinje cells in the cerebellum and partial replacement by fibrous astrocytes. Although the connection between male sterility and congenital quiver is not understood, other neurological defects are known to affect reproductive performance (see Smyth, Jersyk & Montgomery, 1985).

The defect known as faded shaker (Silversides & Smyth, 1986) affects both pigmentation of the skin and the nervous system. The hatchability of eggs is reduced to about two thirds of normal. The weight of the cerebellum is reduced relative to the weight of the cerebral hemispheres. There are no defects in the Purkinje cells but there is a reduction in cerebellar nerve myelination. The melanocytes have normal morphology, but there is lower deposition of melanin pigment within the cells. The melanocytes are also responsible for development of the Schwann cells and myelination in the central nervous system. The faded shaker gene behaves as an autosomal recessive with complete penetrance.

Thus all the genetic defects of the nervous system studied so far are recessive characters, either autosomal or sex-linked. The inheritance of most of them can be explained on the basis of single gene effects. However, the 'arched' trait appears to involve more than one gene, and congenital tremor, originally described by Hutt & Child (1934) as affecting less than 5% of siblings, can be most readily explained on the basis of two pairs of alleles, the defect being autosomal recessive (Sittmann, 1967). Some of the nervous defects are such as to only permit survival for a few days, but others such as epileptiform seizure can survive for a near normal life span.

## 7.3 The skeleton

Like the nervous system the skeletal mutants so far described have arisen spontaneously in domestic flocks. An exception is that of shankless which arose as a result of X-radiation aimed at producing chromosomal rearrangements (see Chapter 5, section 5.3). In this particular case shankless was shown to be due a pericentric inversion in chromosome 2. Many of the skeletal mutants are either lethal or semi-lethal, and these are described in Chapter 8, section 8.3. Although most of the genetic variants are less likely to survive than the 'normal' form, there are some that are specifically selected for by breeders, e.g. the heterozygous form of creeper is present in Scots Dumpies and Japanese bantams, polydactyly (generally 5-toed) is required in the Silkie, Houdan, Favorolle, Sultan and Dorking breeds, and dominant rumplessness in the Rumpless Araucanas. The mutant forms that had been studied before 1949 are described in detail in Hutt's *Genetics of the Fowl* and a comprehensive review including more recent work has been made by Somes (1990). Many of the most studied mutants are those affecting the appendages, but there are also a number affecting the axial skeleton including the head, beak, neck, vertebrae, sternum, and ribs. The limited range of mutants available has not, in general, been particularly useful to the developmental geneticist. Some of the more distal regions of limbs undergo brief periods of rapid growth and these appear more susceptible to mutations during that period. Some of the most well-characterised mutants are discussed below.

### *7.3a Dominant rumplessness*

This character has been known in the fowl for centuries, but at the present day is found principally in the Rumpless Araucanas. Whereas the normal domestic fowl has 16 synsacral vertebrae (vertebrae fused with the pelvis so that individual vertebrae are not easily distinguished), 5 free caudal vertebrae and the pygostyle (see Fig. 7.7), in complete rumplessness two of the sacro-caudal vertebrae are missing, also the pygostyle and caudal vertebrae. In the condition of intermediate rumplessness the sacro-caudal vertebrae are missing but the caudal vertebrae are generally present. Rumpless birds lack a uropygial gland (oil or preen gland) and the number of retrices (tail feathers) is reduced compared with that in a normal bird. The reduction is from 14–16 down to two in extreme examples. Rumplessness can generally be detected by the time of hatching. Dominant rumplessness is due to a single dominant gene, which in some intermediate states is affected by modifier genes.

It is alleged that rumplessness has the selective advantage of facilitating escape from foxes. The lack of an effective preen gland, which is situated at the base of the tail feathers, means that the birds' feathers do not repel water so readily and they are more prone to suffer in damp conditions. Rumpless fowl generally show lower viability and lower fertility when compared with their normal counterparts. The gene for dominant rumplessness has not been mapped.

### *7.3b Polydactyly*

There are a number of variations in the exact form of polydactyly. In most cases it results from duplication or dichotomy of the first toe rather than the

**Fig. 7.7** Examples of skeletal defects in the domestic fowl.

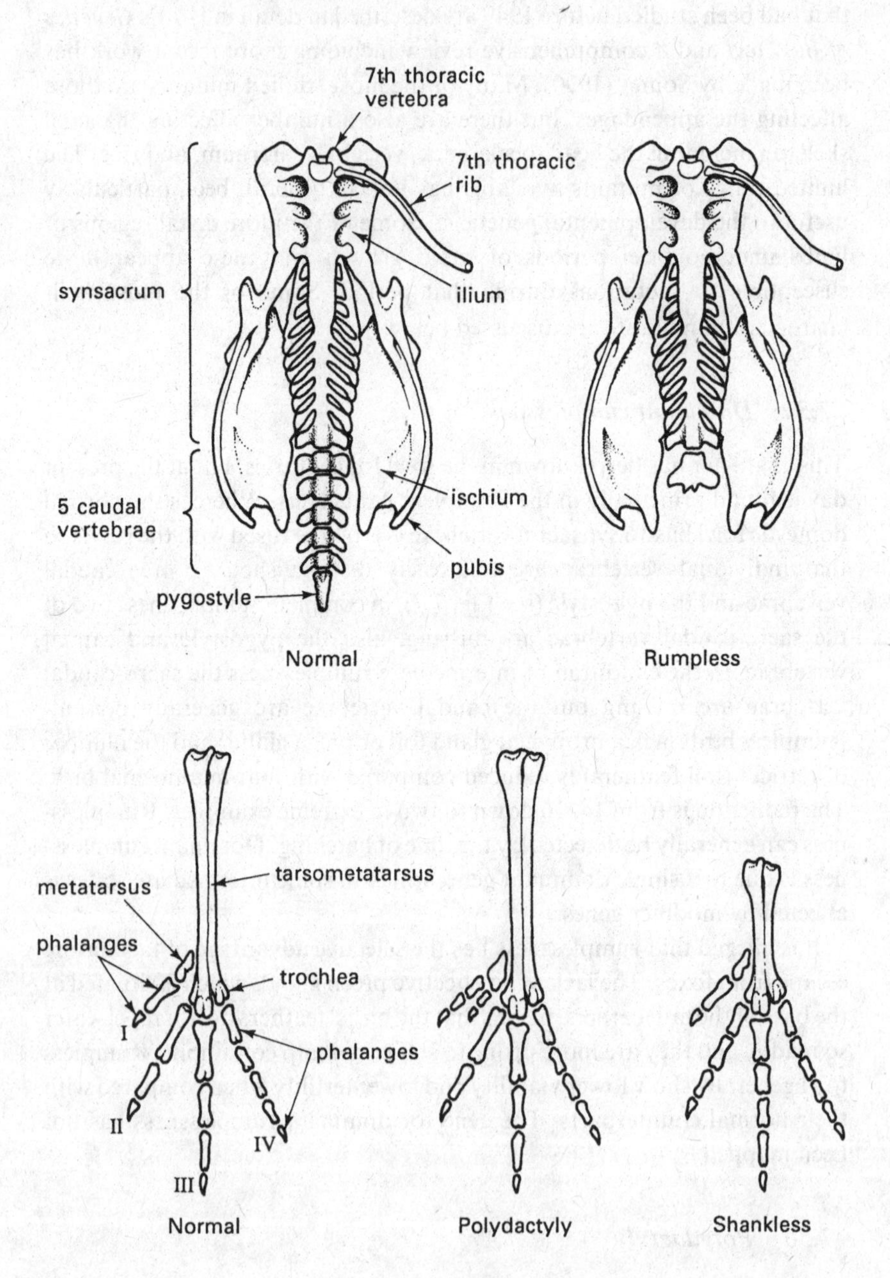

existence of an extra digit (Fig. 7.7). It is discernible at hatching. Polydactyly is caused by a single dominant gene mapped to linkage group IV (see Appendix I), but its expression is irregular. For example, it may be expressed on one foot only, or only the claw may be divided and duplicated. Other factors may thus be concerned with the expression of this gene. A multiple trait involving polydactyly has been discovered by McGibbon & Shackelford (1972). It arose from a cross between a Buttercup male and a Leghorn female. Approximately 17% of the birds affected showed multiple traits: polydactyly, syndactyly (adjacent digits connected) and ptilopody (webbed feet) (symbol: *psp*). The affected birds died within 3 weeks of hatching. The expression of the three traits was variable, particularly in the case of the syndactyly and ptilopody. All chicks had 5 or more toes. The *psp* allele is autosomal and recessive.

### *7.3c Brachydactyly*

Brachydactyly (*By*) is an autosomal, incompletely dominant trait affecting the fourth digit of the foot. It is fully described by Hutt (1949) and so will be described only briefly here. In normal birds the fourth digit is about 10% longer than the second digit, but in brachydactyly the fourth digit is as short as the second. It can be recognised in the embryo at 9–10 days. It occurs most commonly in birds having feathered feet; but this does not seem to be due to linkage of the genes but rather to a process during embryonic development common to both brachydactyly and feathered feet.

### *7.3d Creeper*

Creeper (*Cr*) is an autosomal incompletely dominant trait which is lethal in the homozygote (Chapter 8, section 8.2). The heterozygote *Cr*/*cr* is selected for in breeding Scots Dumpies and Japanese bantams. All the long bones of the legs are shortened; this is most pronounced in the case of the tibia. The fibula is thickened and extends the whole length of the tibia. Creeper is an example of a proximodistal gradient of effects (Langhorst & Fechheimer, 1985). In the normal situation growth of the more distal areas of the limb is concentrated at a specific and brief period during development and appears more susceptible to mutation. This mutation maps on linkage group I, only 0.4 map units from the gene for rose comb (*R*). This means that there is close linkage between the allelic pairs *Cr*,*cr* and *R*,*r*. Thus a breed having *Cr* and *r* genes will show a strong link between creeper and single comb.

### 7.3e *Shankless*

Shankless (*shl*) is a recessive mutant characterised by the absence of shanks. This mutation has been induced by X-radiation which in this instance caused a pericentric inversion to chromosome 2 (Langhorst & Fechheimer, 1985). The inversion can be seen by comparing the karyotype with that of a normal type (see Chapter 5, section 5.3). The phenotype is characterised by complete lack of the tarsometatarsal shaft, extra bones in digits 2, 3, and 4, fusion of the proximal phalanges (Fig. 7.7), and also by a shorter than normal metacarpal region of the wing. It is believed that the mutation occurs within or at the breakpoint of the inverted segment of the chromosome. Many shankless individuals die late in the incubation period, but a few survive to adulthood.

### 7.3f *Diplopodia*

Diplopodia (*dp*) is characterised by micromelia and preaxial polydactyly of both fore and hind limbs, the upper beak being shorter than the lower, and lethality in the final stages of embryonic development. Five different diplopodia mutants have been described. Diplopodia is further discussed in Chapter 8, section 8.3.

### 7.3g *Ametapodia*

Ametapodia (*Mp*) is characterised by lack of the metapodial region of the foot and is due to a single autosomal dominant gene (Cole, 1967). It is phenotypically similar to the shankless mutation. The birds are able to hobble around on the vestiges of the feet. The trait has no adverse effect on the hatchability or viability of the embryos. The gene has been mapped as being on the same linkage group (Group I) as creeper and rose comb.

### 7.3h *Wingless*

Two forms of winglessness have been described, an autosomal recessive (*wg*) (Hutt, 1949) and a sex-linked recessive (*wl*) (Pease, 1962; Lancaster, 1968). Most embryos from both types of winglessness are characterised by complete or partial absence of wings. Expression of the sex-linked form shows variable penetrance. Winglessness can be recognised early in embryonic development by the absence of wing buds. Those less severely affected have rudimentary humeri. In addition there are often other associated deficiencies, such as toes missing or duplicated. The autosomal form *wg* has not been mapped, but the sex-linked form *wl* has been mapped in relation to the genes for silver and for barring.

### 7.3i Other skeletal defects

Other skeletal defects have been discovered, but their genetic basis is only partly understood; they include cleft palate and a related trait, palatal pits. Cleft palate is a condition in which a fissure develops down the centre of the palate, and palatal pits are small pits that may develop on either side of a cleft palate. Although some instances of cleft palate in mammals have been attributed to single gene defects, the majority involve more than one locus. In the instances of cleft palate in the domestic fowl investigated by Juriloff & Roberts (1975), the trait was shown to be semi-lethal and to involve at least two or three recessive loci. The trait is not sex-linked, but in addition to the genetic basis it is also affected by nutrition, namely by riboflavin deficiency. The associated trait 'palatal pits' ($p^t$) is also a recessive trait showing Mendelian inheritance, but only preliminary investigations of its genetic basis have been carried out (Juriloff & Roberts, 1977).

The spontaneous occurrence of syndactylism (partial webbing between the toes) has been studied by Hollander & Brumbaugh (1969). Although it is clear that the extent of webbing between the toes can be increased by breeding, and is therefore genetically controlled, the mode of inheritance has not been resolved.

## References

Arnold, H. H., Lohse, P., Paterson, B. M. & Winter, B. (1988). The regulated expression of chicken muscle genes transfected into myogenic culture cells. In *Mechanisms of Control of Gene Expression*, ed. B.R. Cullen, pp. 85–96. New York: Alan R. Liss.

Asami, Y. Funatsu, T. & Ishiwata, S. (1988). Transition of β-actinin isoforms during development of chicken skeletal muscle. *Journal of Biochemistry*, **103**, 72–5.

Asmundson, V. S. & Julian, L. M. (1956). Inherited muscle abnormality in the domestic fowl. *Journal of Heredity*, **47**, 248–52.

Bandman, E., Matsuda, R. & Strohman, R. C. (1982). Developmental appearance of myosin heavy and light chain isoforms *in vivo* and *in vitro* in chicken skeletal muscle. *Developmental Biology*, **93**, 508–18.

Befus, A. D., Johnston, N., Nielsen, L., Bienenstock, J., Butler, J. & Cosmos, E. (1981). Thymus mast cell deficiency in avian muscular dystrophy. *Thymus*, **3**, 369–76.

Bitgood, J. J., Eutsler, E. P. & Wallace, M. P. (1987). Studies of the pirouette mutation. 1. Lack of linkage association with marked regions of chromosomes 1 and 2. *Poultry Science*, **66**, 38–40.

Breitbart, R. E., Andreadis, A. & Nadal-Ginard, B. (1987). Alternative splicing: a ubiquitous mechanism for the generation of multiple protein isoforms from single genes. *Annual Review of Biochemistry*, **56**, 467–96.

Cerny, L. C. & Bandman, E. (1987). Expression of myosin heavy chain isoforms in regenerating myotubes of innervated and denervated chicken pectoral muscle. *Developmental Biology*, **119**, 350–62.

Chu, Y. & Dietert, R. R. (1989). Monocyte function in chickens with hereditary muscular dystrophy. *Poultry Science*, **68**, 226–32.

Cole, R. K. (1961). Paroxysm – a sex-linked lethal of the fowl. *Journal of Heredity*, **52**, 46–52.

Cole, R. K. (1967). Ametapodia, a dominant mutation in the fowl. *Journal of Heredity*, **58**, 141–6.

Conner, M. H. & Shaffner, C. S. (1953). An arched-neck character in chickens. *Journal of Heredity*, **44**, 223–4.

Cosmos, E., Perey, D. Y. E., Butler, J. & Allard, E. P. (1977). Thymic–muscle interaction: A non-neural influence on metabolic differentiation of anaerobic muscle of normal and dystrophic genotype. *Differentiation*, **9**, 139–45.

Crawford, R. D. (1970). Epileptiform seizures in domestic fowl. *Journal of Heredity*, **61**, 185–8.

Crawford, R. D. (1990). Mutations and major variants of the nervous system. In *Poultry Breeding and Genetics*, ed. R. D. Crawford, pp. 257–72. Amsterdam: Elsevier.

Ebendal, T. & Persson, H. (1988). Detection of nerve growth factor mRNA in the developing chicken embryo. *Development*, **102**, 101–6.

Eppenberger, H. M. (1981). M Portein Myomesin – A specific protein of cross striated muscle cells. *Journal of Cell Biology*, **89**, 185–93.

Firman, J. D. & Beck, M. M. (1984). GABA in brain tissue of paroxysmal (*px*) chick. *Comparative Biochemistry and Physiology*, **79C**, 143–5.

Funatsu, T., Asami, Y. & Ishiwata, S. (1988). β-Actinin: a capping protein at the pointed end of thin filaments in skeletal muscle. *Journal of Biochemistry*, **103**, 61–71.

Godfrey, E. F., Bohren, B. B. & Jaap, R. G. (1953). "Jittery", a sex-linked nervous disorder in the chick. *Journal of Heredity*, **44**, 108–12.

Hastings, E. M. & Emerson, C. P. (1982). cDNA clone analysis of six co-regulated mRNAs encoding skeletal muscle contractile proteins. *Proceedings of the National Academy of Sciences of USA*, **79**, 1553–7.

Hayward, L. J. & Schwartz, R. J. (1986). Sequential expression of chicken actin genes during myogenesis. *Journal of Cell Biology*, **102**, 1485–93.

Hollander, W. F. & Brumbaugh, J. A. (1969). Web-foot or syndactylism in the fowl. *Poultry Science*, **48**, 1408–13.

Hutt, F. B. (1949). *Genetics of the Fowl*. New York: McGraw-Hill.

Hutt, F. B. & Child, G. P. (1934). Congenital tremor in young chicks. *Journal of Heredity*, **25**, 341–50.

Juriloff, D. M. & Roberts, C. W. (1975). Genetics of cleft palate in chickens and the relationships between the occurrence of the trait and maternal riboflavin deficiency. *Poultry Science*, **54**, 334–6.

Juriloff, D. M. & Roberts, C. W. (1977). "Palatal pits" – A new trait in chickens? *Poultry Science*, **56**, 386–8.

Kwiatkowski, R. W., Ehrismann, R., Schweinfest, C. W. & Dottin, R. P. (1985). Accumulation of creatine kinase mRNA during myogenesis: molecular cloning of a B-creatine kinase cDNA. *Developmental Biology*, **112**, 84–8.

Kropp, K., Gulick, J. & Robbins, J. (1986). A canonical sequence organization at the 5′-end of the myosin heavy chain genes. *Journal of Biological Chemistry*, **261**, 6613–8.

Lancaster, F. M. (1968). Sex-linked winglessness in the fowl. *Heredity*, **23**, 257–62.

Langhorst, L. J. & Fechheimer, N. S. (1985). Shankless, a new mutation on chromosome 2 in the chicken. *Journal of Heredity*, **76**, 182–6.

Linkhart, T. A., Yee, G. W., Nieberg, P. S. & Wilson, B. W. (1976). Myogenic defect in muscular dystrophy of the chicken. *Developmental Biology*, **48**, 447–57.

Matsuda, G., Maita, T., Kato, Y., Chen, J.-I. & Umegame, T. (1981) Amino acid sequences of the cardiac, L-2A, L-2B and gizzard 17000–$M_r$ light chains of chicken muscle myosin. *Federation of European Biochemical Societies Letters*, **135**, 232–6.

Markson, L. M., Carnaghan, R. B. A. & Young, G. B. (1959). Familial cerebellar degeneration and atrophy – a sex-linked disease affecting Light Sussex pullets. *Journal of Comparative Pathology*, **69**, 223–9.

McGibbon, W. H. (1973). Crazy – A nervous disorder in Ancona chicks. *Journal of Heredity*, **64**, 91–4.

McGibbon, W. H. (1974). Pirouette: A behavioural mutation in the domestic fowl. *Journal of Heredity*, **65**, 124–6.

McGibbon, W. H. & Shackelford, R. M. (1972). A multiple trait semilethal in fowl. *Journal of Heredity*, **63**, 209–11.

Merrill, G. F., Hauschka, S. D. & McKnight, S. L. (1984). tk Enzyme expression in differentiating muscle cells is regulated through an internal segment of the cellular tk gene. *Molecular and Cellular Biology*, **4**, 1777–84.

Nabeshima, Y., Fujii-Kuriyama, Y., Muramatsu, M. & Ogata, K. (1984).Alternative transcription and two modes of splicing result in two myosin light chains from one gene. *Nature*, **308**, 333–8.

Paterson, B. M. & Bishop, J. O. (1977). Changes in the mRNA population of chick myoblasts during myogenesis *in vitro*. *Cell*, **12**, 751–65.

Pease, M. S. (1962). Wingless poultry. *Journal of Heredity*, **53**, 109–10.

Peberdy, J. F. (1980). *Developmental Microbiology*. Glasgow: Blackie.

Perriard, J. C., Achtnich, U., Cerny, L., Eppenberger, H. M., Grove, B. K., Hossle, H. P. & Schafer, B. (1986). Expression of M-band during myogenesis. In *Molecular Biology of Muscle Development*, pp. 693–707. New York: Alan R. Liss.

Rathbone, M. P., Stewart, P. A. & Vetrano, F. (1975). Dystrophic spinal cord transplants induce abnormal thymidine kinase activity in normal muscles. *Science*, **89**, 1106–7.

Robbins, J., Horan, T., Gulik, J., & Kropp, K. (1986). The chicken myosin heavy chain family. *Journal of Biological Chemistry*, **261**, 6606–12.

Saborio, J. S., Segura, M., Flores, M., Garcia, F. R. & Palmer, E (1979). Differential expression of gizzard actin genes during chick embryogenesis. *Journal of Biological Chemistry*, **254**, 11119–25.

Sanders, B. G., Kline, K. & Morton, C. J. (1980). Serum IgG levels in the Storrs strain of hereditary muscular dystrophic chickens. *Biochemical Genetics*, **18**, 1149–58.

Scott, H. M., Morrill, C. C., Alberts, J. O. & Roberts, E. (1950). The "shaker" fowl. A sex-linked semi-lethal nervous disorder. *Journal of Heredity*, **41**, 254–7.

Sha'afi, R. I, Rodan, S. B., Hintz, R. L., Fernandez, S. M. & Ridan, G. A. (1975). Abnormalities in membrane microviscosity and ion transport in genetic muscular dystrophy. *Nature*, **254**, 525–6.

Shaw, E. M., Guise, K. S. & Shoffner, R. N. (1989). Chromosomal localization of chicken sequences homologous to the β-actin gene by *in situ* hybridization. *Journal of Heredity*, **80**, 475–8.

Silversides, F. G. & Smyth, J. R. (1986). Faded shaker, a lethal pigment and neurological mutation in the chicken. *Journal of Heredity*, **77**, 295–300.

Sittmann, K. (1967). Penetrance of a rare genetic defect. *Genetical Research*, **10**, 229–33.

Smyth, J. R., Jerszyk, M. M. & Montgomery, N. (1985). Congenetical quiver, an inherited neurological defect in the chicken. *Journal of Heredity*, **76**, 263–6.

Somes, R. G. (1990). Mutations and major variants of muscles and skeleton in chickens. In *Poultry Breeding and Genetics*. ed. R. D. Crawford, pp. 209–37. Amsterdam: Elsevier.

Walliman, T., Turner, D. C. & Eppenberger, H. M. (1977). Localization of creatine kinase isoenzymes in myofibrils. I. Chicken skeletal muscle. *Journal of Cell Biology*, **75**, 297–317.

Wilson, B. W., Kaplan, M. A., Merhoff, W. C. & Mori, S. S. (1970). Innervation and the regulation of acetylcholinesterase activity during the development of normal and dystrophic chick muscle. *Journal of Experimental Zoology*, **174**, 39–54.

Wilson, B. W., Abplanalp, H., Buhr, R. J., Entrikin, R. K., Hooper, M. J. & Nieberg, P. S. (1988). Inbred crosses and inherited muscular dystrophy of the chicken. *Poultry Science*, **67**, 367–74.

Zezza, D. J. & Heywood, S. M. (1986). The localisation of a tcRNA102 gene near the 3′ IH terminus of a fast myosin heavy chain gene. *Journal of Biological Chemistry*, **261**, 7455–60.

# 8

# *Lethal genes in domestic fowl*

## 8.1 Introduction

Genes may be transformed by spontaneous mutations, which occur at a low frequency in the range of 1 in every $2 \times 10^{-4}$ to $4 \times 10^{-10}$ meiotic divisions. Agents such as UV light, X-rays and certain chemicals known as mutagens increase the frequency of mutation. Mutations involve changes in the structure of DNA and hence the information coded in the DNA sequences. Different types of changes can occur such as the removal or addition of a base, or the inversion or transposition of a segment of DNA. Mutation is generally thought of as a random process, although certain positions in DNA may be more sensitive to change. Because of the random nature of any change most mutations are detrimental to the organism concerned. It is somewhat analogous to making a random replacement of a component in a computer; it is much more likely that the computer will perform less well, than that it will show additional capabilities. Very occasionally mutants may produce an improved genotype that will have a selective advantage in the environment. This is of great importance in the process of evolution.

Many mutations are **pleiotropic**, that is, they are wide ranging in their effects on the phenotype, and it is in most cases difficult to establish the nature of the primary effect. It is often useful to make a phenotypic classification of mutants by distinguishing the effectiveness of the mutant gene product. Five types are distinguished: (i) **amorphic** – where no active gene product is produced, (ii) **hypomorphic** – where the gene product is less effective than that of the wild type, (iii) **hypermorphic** – where the gene product is more effective than that of the wild type, (iv) **antimorphic** – where the gene product has the opposite effect to that of the wild type, and (v) **neomorphic** – where the mutant expresses a new character. Mutations that are detrimental to the organism vary from those which are incompatible with life to those that only have a slightly adverse effect, and these are sometimes distinguished by the terms **lethal** and **semi-lethal**. For individuals with semi-lethal genes, conditions under which they are nurtured may determine how well they survive. For example, in the sleepy-eyed mutant (Somes, 1968), chicks hatch with their eyelids partly closed and this impairs their vision, making it difficult for them to find food and water. Many of the extreme cases die within 3–4 days of hatching, but if their eyelids are sutured into an open position, mortality decreases dramatically.

Multicellular organisms such as the domestic fowl have very complex developmental programmes which enable the adult to form from a single fertilised egg. The cell formed from fusion of the sperm with the ovum contains all the genetic information required for an individual to complete development. The process of development involves a precise programming, such that particular genes are expressed at given stages of development. This temporal sequence means that mutation of a particular gene will often affect one particular stage in development. Thus many lethal genes terminate life during embryonic development. If the gene is a regulatory gene it will often control a number of other genes in a hierarchical fashion. It is often very difficult to understand the nature of the primary defect from the varied phenotypic effects observed. By contrast, mutations in lower organisms can often be useful in elucidating the programming of the genes. For example, the conversion of a vegetative bacterial cell to its resting spore has been extensively analysed by studying mutants arrested at particular stages of development (Dawes, 1982). Eventually a similar approach may be possible with higher organisms like the domestic fowl.

Lethal mutations probably constitute the most common type of mutation. They may be dominant or recessive; if completely dominant they will not survive, and hence are not introduced into the population. Many lethal mutations will not be observed if they arrest development at a very early stage, or they may even prevent fertilisation from occurring. Lethal mutations are often recessive and thus become evident only in the homozygote. This is because the lethal gene is often of the amorphic type, unable to perform the function of the wild type, rather than because it is of the antimorphic or neomorphic type. Thus in the presence of the normal allele it is not expressed.

Lethal genes are often more evident in domestic animals than in their corresponding undomesticated counterparts. This stems from two factors. Firstly, in a protected environment a semi-lethal mutant is more likely to survive long enough to be observed. Secondly, in order to maintain desirable features in a domestic animal a higher degree of inbreeding may occur; in this way recessive lethals are more likely to become homozygous. Over 100 loci in the domestic fowl have now been shown to have lethal effects (Somes, 1990), some autosomal and some sex-linked. Examples of

Table 8.1. *Lethal genes in the domestic fowl*

| Nature of defect | Symbol | Inheritance | Stage at which lethal | Linkage | Penetrance or expressivity[a] |
|---|---|---|---|---|---|
| Ametapodia | *Mp* | Aut Dom | < 21 days | Group I | |
| Chondrodystrophy | *ch* | Aut Rec | < 21 days | unknown | |
| Coloboma | *co* | SL Rec | | Group V | CP VE |
| Creeper | *Cr* | Aut inc Dom | < 21 days | Group I | |
| Crooked-neck dwarf | *cn* | Aut Rec | 20–21 days | unknown | |
| Diplopodia-1 | *dp-1* | Aut Rec | < 22 days | unknown | VE |
| Diplopodia-2 | *dp-2* | Aut Rec | 15–22 days | unknown | VE |
| Diplopodia-3 | *dp-3* | Aut Rec | | | |
| Diplopodia-4 | *dp-4* | SL Rec | | Group V | |
| Diplopodia-5 | *dp-5* | Aut Rec | 19 days | not Group I, II or III | CP VE |
| Donald Duck | *dd* | Aut Rec | < 22 days | unknown | |
| Ear-tufts | *Et* | Aut Dom | < 21 days | unknown | IP IE |
| Ectrodactyly | *ec* | Aut Rec | 17–20 days | unknown | |
| Faded shaker | *fs* | Aut Rec | 18 days–3 months | unknown | CP |
| Limbless (amelia) | *ame* | Aut Rec | ⩽21 days | unknown | |
| Nanomelia | *nm* | Aut Rec | < 21 days | unknown | |
| Paroxysm | *px* | S-L Rec | *c.* 15 weeks | Group V near *n* | |
| Prenatal sex-linked lethal | *pn* | S-L Rec | 1–7 & 15–22 days | Group V near *S* & *K* | |
| Shaker | *sh-2* | Aut Rec | semi-lethal ⩾14 weeks | unknown | LP |
| Split foot | *sf* | Aut Rec | 17–19 days | unknown | VE |
| Wingless | *wg-2* | Aut Rec | ⩽22 days | unknown | |

[a] CP, complete penetrance; IP, incomplete penetrance; LP, low penetrance; VE, variable expression; IE, irregular expression.
For complete references see Hutt (1949), Somes (1990).

these are given in Table 8.1, some of which are described in more detail in section 8.3, but for a comprehensive list see Somes (1990).

## 8.2 Detection of lethal genes in domestic fowl

Lethal genes often come to light when a high percentage of eggs fails to hatch. Dissecting the unhatched embryos may reveal the defect, especially if it is a gross anatomical defect. In this way a number of lethal genes controlling, for example, nanomelia, split-foot, diplopodia and coloboma became evident. Once an effect has been found it is then necessary to relate it back to its parents. Further breeding experiments should establish whether it is caused by a single gene and how the defect is inherited. Most lethal genes which survive in a population are recessive, and so the commonest way in which a lethal gene will become evident is if both parents are heterozygous for the gene in question. This will lead to the following distribution in the progeny ($a^l$ is a recessive lethal gene).

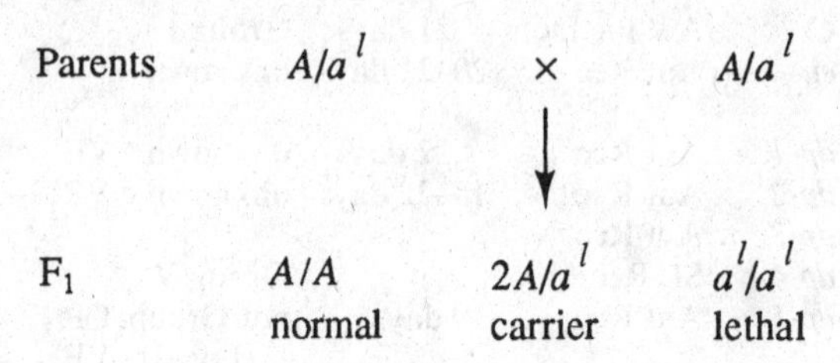


Thus only 75% of the progeny will survive and two thirds of the survivors will be carriers of the lethal gene. If the lethal gene is an incomplete dominant, the outcome will be different. A good example is the Creeper gene (*Cr*), which is present in the heterozygous state in the Scots Dumpy and the Japanese bantams. The homozygous *Cr*/*Cr* is lethal. Thus when Japanese bantams or Scots Dumpies, having the shortened legs, are bred, the following occurs.

| | | | |
|---|---|---|---|
| Parents | *Cr*/*cr* short-legged | × | *Cr*/*cr* short-legged |
| | | ↓ | |
| $F_1$ | *Cr*/*Cr* lethal | 2*Cr*/*cr* short legs | *cr*/*cr* normal legs |

The surviving progeny will have a 2:1 ratio of short-legged:normal legged.

The presence of lethal genes will affect the ratios obtained in a dihybrid cross. The normal dihybrid cross (*AB*/*ab* × *AB*/*ab*) gives the characteristic 9:3:3:1 ratio for pairs of independently assorting alleles. This ratio will be modified if either or both recessive genes are lethal. For example, if *b* were considered a lethal gene the ratio of viable offspring would be 9:3 (*A–B–* :*aaB–*), and if both *a* and *b* were lethal the surviving offspring would be all *A–B–*. If the lethal genes were incompletely dominant the ratios would be again modified. Table 8.2 summarises the main possibilities for a dihybrid

Table 8.2. *Phenotypic ratios for dihybrid crosses involving lethal genes*

| Allelic relationships in dihybid parents | | |
|---|---|---|
| First locus | Second locus | Expected adult phenotypic ratio |
| Dominant-recessive | Lethal rec. | 9:3 A–B–:aaB– |
| Dominant-recessive | Lethal inc. rec. | 3:6:1:2 A–B–:A–Bb:aaB–:aaBb |
| Incomplete dominant | Lethal rec. | 3:6:3 AAB–:AaB–:aaB– |
| Lethal recessive | Lethal inc. rec. | 3:6 AAB–:AaB– |
| Lethal recessive | Lethal rec. | 9:0 A–B– |

A & a are alleles at the first locus, and B & b are alleles at the second locus.
inc, incomplete; rec, recessive.

cross involving two pairs of independently assorting alleles in which one or both loci contain a lethal gene.

## 8.3 Examples of lethal genes in the domestic fowl

Some examples of lethal genes are now described, in order to illustrate their genetics and their expression.

### *8.3a Diplopodia*

A number of abnormalities result in extra digits, some lethal and others not. Polydactyly, in which a fifth toe is present, is found in breeds such as the Dorkings, Silkies and Houdans. It is an autosomal dominant and is not lethal. Talpid is a genetic abnormality in which the wings have several extra digits and are short and palmate, resembling the broad forefeet of a mole (*Talpa*). It is generally lethal at 8–11 days of incubation.

The distinguishing feature of diplopodia is the partial doubling of the structure of the foot, many of the embryos having six toes arranged in two sets of three. Diplopodia was first described in the domestic fowl by Taylor & Gumms (1947). Since then five different mutants have been described and designated *dp-1*, *dp-2*, *dp-3*, *dp-4* and *dp-5*, although *dp-2* now appears to be extinct. All are autosomal recessives except *dp-4*, which is a sex-linked recessive (Abbott & Kieny, 1961). The phenotypes have common features, although some may be more pronounced than others. The main features are preaxial polydactyly, generally resulting in six toes rather than four, micromelia (shortening of the long bones of the limbs), and the upper beak shorter than the lower. Although all forms are lethal, some are more severe and thus result in earlier mortality. In their study Taylor & Gumms (1947) found that *dp-1* chicks only rarely hatched (*c.* 2%) and 98% were dead in shell. The second diplopodia mutant (*dp-2*) described by Landauer (1956) showed similar features to the first, but results suggested that different genes were involved. The third diplopodia mutant (*dp-3*) was discovered in 1957, no chicks ever hatched, although some pipped (Taylor, 1972). Death generally occurred at *c*. 21 days. The severity of morphological expression of characters is considered to decrease in the order *dp-2* > *dp-4* > *dp-1* > *dp-3* ≃ *dp-5* (Taylor, 1972; Somes, 1990). The mutants *dp-1* and *dp-3* are non-allelic. Only a short note has been published on the sex-linked diplopodia (Abbott & Kieny, 1961). The most recent diplopodia mutant to be described is *dp-5* (Olympio, Crawford & Classen, 1983). It shows morphological features distinct from *dp-1*, *dp-2* and *dp-4* and has been shown to be non-allelic with *dp-3*. Linkage tests using the marker genes pea comb, rose comb, dominant white and naked neck have failed to show linkage between these genes and that of diplopodia-5. This suggests that *dp-5* is either not on linkage groups I, II or III, or if it is, it must be some distance from the markers and thus show very weak linkage.

### 8.3b *Ear-tufts*

The presence of ear-tufts is a feature most notable in the Araucana breed of domestic fowl. The ear-tuft consists of a feather-covered epidermal appendage that projects from the side of the head, either at the distal edge of the external ear opening or from the ventral surface of the earlobe (Fig. 8.1). The Araucana is one of the more recent breeds to be accepted into the British Poultry Standards (compare 3rd and 4th editions, 1971, 1982), and there are differences in the British, German and American standards, thus different forms of ear-tufts are preferred by different breeders.

The genetics and the associated lethality of the ear-tufts have been studied by Somes (1978), Somes & Pabilonia (1981) and Pabilonia & Somes

**Fig. 8.1** Ear-tufts in Araucanas.

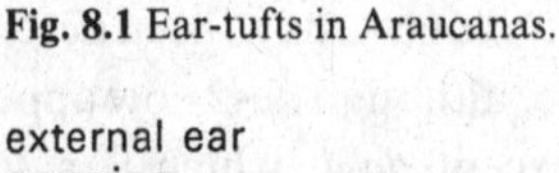

(1983). The ear-tuft trait is associated with abnormalities in ear and throat development. It is inherited as an autosomal dominant which is generally lethal in the homozygous state. The trait shows variable expressivity; there is considerable variation in the size of the ear-tufts, and also a pair may show asymmetry. There is reduced penetrance in the heterozygotes; estimated at between 4 and 14% (Somes & Pabilonia, 1981).

In the homozygous state most embryos die between 17 and 19 days and the few chicks that hatch die within a week. The heterozygotes also show a higher mortality (average *c.* 40%) than non-tufted counterparts. In a detailed examination Pabilonia & Somes (1983) have shown that the abnormalities are caused by incomplete fusion of the hyoid and mandibular arches (Fig. 8.2). This gives rise to a range of shapes of the ear openings, to large holes in the throat or to fissure-like clefts, and is believed to be responsible for the ear-tuft development. From the breeder's point of view there is therefore a fine balance between obtaining birds with well-formed ear-tufts and ones succumbing to the adverse effects of incomplete closure of the hyomandibular arch.

An abnormality leading to ear defects, that develops in a similar way, has been observed in the Japanese quail (Tsudzuki & Wakasugi, 1988). In contrast to the ear-tuft (*Et*) gene, it is an autosomal recessive and it is thought that the gene acts on a different region of the visceral arch.

**Fig. 8.2** Diagram of a four-day-old chick embryo showing the hyoid and mandibular arches. The hyoid and mandibular arches close to form the ear opening during normal embryonic development.

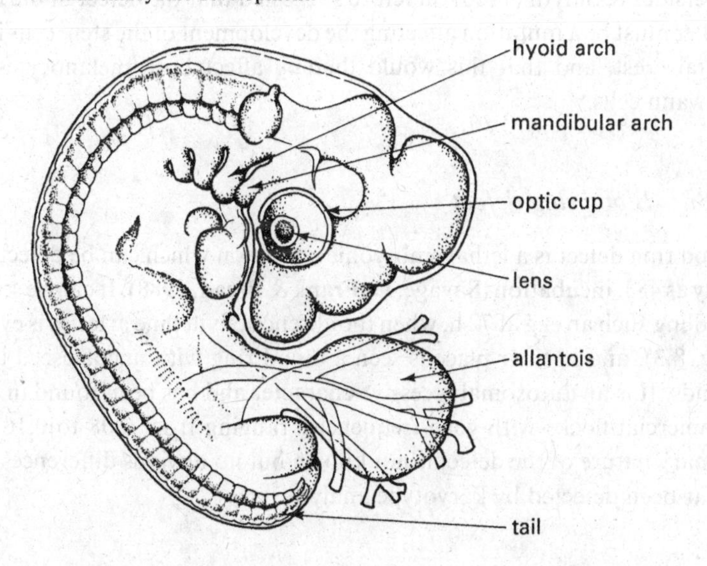

### 8.3c Faded shaker

Faded shaker is a lethal condition, first described in the domestic fowl by Silversides & Smyth (1986), involving both the nervous system and a reduction of feather pigmentation by melanin. Combinations of deficiencies involving pigmentation and the nervous system have been previously described in mammals, but not in the domestic fowl. Chicks having faded shaker die between 18 days of incubation and 3 months after hatching. The characteristic features are a congenital tremor, i.e. a tremor shown at hatching, and a much lighter coloured down than their normal counterparts. In most cases the faded shaker gene behaves as a single autosomal gene (*fs*) with complete penetrance. However, the results of some breeding experiments produced fewer mutants than expected (reduced penetrance). Silversides & Smyth (1986) suggested that this may be due to some modifying gene present in the particular stock used for the breeding experiment. All the birds which were homozygous (*fs*/*fs*) showed both nervous and pigment defects.

Detailed examination showed that the tremor is associated with a deficiency of myelin in the medulla of the cerebellum, and the faded appearance is caused by a deficiency in the amount of melanin deposited in the premelanosome. For details of melanin pigmentation see Chapter 6, section 6.4a. Both melanocytes, responsible for the pigmentation of feathers, and the Schwann cells, responsible for myelin production in the central nervous system, are derived from the neural crest in the embryo. Silversides & Smyth (1986) therefore suggested that the defect in the faded shaker must be a mutation affecting the development of the stem cells in the neural crest, and that this would thereby affect both melanocytes and Schwann cells.

### 8.3d Blood ring defect

Blood ring defect is a lethal embryonic condition which can be detected as early as 48 h incubation (Savage, DeFrank & Brean, 1988). It can be seen by candling such an egg at 72 h, when the absence of vitelline arteries is evident (Fig. 8.3), and in their place is seen a blood ring with uncoalesced blood islands. It is an autosomal recessive character and has been found in some commercial flocks with gene frequencies ranging from 0.08 to 0.16. The primary nature of the defect is not known but no obvious differences have so far been detected by karyotype analysis.

### *8.3e Nanomelia*

Few of the conditions brought about by lethal genes have been investigated at a biochemical level, but this is not the case with nanomelia. Nanomelia is one of seven chondrodystrophies which are embryonic lethals in the domestic fowl (Somes, 1990). Chondrodystrophy is used to describe abnormal development of the limb long bones. In the case of nanomelia there is malformation and shortening of of the limb bones. Embryos die at about 18 days of incubation. The two main constituents of cartilage are collagen and chondroitin sulphate proteoglycan. The chondroitin sulphate proteoglycan is made up of a protein core (*c.* 8–10% of the total material) to which chondroitin sulphate is attached. It is the production of the chondroitin sulphate proteoglycan that is defective in nanomelia. Nanomelic chondrocytes synthesise the core at only about 10% the rate of normal (McKeown-Longo & Goetinck, 1982). Further, the normal protein core has $M_r$ of 370 000, but in the nanomelia condition a smaller core ($M_r$ 300 000) has been detected (O'Donnell *et al.*, 1988). The precise function of the defective gene has yet to be determined.

The examples of lethal conditions described in this section illustrate the point that mutations affecting genes involved in the developmental programme are likely to cause pleiotropic effects on the resulting phenotypes,

**Fig. 8.3** Macroscopic appearance of embryos at 72 h of incubation. *A*, normal, and *B*, blood ring defect (Savage *et al.*, 1988).

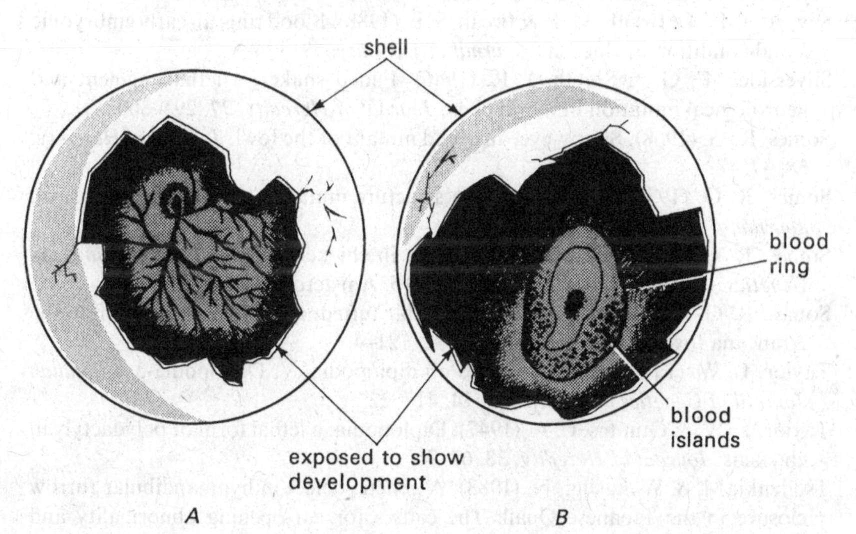

particularly if the gene concerned happens to be at the apex of a bank of other genes. For a comprehensive account of lethal genes in the domestic fowl see Hutt (1949) and Somes (1990).

## References

Abbott, U. K. & Kieny, M. (1961). Sur la croissance *in vitro* du tibiotarse et du perone de l'embryon de poulet "diplopode". *Comptes Rendus Academie des Sciences*, **252**, 1863–5.

*British Poultry Standards*, 3rd edn, ed. C. G. May (1971), 4th edn, ed. D. Hawksworth (1982). London: Butterworths.

Dawes, I. W. (1982). Differentiation: sporogenesis and germination. In *Biochemistry of Bacterial Growth*, 3rd edn, ed. J. Mandelstam, K. McQuillen and I. W. Dawes, pp. 379–84. Oxford: Blackwell Scientific Publications.

Hutt, F. B. (1949). *Genetics of the Fowl*. New York: McGraw-Hill.

Landauer, W. A. (1956). A second diplopod mutation of the fowl. *Journal of Heredity*, **47**, 57–63.

McKeown-Longo, P. J. & Goetinck, P. F. (1982). Characterization of the tissue-specific proteoglycans synthesized by chondrocytes from nanomelic chick embryos. *Biochemical Journal*, **201**, 387–94.

O'Donnell, C. M., Kaczman-Daniel, K., Goetinck, P. F. & Vertal, B. M. (1988). Nanomelic chondrocytes synthesize a glycoprotein related to chondroitin sulfate proteoglycan core protein. *Journal of Biological Chemistry*, **263**, 17749–54.

Olympio, O. S., Crawford, R. D. & Classen, H. L. (1983). Genetics of the diplopodia-5 mutation in domestic fowl. *Journal of Heredity*, **74**, 341–3.

Pabilonia, M. S. & Somes, R. G. (1983). The embryonic development of ear-tufts and associated structural head and neck abnormalities of the Araucana fowl. *Poultry Science*, **62**, 143–52.

Savage, T. F., DeFrank, M. P. & Brean, S. E. (1988). Blood ring: an early embryonic lethal condition in chickens. *Journal of Heredity*, **79**, 124–8.

Silversides, F. G. & Smyth, J. R. (1986). Faded shaker, a lethal pigment and neurological mutation in the chicken. *Journal of Heredity*, **77**, 295–300.

Somes, R. G. (1968). Sleepy-eye, an eyelid mutant of the fowl. *Journal of Heredity*, **58**, 375–8.

Somes, R. G. (1978). Ear-tufts: a skin structure mutation of the Araucana fowl. *Journal of Heredity*, **69**, 91–6.

Somes, R. G. (1990). Lethal mutant traits in chickens. In *Poultry Breeding and Genetics*, ed. R. D. Crawford, pp. 293–315. Amsterdam: Elsevier.

Somes, R. G. & Pabilonia, M. S. (1981). Ear tuftedness: a lethal condition in the Araucana fowl. *Journal of Heredity*, **72**, 121–4.

Taylor, L. W. (1972). Further studies on diplopodia. V. Diplopodia-3. *Canadian Journal of Genetics and Cytology*, **14**, 417–22.

Taylor, L. W. & Gumms, C. A. (1947). Diplopodia: a lethal form of polydactyly in chickens. *Journal of Heredity*, **38**, 67–76.

Tsudzuki, M. & Wakasugi, N. (1988). A genetic defect in hyomandibular furrow closure in the Japanese Quail: The causes for ear-opening abnormality and formation of an ear tuft. *Journal of Heredity*, **79**, 160–4.

# 9

# Quantitative genetics

## 9.1 Introduction

The characters discussed in Chapters 3–8 are generally discrete in nature. There is a marked distinction between different alleles, and inheritance occurs in straightforward Mendelian fashion. The varieties of plumage, the type of comb, the feathering of the legs, polydactyly, and the colour of skin are all differences of kind. They are the types of character that both poultry fancier and geneticist are generally most interested in: the former because they represent many of the attractive features of the birds and they give them their individuality, and the latter because they are the more satisfactory traits to analyse in genetic terms.

However, most of the characters in which the commercial breeder is interested show continuous variation, e.g. body weight, proportion of body fat to muscle, size of egg, rate of growth and rate of egg laying. With these, there are not two alternative phenotypes which are easily distinguished, but continuous variations between two extremes. Most of these characters are quantitative and easily measured.

Characters that exhibit continuous variation are more difficult to analyse and, in fact, were a puzzle to early geneticists, until Nilsson-Ehle, a Swedish geneticist, showed in 1910 that it was possible to account for the colours of wheat kernels, which ranged from red, through medium-red, light red and very light red, to white, by assuming that two pairs of alleles were responsible for the colour, and that each allele had an additive effect. Thus if we start with a homozygous red ($R_1R_1R_2R_2$) and cross with a homozygous white ($r_1r_1r_2r_2$) and then allow the $F_1$ to self-pollinate to form the $F_2$ generation, the resultant phenotypes occur in the following proportions:

| Phenotype | Proportions | Genotype |
|---|---|---|
| red | 1/16 | *RRRR* |
| medium red | 4/16 | *RRRr* |
| light red | 6/16 | *RRrr* |
| very light red | 4/16 | *Rrrr* |
| white | 1/16 | *rrrr* |

With just two pairs of alleles we can account for five colours, but if the same treatment is applied to three pairs of alleles each having an additive effect on the colour then seven shades would be possible, and with four pairs of alleles nine shades of colour; with $n$ pairs, $2n+1$ phenotypes are possible (Fig. 9.1). The number of shades increases with the number of alleles involved; the phenotypic ratio has the form $(a+b)^{2n}$, where $a$ and $b$ represent the allelic pairs. In making these deductions it has been assumed that each pair of alleles contributes equally to the character, e.g. $R_1$ makes the same as $R_2$, and $r_1$ makes the same as $r_2$. If they contribute to differing extents then the number of possibilities becomes greater. As the number of pairs of alleles governing a character increases, the transitions between each genotype become less distinct, so that ultimately there is a continuum between the two extremes, and the distribution between each genotype fits a smooth curve, which is known as **normal distribution** (Fig. 9.2).

The 'quantitative characters' (characters that can be measured or quantified) thus arise because they are **polygenic**. In the early work on egg

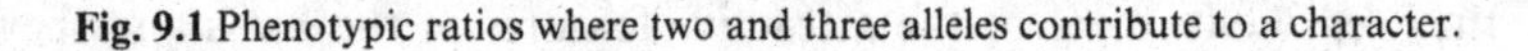

**Fig. 9.1** Phenotypic ratios where two and three alleles contribute to a character.

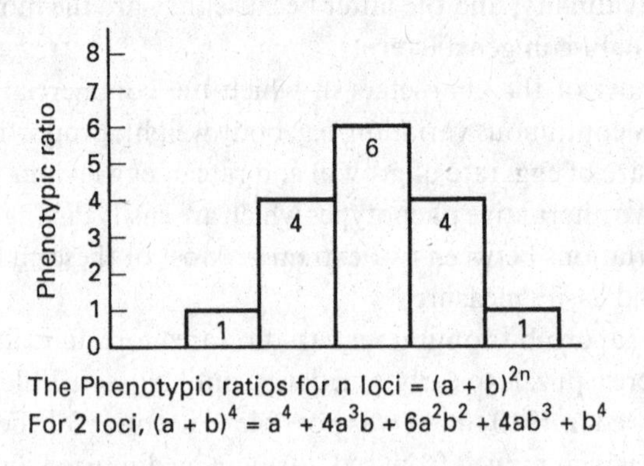


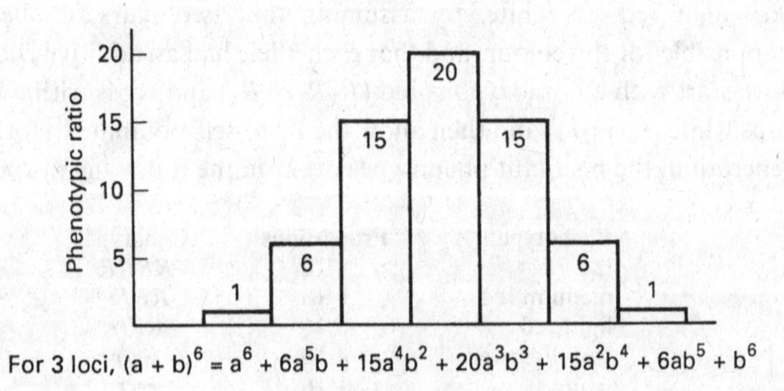

production attempts were made to fit results to those expected from a simple Mendelian relationship, but it was not until 1936 that it was proposed that egg production was a polygenic character (see Fairfull & Gowe, 1990).

The number of genes accounting for a polygenic trait is often unknown. An approximate estimate of the number of genes can sometimes be made in simpler cases by determining the proportion in the $F_2$ generation that have the extreme phenotype. For example, where two pairs of alleles are involved the extremes will represent 1/16 each of the total; if there are three pairs of alleles the extremes will be 1/64, and four alleles 1/256. The extremes in the example given earlier would be $R_1R_2R_3R_4$ and $r_1r_2r_3r_4$. The extremes are $(1/4)n$ of the population, where $n$ is the number of alleles involved. The study of polygenic characters is generally referred to as **quantitative genetics**. Polygenic characters are of particular importance in breeding, and also in explaining natural selection. In a population showing a normal distribution for a particular character, natural selection may allow the best fitted extreme to reproduce more successfully and hence shift the overall distribution of the genotypes.

The analysis of quantitative characters is more complex than that of qualitative characters, and any detailed treatment requires a knowledge of statistics. In this chapter an outline of basic principles is given in order to provide sufficient background to understand the parameters used, and this is followed by a description of the quantitative characters that are of most importance to the poultry breeder, and which have resulted in fundamental changes in the commercial production of broilers and layers over the last 30 years.

**Fig. 9.2** Normal distribution. $\mu$, mean; $\sigma$, standard deviation.

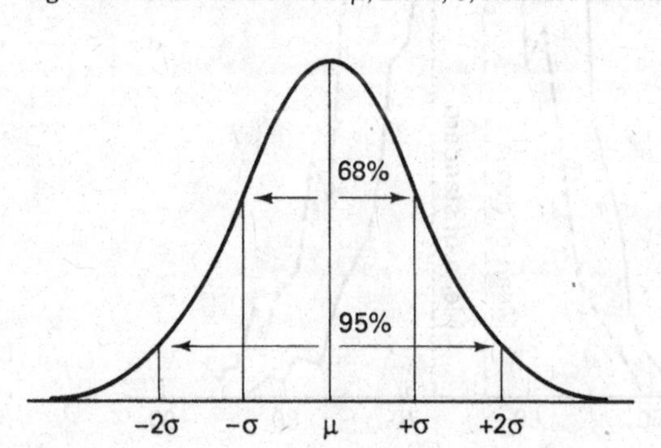

## 9.2 Mean, standard deviation, variance, covariance and correlation coefficient

Fig. 9.3 compares the egg weights from standard-sized breeders and dwarf broiler breeders. Both groups consisted of hens of similar age. The egg weights from both groups show normal distribution, like that in Fig. 9.2. Two features of these egg weights are easily apparent from the figure: (i) those from the standard-sized breeders are on average heavier than those from the dwarf breeder, and (ii) the spread of weights, or distribution, is similar in both cases. These features are expressed as the **mean** weight, and either the **variance** or the **standard deviation**, respectively.

The arithmetic mean is simply the average of all the weights in the sample, and is expressed as follows:

$$\bar{X} = \sum X_i / N$$

**Fig. 9.3** Egg weight frequency distributions of similarly aged standard sized and dwarf broiler breeder females. The means ($\bar{X}$) ± standard deviations (s) are 67.48 ± 5.47 g for the standard sized breeder, and 62.84 ± 5.93 g for the dwarf breeder (Whiting & Pesti, 1983).

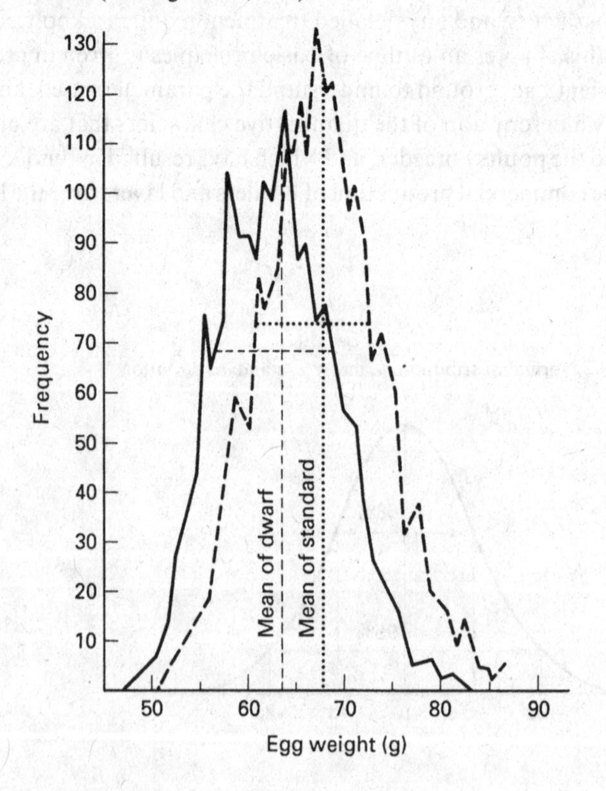

where $\bar{X}$ is the mean, N is the number of weights, and $\sum X_i$ is the sum of the weights. Where the mean of the population, as opposed to a sample of the population, is estimated the symbol $\mu$ is used, instead of $\bar{X}$.

The distribution of the weights can be expressed as the variance, which is defined as follows:

$$V = \sum(X_i - \bar{X})^2/n - 1$$

where V is the variance, given by the sum of the differences of the individual values from the mean, all squared, divided by $n-1$, where $n$ is the sample number.

Alternatively, the distribution can be expressed as the standard deviation, s, which is $\sqrt{V}$. As with the mean, different symbols are used to represent the sample standard deviation (s) and the standard deviation of a population, which is given the Greek letter $\sigma$. In some ways the standard deviation is more easily comprehended than variance, since it is expressed in the same units as the mean. Thus for the two sets of egg weights given in Fig. 9.3, the egg weights from the standard-sized breeder have a mean ± standard deviation $(\bar{X} \pm s) = 67.48 \pm 5.47$ g, and for the egg weights from the dwarf sized breeder $\bar{X} \pm s = 62.84 \pm 5.93$ g. With a normal distribution 68% of a sample lies within one standard deviation of the mean, and 95% within two standard deviations (see Fig. 9.3: 68% of the standard-sized breeders have egg weights between 62.01 and 73.41 g). The term variance is expressed in units which are the square of the mean. The advantage of using variance is that if it is caused by more than one factor, then the variances due to the separate factors are additive. The importance of this is shown in section 9.3 in connection with heritability.

It is often important to determine whether or not two variables are related. For example, is the weight of abdominal fat in the domestic fowl related to the total body weight? The usual measures for this are either the **covariance (cov)** or the **correlation coefficient (r)**. The covariance between two variables $x$ and $y$, e.g. weight of abdominal fat and total body weight, is defined as follows:

$$\text{cov}_{xy} = 1/N\sum(x_i - \bar{x})\,(y_i - \bar{y})$$

The difference between each $x$ term $(x_i)$ and the mean $\bar{x}$ is multiplied by the corresponding $y$ term $(y_i)$ minus its mean $\bar{y}$. The sum of these products is divided by the number of pairs of observations. An equivalent way of expressing covariance is:

$$\text{cov}_{xy} = 1/N\sum x_i y_i - \bar{x}\bar{y}$$

The latter is generally easier to evaluate on a calculator, and is exactly equivalent mathematically.

Covariance may have a positive or negative value or may be zero. Positive values signify a positive correlation, negative values a negative correlation, and zero no correlation. Since the value of covariance depends on the units and magnitude of the measurements, it is often useful to standardise by determining the correlation coefficient r, which is obtained by dividing the covariance by the product of the two standard deviations:

$$r_{xy} = \text{cov}_{xy}/s_x \times s_y$$

The correlation coefficient varies between +1 and −1; the nearer to +1 the stronger the positive correlation, whereas the nearer to −1 the stronger the negative correlation. Values close to zero show no correlation.

An example from data from Abplanalp *et al.* (1984a) illustrates how a correlation coefficient is determined. The data compare body weight and feed intake for some inbred lines (numbered 054 to 082) of domestic fowl.

| **Inbred line** | **Body weight (kg)** | **Feed intake/day (g)** | ***xy*** |
|---|---|---|---|
| 054 | 1.347 | 94.7 | 127.56 |
| 056 | 1.233 | 88.1 | 108.63 |
| 058 | 1.280 | 101.1 | 129.63 |
| 070 | 1.215 | 93.6 | 113.72 |
| 080 | 1.381 | 100.8 | 139.2 |
| 082 | 1.146 | 80.9 | 92.7 |
| Mean | 1.267 | 93.2 | |
| Standard deviation | 0.0872 | 7.74 | $\sum xy = 711.23$ |

$$\sum \bar{x}\bar{y} = 1.267 \times 93.2 \times 6 = 708.5$$

$$\text{cov}_{xy} = 1/6 \times 711.23 - 708.5 = 0.454$$

$$r_{xy} = 0.454/(0.0872 \times 7.74) = 0.67$$

There is thus a strong correlation between body weight and food intake.

## 9.3 Heritability

If the range of egg weights from the two strains of hen shown in Fig. 9.3 is considered, the variation in weights arises from two main causes: genetic and environmental. The size of an egg will depend both on the genetic makeup of the hen, and also on environmental factors such as nourishment. The breeder is continually faced with the question of how to improve his flock, and in particular whether it is possible to improve it by genetic means. Geneticists have devised a measurement called **heritability** which enables this to be roughly quantified.

Bearing in mind that variance arising from different causes is additive, the phenotypic variance that is observed is the sum of genetic and environmental variances.

$$V_P = V_G + V_E$$

where $V_P$, $V_G$ and $V_E$ are phenotypic, genetic and environmental variances, respectively.

The genetic variance can be divided into a number of different factors; the main ones usually considered are (i) the additive effect of genes, (ii) the dominance effects of genes, and (iii) the interaction of genes, by processes such as epistasis. We can thus state:

$$V_G = V_A + V_D + V_I$$

Where $V_G$, $V_A$, $V_D$ and $V_I$ are genetic, additive, dominance and interaction variances, respectively. These factors will be explained in turn.

Suppose that there were two pairs of alleles, $A_1$, $A_2$, $B_1$ and $B_2$ that control egg weight (in fact there are many more, but it is simpler to consider two, in order to illustrate the additive effect). Consider also that there is no dominance between $A_1$ and $A_2$, and also none between $B_1$ and $B_2$. For this example we will assume that the presence of genes $A_1$, $A_2$, $B_1$ and $B_2$ contribute 1, 2, 1.5 and 3 grams respectively to the basic weight of the egg. There are nine possible genotypes, and the egg weights above the basic that would be expected if the genes have additive effects are:

$A_1A_1B_1B_1 = 2.5$ $A_1A_2B_1B_1 = 3.0$ $A_2A_2B_1B_1 = 3.5$
$A_1A_1B_1B_2 = 3.25$ $A_1A_2B_1B_2 = 3.75$ $A_2A_2B_1B_2 = 4.25$
$A_1A_1B_2B_2 = 4.0$ $A_1A_2B_2B_2 = 4.5$ $A_2A_2B_2B_2 = 5.0$

If the above pattern is modified to take account of dominance, the distribution of phenotypes would differ. Supposing that $A_2$ shows complete dominance over $A_1$, and $B_2$ complete dominance over $B_1$, the heterozygote phenotypes would have the same value as the homozygous dominants, e.g. $A_1A_1 = 1$, $A_1A_2 = 2$, and $A_2A_2 = 2$, and so the overall pattern would be as follows:

$A_1A_1B_1B_1 = 2.5$ $A_1A_2B_1B_1 = 3.5$ $A_2A_2B_1B_1 = 3.5$
$A_1A_1B_1B_2 = 4.0$ $A_1A_2B_1B_2 = 5.0$ $A_2A_2B_1B_2 = 5.0$
$A_1A_1B_2B_2 = 4.0$ $A_1A_2B_1B_2 = 5.0$ $A_2A_2B_2B_2 = 5.0$

The third genetic factor contributing to genetic variance is the interaction variance that arises through epistasis. For example, if the A genes were epistatic to the B genes the pattern would be further modified.

Genetic variance therefore arises from a combination of the factors described above. The heritability is then defined as the proportion of the phenotypic variance which can be assigned to genetic variance.

$$\text{Heritability (broad sense)} = H^2 = V_G/V_p$$

This is referred to as broad sense heritability, since it relates to all types of genetic factors. It is given the symbol $H^2$ rather than H, since it relates directly to variance rather than standard deviation and its value varies between 0 and 1. If the phenotypic variance is due entirely to genetic factors and is not influenced by the environment, heritability has a value of 1. At the other extreme, if it is entirely due to environmental factors, it will have a value of zero.

Of the three genetic factors mentioned above the only one which can be readily selected for in a breeding programme is the additive variance. The reasons for this are well explained by Suzuki *et al.* (1989). For these reasons heritability is more usually defined in the narrow sense, to include only the additive genetic variance, and is given the symbol $h^2$.

$$h^2 = V_A/V_P$$

Whenever heritability is mentioned later in this chapter, it refers to $h^2$ (narrow sense). The main application of $h^2$ is to be able to predict whether a given flock or population can be improved with respect to a particular trait by breeding. Although not rigidly defined, high heritability is usually reckoned as $>0.5$, medium heritability as 0.2–0.5, and low heritability as $<0.2$.

The heritability for a particular character is not a fixed quantity for a particular species or even strain. It will differ from flock to flock. For example, if a strain has been bred selectively for high egg laying then the value of $h^2$ will almost certainly be lower than that for one which has not been selectively bred for egg production. The measurement of $h^2$ in the former will indicate whether there is any scope for further improvement through breeding.

There are a number of methods of determining heritability ($h^2$) and these are more fully described by Stansfield (1983), Falconer (1989) and Suzuki *et al.* (1989). Many of the methods entail the measurement of phenotypic variance in a group of organisms whose genetic relationships are known and which have experienced the same environment. Related individuals must have a higher proportion of genes in common with each other than with the population at large: parents and offspring have at least 50% genetic identity with each other, full sibs (i.e. offspring of the same parents) will have at least 50% genetic identity, and half sibs at least 25% genetic identity. The figures of 50% and 25% are the minimum in each instance, since there will in addition almost certainly be some genes in common between both parents.

It can be shown that the extra similarity between, for example, parent and offspring compared with the population at large can be expressed as a covariance (Fincham, 1983; Falconer,1989):

$$\mathrm{cov}_{(\text{parent–offspring})} = \tfrac{1}{2}V_A$$

The dominance component does not in this case enter the equation, since in a parent–offspring relationship there is an equal chance that it may enhance or reduce the covariance and hence cancel out. With full sibs there is a dominance component since there is a quarter chance that an offspring may inherit a particular allele from one parent together with a particular allele from the other (thus: $\frac{1}{2} \times \frac{1}{2} = \frac{1}{4}$) and in this case covariance is:

$$\mathrm{cov}_{(\text{full sibs})} = \tfrac{1}{2}V_A + \tfrac{1}{4}V_D$$

With half sibs the covariance is:

$$\mathrm{cov}_{(\text{half sibs})} = \tfrac{1}{4}V_A$$

Since the parent–offspring and the half sib relationships have no dominance term in the covariance, they are more useful in determining heritability ($h^2$). Thus, if the phenotypic variance $V_P$ of the group is determined, together with the appopiate covariance, the additive variance $V_A$ can be calculated and hence $h^2$:

$$h^2 = V_A/V_P$$

## 9.4 Selection in principle

A knowledge of $h^2$ for a particular trait helps to predict the likely success or failure of selection. Artificial selection effectively changes the gene frequencies in such a way that the group selected approaches homozygosity for the genes concerned and this is accompanied by a reduction in variance. With quantitative traits a reduction in the additive genetic component is that which is principally selected for. Suppose that the heritability of a trait has been found to be near to unity, artificial selection can be expected to bring about a significant improvement within a few generations.

Consider the selection process for a particular trait having $h^2 = 1$. If the mean value of this trait is M, and only those individuals having a value of $\geqslant M+S$ (selection differential) are selected for further breeding, then the mean value of the next generation would be expected to be $M+S$. If, on the other hand, $h^2=0$, then selecting those individuals having a value of $\geqslant M+S$ and breeding from these would still give a mean value in the next generation of M; i.e. in this case there would be no improvement by selection. In practice, in the vast majority of cases, the result will lie between the two extremes. Suppose this intermediate value was $M+R$ (R = Response) then the relationship to heritability is $h^2=R/S$. The $h^2$ estimated by this method is referred to as **realised heritability** (Falconer, 1989). It should be noted that it is equally possible to select against a trait by selecting those having a value $\leqslant M-S$.

A limitation of heritabilities as a predictive index is that the value applies strictly only for the first generation. One would expect a progressive decline in the value of $h^2$ in subsequent generations as the ability to improve by genetic means declines. It is found that the heritability value is usually maintained for four or five generations, e.g. selection in a random-bred line of meat-type domestic fowl for body weight for four generations gave $h^2=0.47$, and continued selection over seven further generations gave $h^2$ values between 0.21 and 0.29 (see Marks, 1985). If the value of $h^2$ is very low, then it suggests that a change of environment or husbandry practice could be more usefully tried. In breeding programmes it is rare that only one trait is being selected for, and this means that it is often necessary to give weightings to the different traits to be selected, by devising a selection index (see Stansfield, 1983). For example, Lamont *et al.* (1987) used a selection index to select positively for body weight and egg mass, but negatively for feed consumption. The index is expressed as an equation:

$$\text{Index} = b_1\text{BW32} + b_2\text{EM1} - b_3\text{FC1}$$

where BW32 is the body weight in grams at 32 weeks, EM1 the mean weight of eggs per day in grams, and FC1 mean feed consumed per day in grams from 30 to 34 weeks of age, and $b_1$, $b_2$ and $b_3$ are weighting coefficients.

A number of methods of selection are practised: (i) mass selection, (ii) family selection, (iii) pedigree selection, and (iv) selection by progeny testing. In poultry breeding, mass selection and progeny testing are the most important. Mass selection simply means that a group within the population is selected having the necessary selection differential, and this is used for breeding; similar selection is then applied to subsequent generations. Mass selection is useful in the initial selection procedures, but to gain the optimum improvement a more refined method is necesary. Progeny testing is more laborious and requires waiting until sexual maturity. It is most useful for traits that can be expressed only in one sex, e.g. egg laying, or that cannot be measured until after slaughter, e.g. abdominal fat, or that have a low $h^2$ and cannot otherwise be selected for very accurately. It is usually impractical to carry out progeny testing on the whole population and therefore those which are tested have usually undergone a preselection.

A progeny test is a method of determining the breeding value of an animal by the performance or phenotypic value of its offspring. The **breeding value** is a measure of an animal's value in a breeding programme. It is a measure of the animal's expected performance or phenotype in relation to the population mean. It is quantified as twice the difference of the expected progeny mean from the population mean. The reason that it is twice the difference is that the progeny carry only half of the parent's genes (Falconer, 1989).

## 9.5 Selection in practice

It can be seen from Table 9.1, which summarises the heritabilities of some important traits, that, in general, traits which are important for survival have low values, e.g. fertility and hatchability, whereas those not essential for survival, such as body weight and abdominal fat, have high heritabilities.

The principles of quantitative genetics have been applied in the poultry industry since the 1950s, and there have thus been over 30 generations in which to effect improvement. There have been some dramatic improvements in 'performance', for example, today's broiler reaches a market weight of 1800–2000 g in 49 days compared with 100 days in the 1950s, whilst consuming about half of the food (Siegel & Dunnington, 1985). It is also clear that the limits of improvement by selection have not yet been reached. The main developments have been in various aspects associated with growth in broilers, and all aspects associated with egg laying in layers. In addition, resistance to disease is particularly important.

The various aspects of selection in practice will be considered in turn. In most cases selection for more than one trait is necessary, and thus correlations of genetic traits will also be discussed. There is now a large body of information on selection for meat, egg laying and disease resistance, but much of this is difficult to summarise, since there are so many parameters that have been varied in different studies, e.g. the period over which growth or egg laying has been measured, the breed or strain used, and the management conditions used. For detailed compilations of results on growth and meat production, egg laying and disease resistance the reader is

Table 9.1. *Examples of heritability in the domestic fowl*

| Trait | Heritability estimate | Reference |
|---|---|---|
| A. Body weight and composition | | |
| Body weight | 0.39–0.91 | Bowen & Washburn, 1984; Chambers *et al.*, 1984; Siegel, 1962; Chambers, 1990 |
| Weight gain | 0.11–0.81 | Chambers *et al.*, 1984; Kim *et al.*, 1987; Chambers, 1990 |
| Abdominal fat | 0.6–1.0 | Chambers *et al.*, 1984; Ricard & Rouvier, 1969; Becker *et al.*, 1984; Chambers, 1990 |
| B. Eggs and egg laying | | |
| Hen housed egg production | 0.11–0.44 | Fairfull & Gowe, 1990 |
| Survivor egg production | 0.13–0.28 | Fairfull & Gowe, 1990 |
| Age at 1st egg | 0.15–0.3 | Bowman, 1968 |
| Egg size | 0.4–0.5 | Bowman, 1968 |
| Egg shape | 0.25–0.50 | Bowman, 1968 |
| Shell thickness | 0.25–0.60 | Bowman, 1968 |
| Shell strength | 0.09–0.57 | Washburn, 1990 |
| Albumen quality | 0–0.66 | Washburn, 1990 |
| Yolk cholesterol | 0.11–0.25 | Washburn, 1990 |
| C. Reproduction | | |
| Fertility | 0.0–0.05 | Bowman, 1968; Chambers, 1990 |
| Total mortality (average) | 0.07 | Gavora, 1990 |
| Hatchability | 0.09–0.15 | Bowman, 1968; Chambers, 1990 |
| Mating frequency (♂) | 0.18–0.20 | Dunnington & Siegel, 1983 |
| Sexual maturity | 0.25–0.39 | Chambers, 1990 |
| Broodiness | 0.11 | Saeki, 1957 |
| D. Feed, feed efficiency and metabolism | | |
| Feed consumption | 0.39 | Chambers *et al.*, 1984 |
| Feed efficiency | 0.03–0.48 | Chambers *et al.*, 1984; Thomas *et al.*, 1958 |
| Residual feed consumption | 0.50 ± 0.22 | Hagger & Marguerat, 1985 |
| Fasting metabolic rate | 0.23–0.43 | Damme *et al.*, 1986 |
| E. Disease resistance | | |
| $\gamma$ globulin levels | 0.26 | Garnett & Roberts, 1972 |
| Haemagglutinin titres | 0.11–0.57 | Kim *et al.*, 1987; van der Zijpp, 1983 |
| Specific disease resistance (average) | 0.25 | Gavora, 1990 |
| F. Others | | |
| Heat stress survival time | 0.37 ± 0.12 | Bowen & Washburn, 1984 |
| Plasma thyroxine levels | 0.22 ± 0.10 | Bowen & Washburn, 1984 |
| Spur incidence (in ♀) | 0.21 | Fairfull & Gowe, 1986 |
| Spur length (in ♀) | 0.42 | Fairfull & Gowe, 1986 |
| Imprinting | 0.09 | Fischer, 1969 |
| Feather pecking | 0.07 | Bessei, 1985 |

For more detailed information, see Fairfull & Gowe (1990).

referred to Chambers (1990), Fairfull & Gowe (1990) and Gavora (1990), respectively. In any study in which selection for these polygenic traits is made, it is important to have a suitable control strain. This is particularly so in long term studies carried out over a number of years to allow for fluctuations in environment or in management conditions. The different types of control strains are discussed in detail by Gowe & Fairfull (1990).

### *9.5a Selection for meat*

Prior to the 1930s poultry were reared together commercially both for meat and egg laying and 'pure' breeds were used. Since the 1930s in America, and more recently in the United Kingdom, poultry have been reared separately for meat (broilers) and for egg laying, and they have been selectively bred for one or other purpose. Pure breeds have been replaced by hybrids from two, three or four way crosses, and this has enabled the breeder to take advantage of **heterosis** or hybrid vigour (see section 9.6). The selection in modern broilers has resulted in an eightfold increase in breast muscle mass at 7 weeks of age compared with layer strains.

In order to optimise breeding for meat production, selection has to be made for a number of characteristics, including growth rate, feed efficiency and carcass quality. However, in order to produce large numbers of broilers attention has also to be paid to hatchable egg production. Viability is important whether selection is for broilers or layers, but it has a very low heritability ($h^2 = 0.05$) and is largely dependent on incubation conditions. A practice that is often adopted is to produce separately a male strain and a female strain; the male strain is selected for growth rate and carcass quality, and the female strain for egg production and growth rate, the broiler being the hybrid progeny of the two strains. Many of the characteristics required of the broiler are ones having medium or high heritabilities.

Many studies have been made of growth rates. In 176 published heritability estimates of body weights of chickens of 6–12 weeks age a median value of 0.41 was obtained (Siegel, 1962). Measurements of body weights over groups of ages ranging from 4 weeks to mature adults (greater than 24 weeks) also showed medium to high heritability (Kinney, 1969). In some highly selected strains the heritability for growth rate is very much lower. Body weight is correlated with a number of other parameters (see Marks, 1985). Body weight at 8 weeks is positively correlated with the initial egg weight ($r = 0.26$), and with adult body weight ($r = 0.6$), but negatively correlated with the age of the first egg and egg production ($r = -0.12$). The mechanisms by which the higher body weight and growth rates are achieved are still uncertain, but there are a number of indications. Higher body weights are linked with better food efficiency and low oxygen consumption. Selection for body weight gain is also linked both with efficiency of food utilisation (r = +0.71) (Pym & Nicholls, 1979) and increased appetite (see Marks, 1985). Neurological and endocrinological mechanisms are considered to be involved. When the level of the enzyme ornithine decarboxylase is compared in the muscle of broilers and in layer strains, at the age of one week, the former is found to have about 20-fold higher activity (Bulfield, Isaacson & Middleton, 1988). Ornithine decarboxylase is an enzyme which signals the onset of proliferation in cells (Tabor & Tabor, 1984) and so may signal a growth spurt.

Besides growth rate, carcass quality is very important. The latter term encompasses a large number of factors, some of which can be measured only after slaughter, which makes their selection more difficult and costly. The type of meat and the proportion of body fat are important factors. The breast muscles are generally regarded as the choice morsels. The measurements most often used in this connection are breast angle or breast width and cross sectional area. The breast angle has a heritability of 0.4 (Bowman, 1968). A genetic correlation between growth and abdominal fat has been demonstrated in a number of studies with values of r = 0.2 to 0.76 (Chambers, 1990). The abdominal fat is generally assessed after slaughter, although methods which can be applied to the live bird have been devised. These include using calipers as an abdominal fat probe (Pym & Thompson, 1980) and assaying plasma lipoproteins as an indication of fatness (Griffin & Whitehead, 1982).

Much work has been carried out to see whether adipose tissue can be selected against, without a general reduction in lean meat. Abdominal fat has a medium to high heritability (Ricard & Rouvier, 1969; Ricard, 1974; Leclercq, Blum & Boyer, 1980; Becker *et al.*, 1984) and is positively correlated with body weight and growth rate (Abplanalp, Tai & Napolitano, 1984). It is therefore quite feasible to select successfully against abdominal adipose tissue (Leclerq, 1983). Whitehead & Griffin (1985) have selected on the basis of plasma triglyceride estimates, and have obtained broilers with 20% less fat but of comparable weights; also the feed conversion was 4% better. Grunder & Chambers (1988) have shown that there is a good correlation (r = *c*.1) between abdominal fat weight and plasma very low density lipoproteins (VLDL). It is thus possible to select against the proportion of abdominal fat by indirect selection for low plasma VLDL. There is also evidence which suggests that the low plasma glucose concentration relates to the tendency to high abdominal fat, although direct selection on the basis of body fat is preferable (Leclercq, Simon & Ricard, 1987). Cahaner, Nitsan & Nir (1986) have shown that when fowl are selected for low abdominal fat, there is also a reduction in other adipose tissues associated with the gizzard, sartorius, neck and mesentery, and also a reduction of lipids in the liver and blood plasma.

An alternative method of selecting against abdominal fat is by regulation of the salt content of the diet. It is well established that there is an inverse relationship between carcass lipid and carcass moisture (Twinning, Thomas & Bossard, 1978). Marks & Washburn (1983) have shown high feed:water ratios in broilers may be associated with a reduction in abdominal fat. Marks (1987) fed chickens in two groups on a regular (0.4%) and a high (1.6%) salt diet and selected from each group on the basis of high body weight at 8 weeks. He found that the high salt diet selection resulted in females with lower abdominal fat levels, whereas the males were unaffected.

Selection of broilers for rapid growth rate and high body weight is not without its problems. Siegel & Dunnington (1985) pointed out that heritability estimates for growth rate in many broilers are still moderately high, and that therefore selection can, in theory, still produce further gains, but that other problems will become more pronounced. The reproductive capabilities of broiler breeders become adversely affected when they are selected on the basis of food consumption and it also leads to obesity. It is also found that overweight is associated with an increased number of leg disorders. Mercer & Hill (1984) have shown that the skeletal defects are heritable and they correlate positively with body weight.

Besides the obvious nutritional value of the broiler, there are also the aesthetic considerations, e.g. the appearance of the carcass for marketing. Selection is made for rapid feathering, since this reduces the tendency to form unattractive blisters on the breast of the carcass. The occurrence of breast blisters has a heritability of *c.*0.2. In the UK the preference is for white skinned birds, and these are therefore selected for. Birds must also have white feathers, since the remains of dark coloured pin feathers detract from the value (Bowman, 1968).

### *9.5b Selection for egg-laying*

Since the 1940s there has been a considerable improvement in the performance of layers. Part of this is attributable to improvement in nutrition and management, but also very significantly to genetic factors. The most important improvements have been in (i) the rate of laying, (ii) the quality and size of egg, and (iii) the feed efficiency. The last of these is now particularly important, since egg production is very competitive and a large part of the production costs is in the feeding.

Several studies have been carried out aimed at selection for improved egg laying. To carry out these studies successfully often requires operating on comparatively large scales over periods of several years. Some have effectively been national studies carried out over 20 years or more. Improvements in various aspects of egg production have been achieved by (i) selection between strains and strain replacement, (ii) the formation and screening of new synthetic strains, and (iii) selection within strains (Bulfield & McKay, 1987). Since many of the characters selected for are interdependent, e.g. egg number and egg weight, selection for one character often affects others as well.

There are several methods for measuring the rate of laying (see Fairfull & Gowe, 1990); that most used is **hen housed egg production**. It is defined as the number of eggs laid by a set of housed birds within a given period divided by the number of birds alive at the start of the recording period. This combines a measure of three traits: (i) age at the first egg, (ii) rate of egg production from the start of lay, and (iii) viability. Any birds that die within the test period will adversely affect the hen housed egg production. Long term selection for egg production has been carried out at several centres, e.g. California (Abplanalp *et al.*, 1964), Cornell (Hutt & Cole, 1947), Virginia (Dunnington & Siegel, 1985), Zürich (Hagger & Marguerat, 1985), Australia (see Gowe & Fairfull, 1985) and Canada (Gowe & Fairfull, 1985).

The Canadian study (Gowe & Fairfull, 1985) is used here as an illustration. The study was begun in 1950 using strains of White Leghorns. Strains were selected for hen housed egg production up to 273 days of age. It can be seen in Fig. 9.4, which is a extract from the data of Gowe & Fairfull (1985), that there has been a steady increase in the number of eggs laid by the selected strain compared with the unselected strain. In this particular study the age of laying of the first egg decreased from around 180 days to around 140 days over the 30 year period (Fig. 9.5). The mean mortality also declined from about 30% in 1951 to 6% in 1980, but this was largely attributable to improved management and disease control rather than to any genetic changes. This point highlights the importance of having control strains.

The changes brought about by selection in this study were used to calculate the genetic gain. The genetic gain per generation is obtained by calculating the change in the selected strain compared with the control strain mean regressed over the generations for which the measurements were made. Over the 30 year period the genetic gains were 0.8 and 1.3% for hen housed egg production at 273 days, 0.04 and 0.29% for mortality at 497 days, 0.2 and 0.25% for the rate of egg production to 497 days, and 0.26 and 0.3% for egg weight at 225 days. Two figures are given in each case since two selected strains were compared with one unselected strain. Egg quality, which was measured only after 1969 in this study, also showed gains of between 0.29 and 0.55 Haugh units. (One measure of egg quality is to measure the viscosity of egg albumen. A Haugh unit is a measure of the logarithm of albumen height corrected for egg weight. It is strongly dependent on the freshness of the egg.) There have thus been genetic gains in rate of lay, size and quality of eggs, and a slight decrease in the mortality of the layers over the 30 year period of study. A further point of interest is that the mean heritability for hen housed egg production at 273 days over the period 1953–80 was 0.35 for the dam and 0.105 for sire, and for the earlier part of the study (1953–64) it was 0.406 for the dam and 0.078 for the sire. This shows clearly that greater genetic gains for egg production can be obtained by selection of the dam than of the sire.

Gowe & Fairfull (1985) reported a number of genetic correlations: positive correlations between egg size and age at the first egg (the age at which the first egg is laid), sexual maturity and egg production, and a negative correlation between rate of egg production and egg weight. There was no evidence in this study that selection had caused a decrease in genetic

**Fig. 9.4** Hen-housed egg production to 273 days for a control and a selected strain. The fall in production in 1969 was due to an outbreak of Marek's disease. Data are from Gowe & Fairfull (1985).

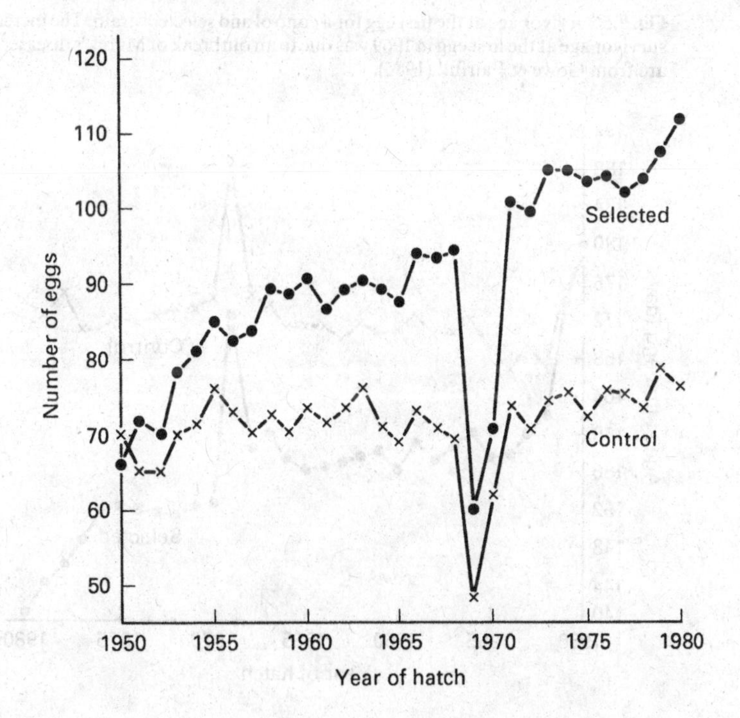

variance or in heritability over 30 years, and there is no evidence from commercial flocks in general of upper limits to selection (Fairfull & Gowe, 1990). Thus further improvement by breeding still appears possible. Both genetic and environmental variance increased with the age of the birds, and Liljedahl *et al.* (1984) suggest that this reflects a reduced ability to cope with environmental stress.

Much work has been carried out on the **ovulation cycle** and how it is controlled (Fairfull & Gowe, 1990). The pattern of egg laying is governed by the interaction of two synchronous biological cycles (see Naito *et al.*, 1989). One is a circadian oscillator that controls oviposition, which shows 23–26 h cycles, and is set off by an environmental time setter, sometimes referred to by the German 'zeitgeber'. This oscillator is particularly affected by light–dark cycles close to 24 h. The other cycle has a slightly longer period (between 24 and 27 h) and is concerned with follicular maturation during which ovulation occurs in response to increased concentrations of luteinising hormone released by the anterior pituitary gland. This release occurs in an 8–9 h period of the normal 24 h cycle of follicular maturation (Williams & Sharp, 1978).

Under a normal light–dark cycle of 24 h, oviposition starts early in the day. Each successive day laying occurs later until the sequence ends with

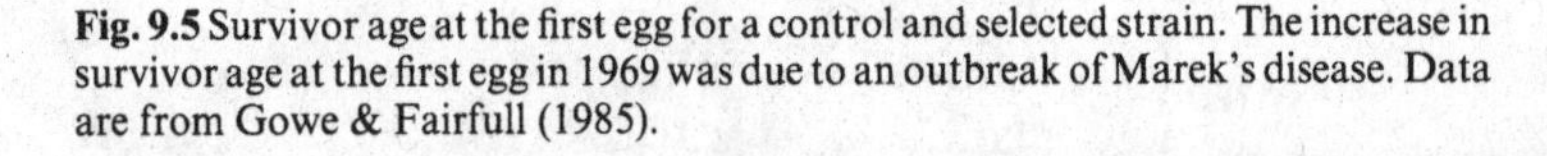

**Fig. 9.5** Survivor age at the first egg for a control and selected strain. The increase in survivor age at the first egg in 1969 was due to an outbreak of Marek's disease. Data are from Gowe & Fairfull (1985).

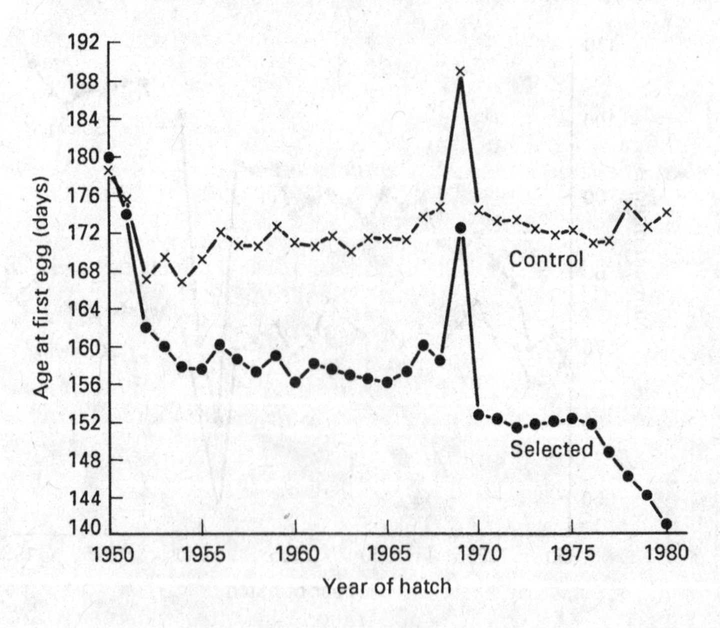

oviposition in the afternoon. At least one day intervenes before a new sequence begins with oviposition early in the day. Various changes to the light–dark cycle, including that of continuous lighting, have been tested to see whether this will alter the rate of ovulation and oviposition, and whether any increases in egg laying can be obtained by selecting over a period of generations.

With light–dark periods of between 21 and 28 h (lighting schemes of light–dark cycles of less than 24 h are often referred to as ahemeral lighting schemes) successive ovipositions never occur consistently within shorter intervals, and within this range there is an upper limit possible on the basis of one egg per cycle, e.g. for 28 h, 86 eggs per 100 hens per day, and for 21 h, 114 eggs per 100 hens per day (Foster, 1985). However, Foster (1985) studied the response of selection on ovulation frequency of 22 and 23 h regimes (Fig. 9.6*A*). It can be seen that selection improves the rate of ovulation on 23 and 22 h light–dark regimes, but for an increased rate of ovulation to be useful it has to lead to increased oviposition. It can be seen (Fig. 9.6*B*) that a limit of about 90 eggs per 100 hens per day is achieved with no improvement on a 23 h light–dark cycle than on a normal regime.

**Fig. 9.6** Changes in the ovulation frequency (*A*), and in oviposition frequency (*B*) in response to selection for an increase in ovulation frequency, under a normal (24 h) lighting regime ( × ), and under 23 h (■) and 22 h (●) regimes. Data collected over 10–14 generations (Foster, 1985).

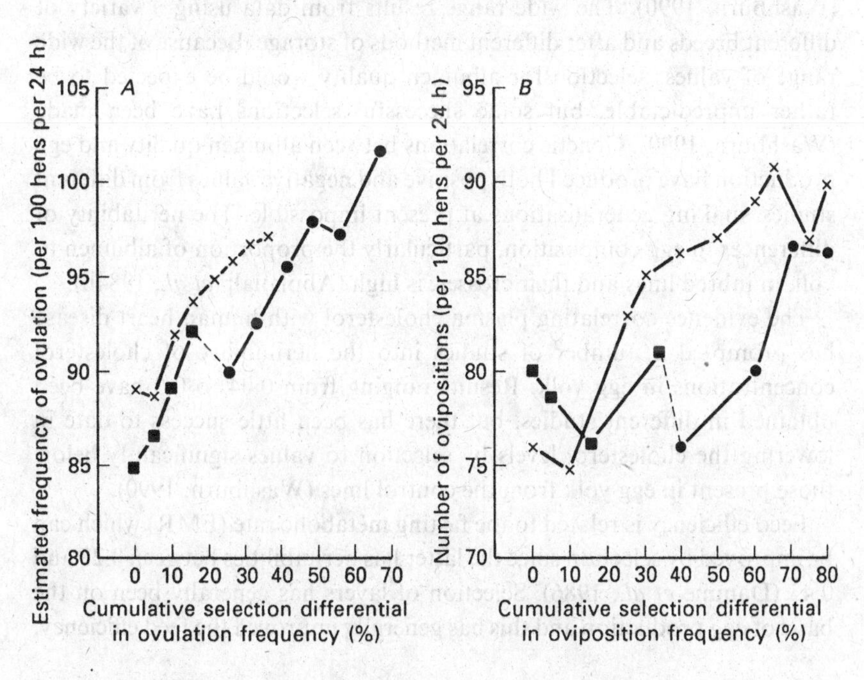

Similar studies were carried out by Naito *et al.* (1989), who selected for increased rate of lay under 23 and 24 h light–dark cycles. The realised heritability for rate of lay was 0.25 ± 0.04 in the 23 h regime and 0.15 ± 0.05 in the 24 h regime. They suggested that a 23 h regime might improve the laying performance of a population approaching a plateau of egg production. Yoo, Sheldon & Podger (1988) estimated the heritability of oviposition interval on a normal light–dark regime for a population of White Leghorns to be 0.54. They suggested that oviposition interval rather than the laying intensity might be a better criterion for selection. Laying intensity includes the interval between sequences as well as the oviposition interval.

Egg quality comprises a number of factors including shape, shell quality and proportion of albumen to yolk (for a review, see Washburn, 1990). Shape is generally judged by measurements of length and breadth, the aim generally being to produce a completely regular shape some way between an ovoid and pear shaped. Assessment is thus rather difficult to quantitate and this is reflected in the heritability measurements which range between 0.1 and 0.6. Shell quality can be judged by thickness, ability to withstand breakage, or the specific gravity of the egg. Values for $h^2$ from different studies using one of these methods have ranged from 0.09 to 0.57 (Washburn, 1990). The quality of albumen, and the proportion of albumen to yolk can be also selected for. Measurements of $h^2$ for albumen quality as Haugh units have yielded a wide range of results from zero to 0.66 (Washburn, 1990). The wide range results from data using a variety of different breeds and after different methods of storage. Because of the wide range of values, selection for albumen quality would be expected to be rather unpredictable, but some successful selections have been made (Washburn, 1990). Genetic correlations between albumen quality and egg production have produced both positive and negative values from different studies, making generalisations at present impossible. The heritability of differences in egg composition, particularly the proportion of albumen to yolk in inbred lines and their crosses, is high (Abplanalp *et al.*, 1984b).

The evidence correlating plasma cholesterol with human heart disease has prompted a number of studies into the heritability of cholesterol concentrations in egg yolk. Results ranging from 0.04 to 0.3 have been obtained in different studies, but there has been little success to date in lowering the cholesterol levels by selection to values significantly below those present in egg yolk from the control lines (Washburn, 1990).

Feed efficiency is related to the fasting metabolic rate (FMR) which can be improved by selection since the latter has heritabilities between 0.23 and 0.43 (Damme *et al.*, 1986). Selection of layers has generally been on the basis of egg production and this has generally improved the feed efficiency, since egg production has increased while the feed required for maintainence has remained nearly constant. An index which is generally used in assessing feed efficiency is **residual feed consumption**. It is defined as the observed feed consumption minus the expected feed consumption. The expected feed consumption is a complex function of body weight and egg mass in a given period of time. Hagger & Marguerat (1985) determined the heritability of residual feed consumption as $0.52 \pm 0.22$. On the basis of this medium value for heritability they suggested that improved feed efficiency is likely to be more successful if selection is on the basis of residual feed consumption than on the basis of egg production. It is not yet clear what is the basis of the different feed efficiencies in their experiments. It is not due to different feather cover which might cause differential heat losses, since similar values are obtained in both well and poorly feathered birds. They suggest that physiological or behavioural factors may be involved.

### 9.5c *Selection for disease resistance*

With the intensive rearing of both layers and broilers, resistance to disease becomes important. Selecting for disease resistance can be a difficult and a long term process. The heritability is generally low and it requires exposing test populations to the disease in order to select the better strain, which is a costly process. Much effort was expended on trying to produce strains of fowl resistant to Marek's disease, but once a successful vaccine had been found interest in producing resistant strains diminished. Nevertheless improvements in disease resistance have been achieved and also a much greater understanding of the controlling genes. The latter is discussed in Chapter 11, on immunogenetics.

Some of the earliest experiments on selection for disease resistance were carried out in the 1930s when strains were selected for resistance to *Salmonella* (see Hartmann, 1985). In the late 1950s Marek's disease reached epidemic proportions and this prompted increased work on the disease. Work on the blood groups in domestic fowl began in the late 1940s (see Briles, 1984), but it was not until 1961 that Schierman and Nordskog recognised that the blood group B locus was also that of the Major Histocompatibility Complex (MHC), the latter playing a major role in the immune response. The MHC is described in more detail in Chapter 11, section 11.7. Briles, Stone & Cole (1977) showed that strains resistant to Marek's disease had a higher frequency of certain alleles.

Several laboratories that have studied fowl genetics for many years have maintained strains resistant or susceptible to certain diseases. At Cornell (USA), Hutt and Cole (1947) have maintained two lines resistant (K and

C), and one susceptible (S) to the 'Leukosis Complex', a group of viruses which includes Marek's disease. At East Lansing (USA), Waters (1945) maintained inbred lines resistant and susceptible to neural and visceral types of leukosis.

Some of the important questions about disease resistance are: (i) how successfully can disease resistance be selected for, (ii) how specific is any resistance so obtained, (iii) does disease resistance correlate closely with other genetic characters, and (iv) what role does the B blood-group locus play in disease resistance? The diseases that have been studied in detail are Marek's disease, tumours induced by Rous Sarcoma Virus (RSV), lymphoid leukosis, and caecal coccidiosis.

The blood group B is important in the immune response. Twenty-seven haplotypes (for further explanation see Chapter 11, section 11.7) have been identified at the B locus by serotyping (Briles & Briles, 1982). Hansen, Law & Van Zandt, (1967) showed that carrying the $B^{21}$ allele caused better viability and fewer losses from Marek's disease. As a result of further work, $B^2$, $B^6$ and $B^{21}$ have been shown to be alleles conferring moderate resistance, whereas $B^3$, $B^5$, $B^{13}$, $B^{15}$, $B^{19}$ and $B^{27}$ are highly susceptible (Hartmann, 1985). Gavora *et al.* (1986) selected separate strains of Leghorns, two on the basis of increased egg production and one on the basis of resistance to Marek's disease. They then compared the B haplotypes present in these strains with those in an unselected control strain. They found that there was an increase in the frequency of $B^{21}$ and $B^2$ in all three selected strains, compared with a control strain. $B^2$ haplotype is also associated with an ability to cause regression of tumours induced by Rous Sarcoma Virus. Thus there appears to be a correlation between egg production and resistance to Marek's disease. Gavora *et al.* (1986) suggested that in commercial flocks it is probably best to produce heterozygotes at the B locus, with one of the alleles resistant to Marek's disease. Since the B alleles are codominant, it means that the other B allele could be chosen to confer resistance to other diseases.

The B haplotypes are not only linked with disease resistance, but may also be linked with certain 'economic traits'. Lamont *et al.* (1987) selected from a line of White Leghorns using two selection indices, one based on body weight and egg mass, the other based on body weight, egg mass and feed consumption. In these particular strains $B^2$ and $B^{13}$ made up $>75\%$ of the gene pool in each line. Using either of the selection indices $B^2$ increased at the expense of $B^{13}$. Another of the haplotypes, $B^{14}$, increased under the second index but not the first. In a similar study by another group (Simonsen *et al.*, 1982), where the strains had predominantly haplotypes $B^{15}$ and $B^{19}$, $B^{15}$ increased at the expense of $B^{19}$.

A haplotype that confers resistance to one disease does not necessarily confer resistance to another. This is illustrated by studies on selection for resistance and susceptibility to acute caecal coccidiosis (Johnson & Edgar, 1986). Two blood group loci are involved, B and C. The resistant line has predominantly $B^5$, $B^r$ and $C^4$, the susceptible line has $B^1$, $B^3$, $B^4$, $B^6$, $C^2$ and $C^3$, and both lines share $B^2$ and $C^1$. (N.B. $B^r$ is a recombinant of $B^2$ and $B^5$). Therefore $B^5$, which confers resistance to coccidiosis, renders the strain susceptible to Marek's disease. Thus it is possible that the gene frequencies oscillate in a population as it becomes exposed to different diseases.

The ability to cause the regression of tumours is partly a function of the immunological system and thus involves the major histocompatibility complex (see Chapter 11, section 11.7). The effect of B haplotypes on the regression of Rous Sarcoma Virus is reviewed by Schierman & Collins (1987).

### *9.5d Selection for hormone levels*

Many of the changes in growth rate and feed efficiency are regulated by hormones. The thyroid hormones, thyroxine and tri-iodothyronine are the principal hormones that govern the basal metabolic rate and are thus important in regulation of body temperature. Susceptibility to heat stress is important when chickens are being intensively reared. Heat stress survival time has a low to moderate heritability but selection for improved heat stress survival time alone would be unsatisfactory since it has a negative correlation with body weight. Bowen & Washburn (1984) therefore investigated if it might be possible to select for other physiological parameters, in particular thyroid function, in order to obtain a higher heat stress survival time without reducing body weight. They used thyroid stimulating hormone (TSH) injected into newly hatched chicks and investigated the response. There was a good correlation in the reponse of circulating thyroxine to TSH injection. However, those groups selected for either good or poor response in the levels of thyroxine did not differ in their heat tolerance. Thus one would not expect an improvement in heat tolerance after selecting for lower thyroxine levels after TSH injection.

Strains that have been selected for broilers (i.e. having good growth rates) have been compared with non-selected strains for levels of various hormones. Circulating growth hormone levels were found to be lower in the selected strain. Prolactin levels also tended to be lower in the selected strains, but tri-iodothyronine had higher circulating levels in the selected strain (see Marks, 1985).

### 9.5e *Selection for other characters*

In a series of experiments in which male chickens were selected for 23 generations for the number of completed matings, selection caused a steady increase in the cumulative number of matings with a heritability of 0.18 ± 0.02 (Dunnington & Siegel, 1983).

The presence and length of spurs in hens can be a nuisance in intensive rearing. The heritability and correlation with other traits was investigated by Fairfull & Gowe (1986). Reduction of the incidence and size of spurs in hens can be brought about by mass selection; their heritabilities are 0.21 and 0.4–0.44, respectively. They show a low genetic correlation with economic traits and thus can be reduced without harm.

### 9.5f *Genotype × Environment interactions*

An important aspect of selection for the commercial breeder is how well a particular breed or strain performs in a given environment. Nowadays strains developed in one country may be distributed to other countries, where climate and management methods differ. It is clear that certain genotypes perform differently under different environmental conditions. Genotype × Environment is the term used to encompass this area of study. The potential importance of this in poultry breeding was first recognised by Munro (1936).

A detailed list of environmental factors is given by Sheridan (1990) including husbandry/management, economic factors, housing, and geographical location. These interactions pose a number of questions for the breeder, e.g. is it best to select under optimal environmental conditions, or under the conditions in which the strain will have to perform? The present practice with respect to disease resistance is generally to develop under optimal environmental conditions, i.e. in the absence of pathogens, whereas with respect to temperature development is more usually at the temperature at which they will subsequently have to perform.

## 9.6 Breeding methods

Once the characteristics of a quantitative trait have been determined (e.g. the heritability, any positive or negative correlations with other traits, whether the genetic variance is additive, and whether it shows dominance), the method of breeding to achieve the desired aims has to be decided. This entails selecting the individuals from within a group and the subsequent mating system to be used.

**Random mating (panmixis)** describes mating in which no restraints are imposed on mating within the experimental group or population. This method generally maintains the genetic constitution of the group or population and also maintains greatest diversity of progeny. **Positive assortive mating** is mating in which like individuals within a group are mated. They may be alike phenotypically, e.g. the heaviest individuals within a flock, or alike genotypically, e.g. full sibs. The latter constitutes inbreeding and causes the same gene to be inherited by an individual from each of its parents as a consequence of common ancestry. This is referred to as inheritance by descent. An individual may be homozygous at a particular locus, either because of inbreeding, or simply by chance because of the frequency of that gene in the population. As an example of the latter if a particular gene had a frequency of 40% in the population, there is a 16% ($0.4 \times 0.4 = 0.16$) chance of an individual being homozygous for that gene in the progeny.

The extent of inbreeding is expressed as the **inbreeding coefficient**, which is defined either on an individual basis, or on a population basis. On an individual basis it indicates the probability that two alleles at any one locus are identical by descent.

$$\text{Inbreeding coefficient } (F_x) = \tfrac{1}{2}R_{SD}$$

where $R_{SD}$ is the relationship between the two parents.

For example, for the progeny of full sibs, as shown in Fig. 9.7: the probability is that C will have $\frac{1}{2} \times \frac{1}{2} = \frac{1}{4}$ and B will have $\frac{1}{2} \times \frac{1}{2} = \frac{1}{4}$, and so in the progeny $R_{BC} = \frac{1}{4} + \frac{1}{4} = \frac{1}{2}$, and thus $F_x = \frac{1}{2}R_{BC} = 0.25$.

On a population basis the inbreeding coefficient indicates the percentage of all loci which were heterozygous in the base population that now have probably become homozygous as the result of inbreeding. The inbreeding

**Fig. 9.7** Inbreeding coefficient of full sibs.

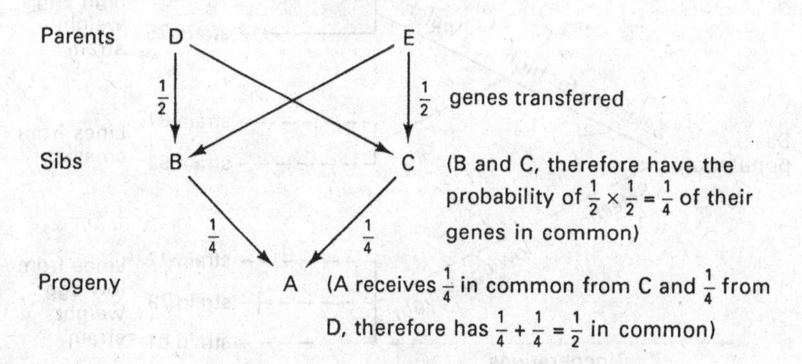

coefficient is thus a relative term, and relates to a particular base population. It can be estimated as follows. The base generation is given an inbreeding coefficient, $F_0 = 0$. For one generation of inbreeding the coefficient of inbreeding, $F_1 = 1/2N$, where $N$ is the size of the population.

For a second generation of inbreeding, homozygotes will be formed from the remaining heterozygotes and also from the homozygotes formed in the first generation, giving an accumulative inbreeding coefficient:

$$F_2 = 1/2N + (1 - 1/2N)F_1$$

Similar accumulations will occur for subsequent generations; thus for any generation

$$F_n = 1/2N + (1 - 1/2N)F_{n-1}$$

The inbreeding coefficient can vary between 0 and 1. A value of zero implies no homozygosity relative to the base population, whereas a value of unity implies complete homozygosity. For example, inbreeding experiments at Iowa State College were begun with 23 lines of White Leghorns and by 1946 some lines had an inbreeding coefficient of 0.85. Seven highly inbred lines were developed by the Swiss Federal Institute of Technology (Hagger, 1985). The base population was used to select two lines, one for low egg weight and one for high egg weight, and these were inbred and subsequently divided into seven separate inbred lines as shown in Fig. 9.8.

Over a period the average inbreeding coefficient, which can be calculated if the pedigrees are known, gradually rose in the following way.

| Year | 1974 | 1975 | 1976 | 1977 | 1978 | 1979 | 1980 | 1981 |
|---|---|---|---|---|---|---|---|---|
| Inbreeding coefficient | 0.75 | 0.80 | 0.84 | 0.86 | 0.89 | 0.91 | 0.93 | 0.94 |

**Fig. 9.8** Population structure of inbred lines from the Swiss Federal Institute of Technology (Hagger, 1985).

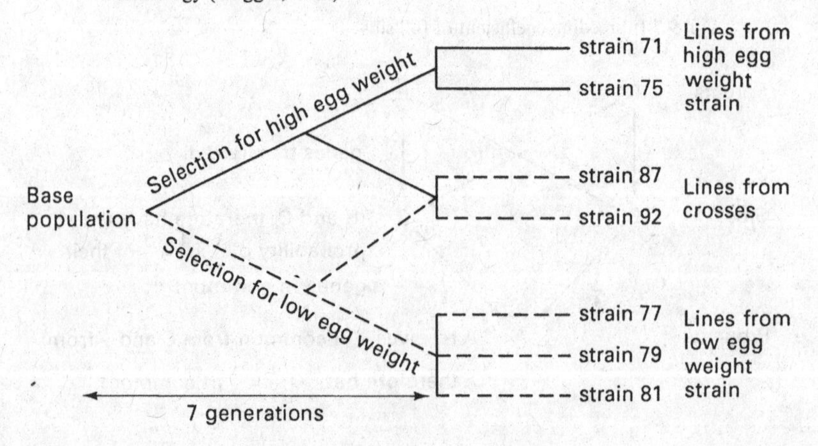

These inbred lines have been used in a variety of breeding studies including one on heterosis (see below). A detailed analysis of inbreeding and it importance in poultry genetics is given by Abplanalp (1990).

The opposite of positive assortive mating is **negative assortive mating**. In the latter, individuals more distantly related either genotypically or phenotypically are mated. An example of phenotypic dissimilarity would be mating two groups within a population that have the highest and the lowest body weights, respectively. Negative assortive mating based on genetic unrelatedness is of greatest importance in poultry breeding when separate strains have been inbred and selected for particular characters for a number of generations and are then crossed with one another. The inbreeding for several generations often causes **inbreeding depression**, which is thought to be due to a number of recessive genes which have detrimental qualities becoming homozygous in the strain. Bringing together two such strains often leads to an improvement in vigour, known as **hybrid vigour** or **heterosis.**

One hypothesis used to explain heterosis is that it arises from the action and interaction of dominant genes for growth or fitness. The cross will generally increase the variety of dominant genes in the progeny, and these may interact to increase vigour. The second hypothesis is that it is due to overdominance, i.e. the heterozygote is superior to either homozygote for a particular character, e.g. $a/a < a/A > A/A$. Both explanations may be correct for different loci. Epistasis is now also thought to contribute (Fairfull, 1990).

Heterosis is usually expressed quantitatively as the difference between the value of a trait in the hybrid and the mean of the values in the parent strains. For example, suppose that two strains have been inbred for a number of generations, and are then cross mated. We designate the strains *AA* and *BB*, and the average value of the trait considered $\bar{A}\bar{A}$ and $\bar{B}\bar{B}$. We can cross breed in two reciprocal ways: either with a male from *AA* to a female from *BB*, or with a male from *BB* to a female from *AA*. The convention is to indicate the male first. Thus $\bar{A}\bar{B}$ would be the average value of the trait in the group obtained from crossing male of *AA* with females of *BB*.

The heterosis is then expressed either as:

$$H = (\bar{A}\bar{B}/2 + \bar{B}\bar{A}/2) - (\bar{A}\bar{A}/2 + \bar{B}\bar{B}/2)$$

or as percentage heterosis:

$$\%H = \frac{(\bar{A}\bar{B}/2 + \bar{B}\bar{A}/2) - (\bar{A}\bar{A}/2 + \bar{B}\bar{B}/2) \times 100\%}{(\bar{A}\bar{A}/2 + \bar{B}\bar{B}/2)}$$

The latter form (%H) is more useful when comparing the heterosis for different traits that generally will be expressed in different units.

A number of studies of heterosis have been made in the domestic fowl, most concerned with characters which are important in layers and broilers. Hagger (1985) studied heterosis for a number of traits when the seven inbred lines of White Leghorns described in Fig. 9.8 were crossed. It is generally more common to find positive heterosis than negative heterosis, and the precise value for any trait will depend on the inbreeding pedigree of the strains before crossing. Hagger (1985) found positive heterosis for the following traits: average weight of eggs up to 40 weeks (+5.8%), body weight at 40 weeks (+7.8%), egg production up to 40 weeks (+45.1%) and egg production from 41 to 60 weeks (+35.8%), but negative heterosis for age at which the first egg was laid (−11.3%).

Zelenka, Siegel & Van Krey (1986) studied the relationship between the formation of double yolks and the duration of the phase of rapid growth of yolk, in crosses between two inbred strains of White Plymouth Rocks that had been selected for high and low body weights. They found positive heterosis for the number of eggs laid during the first 30 days of lay (+12%), but negative heterosis for the incidence of double-yolked eggs (−9%) and a small negative heterosis for the duration of the period of rapid growth of yolk (−1%). Abplanalp *et al.* (1984b) studied heterosis for the quality of eggs in crosses between six inbred lines of Leghorns. They found positive heterosis for egg weight and for the proportion of yolk, negative heterosis for the proportion of albumen, and no significant heterosis for the proportion of shell.

Barbato, Siegel & Cherry (1983) studied the inheritance of body weight and associated traits in young chickens from two inbred lines which had been selected for low and high body weight. The degree of heterosis of eleven traits was measured on crossing the strains. Heterotic effects were generally lacking (i.e. values between ±2%) for overall nutrient intake and gross morphological measures of growth such as body weight and shank length, but in contrast, traits involved with the efficiency of the organism to allocate nutrient intake into specific growth components showed moderate to high degrees of heterosis. For a more comprehensive set of data, see Fairfull (1990).

This chapter has aimed to be an introduction to the principles of quantitative genetics and their application to the domestic fowl; a list of further reading on both the principles and the application to the domestic fowl is given at the end of the reference section.

## References

Abplanalp, H. (1990). Inbreeding. In *Poultry Breeding and Genetics*, ed. R. D. Crawford, pp. 955–84. Amsterdam: Elsevier.

Abplanalp, H., Lowry, D. C., Lerner, I. M. & Dempster, E. R. (1964). Selection for egg number with X-ray induced variation. *Genetics*, **50**, 1083–100.

Abplanalp, H., Okamoto, S., Napolitano, D. & Len, R. E. (1984a). A study of heterosis and recombination loss in crosses of inbred Leghorn lines derived from a common base population. *Poultry Science*, **63**, 234–39.

Abplanalp, H., Peterson, S. J., Okamoto, S. & Napolitano, D. (1984b). Heterosis, recombination effects and genetic variability of egg compsition in inbred lines of White Leghorns and their crosses. *British Poultry Science*, **25**, 361–7.

Abplanalp, H., Tai, C. & Napolitano, D. (1984). Genetic correlations of abdominal fat with production traits of Leghorn hens based on inbred line averages. *British Poultry Science*, **25**, 343–7.

Barbato, G. F., Siegel, P. B. & Cherry, J. A. (1983). Inheritance of body weight and associated traits in young chickens. *Zeitschrift für Tierzuchtung und Zuchtungsbiologie*, **100**, 350–60.

Becker, W. A., Spencer, J. V., Mirosh, L. W. & Verstrate, J. A. (1984). Genetic variation of abdominal fat, body weight, and carcass weight in a female broiler line. *Poultry Science*, **63**, 607–11.

Bessei, W. (1985), Pecking and feather loss – genetics aspects. *2nd European Symposium on Poultry Welfare, World Poultry Science Association*, Celle, pp. 211–18.

Bowen, S. J. & Washburn, K. W. (1984). Genetics of heat tolerance and thyroid function in Athens-Canadian randombred chickens. *Theoretical and Applied Genetics*, **69**, 15–21.

Bowman, J. C. (1968). Production characters in poultry. In *Genetics and Animal Breeding*, ed. I. Johansson and J. Rendel, pp. 340–86. Edinburgh: Oliver & Boyd.

Briles, W. E. (1984). Early chicken blood group investigations. *Immunogenetics*, **20**, 217–26.

Briles, W. E. & Briles, R. W. (1982). Identification of haplotypes of the chicken major histocompatibility complex. *Immunogenetics*, **15**, 449–59.

Briles, W. E., Stone, H. A. & Cole, R. K. (1977). Marek's disease: Effects of *B* histocompatibility alleles in resistant and susceptible chicken lines. *Science*, **195**, 193–5.

Bulfield, G., Isaacson, J. H., & Middleton, R. J. (1988). Biochemical correlates of selection for weight-for-age in chickens: twenty-fold higher ornithine decarboxylase levels in modern broilers. *Theoretical and Applied Genetics*, **75**, 432–7.

Bulfield, G. & McKay, J. C. (1987). Genetic manipulation of egg quality. In *Egg Quality – Current Problems and recent Advances*, ed. R. G. Wells and C. G. Belyavin, pp. 195–202. London: Butterworths.

Cahaner, A., Nitsan, Z. & Nir, I. (1986). Weight and fat content of adipose and nonadipose tissues in broilers selected for or against abdominal adipose tissue. *Poultry Science*, **65**, 215–22.

Chambers, J. R. (1990). Genetics of growth and meat production in chickens. In *Poultry Breeding and Genetics*, ed. R. D. Crawford, pp. 599–643. Amsterdam: Elsevier.

Chambers, J. R., Bernon, D. E. & Gavora, J. S. (1984). Synthesis and parameters of new populations of meat-type chickens. *Theoretical and Applied Genetics*, **69**, 23–30.

Damme, K., Pirchner, F., Willeke, H. & Eichinger, H. (1986). Fasting metabolic rate in hens. 2. Strain differences and heritability estimates. *Poultry Science*, **65**, 616–20.

Dunnington, E. A. & Siegel, P. B. (1983). Mating frequency in male chickens: Long-term selection. *Theoretical and Applied Genetics*, **64**, 317–23.

Dunnington, E. A. & Siegel, B. P. (1985) Long term selection for 8-week body weight in chickens – direct and correlated responses. *Theoretical and Applied Genetics*, **71**, 305–13.

Fairfull, R. W. (1990). Heterosis. In *Poultry Breeding and Genetics*, ed. R. D. Crawford, pp. 913–34. Amsterdam: Elsevier.

Fairfull, R. W. & Gowe, R. S. (1986). Genotypic and phenotypic parameters of spur incidence and length in White Leghorn hens. *Poultry Science*, **65**, 1995–2001.

Fairfull, R. W. & Gowe, R. S. (1990). Genetics of egg production in chickens. In *Poultry Breeding and Genetics*, ed. R. D. Crawford, pp.705–60. Amsterdam: Elsevier.

Falconer, D. S. (1989). *Introduction to Quantitative Genetics*, 3rd edn. Edinburgh: Longman.

Fincham, J. R. S. (1983). *Genetics*. Bristol: Wright.

Fischer, G. J. (1969). Heritability in the following response of White Leghorns. *Journal of Genetic Psychology*, **114**, 215–17.

Foster, W. H. (1985). The genetics of ovulation efficiency. In *Poultry Genetics and Breeding*, ed. W. G. Hill, J. M. Manson and D. Hewitt, pp. 157–68. Harlow: British Poultry Science Ltd.

Gavora, J. S. (1990). Disease Genetics. In *Poultry Breeding and Genetics*, ed. R. D. Crawford, pp. 805–46. Amsterdam: Elsevier.

Gavora, J. S., Simonsen, M., Spencer, J. L., Fairfull, R. W. & Gowe, R. S. (1986). Changes in the frequency of major histocompatibility haplotypes in chickens under selection for both high egg production and resistance to Marek's disease. *Journal of Animal Breeding and Genetics*, **103**, 218–26.

Gowe, R. S. & Fairfull, R. W. (1985). The direct response to long-term selection for multiple traits in egg stocks and changes in genetic parameters with selection. In *Poultry Genetics and Breeding*, ed. W. G. Hill, J. M. Manson and D. Hewitt. pp.125–46. Harlow: British Poultry Science Ltd.

Gowe, R. S. & Fairfull, R. W. (1990). Genetics controls in selection. In *Poultry Breeding and Genetics*, ed. R. D. Crawford, pp. 935–54. Amsterdam: Elsevier.

Griffin, H. D. & Whitehead, C. C. (1982). Plasma lipoprotein concentration as an indicator of fatness in broilers: development and use of a simple assay for plasma very low density lipoproteins. *British Poultry Science*, **23**, 307–13.

Grunder, A. A. & Chambers, J. R. (1988). Genetic parameters of plasma very low density lipoproteins, abdominal fat lipase, and protein, fatness, and growth traits of broiler chickens. *Poultry Science*, **67**, 183–90.

Hagger, C. (1985). Line crossing effects in a diallel mating system with highly inbred lines of White Leghorn chickens. *Theoretical and Applied Genetics*, **70**, 555–60.

Hagger, C. & Marguerat, C. (1985). Relationship of production traits and egg composition to feed consumption and the genetic variability of efficiency in laying hens. *Poultry Science*, **64**, 2223–9.

Hansen, M. P., Law, R. J. & Van Zandt, J. N. (1967). Differences in susceptibility to Marek's disease in chickens carrying two different *B* locus blood group alleles. *Poultry Science*, **46**, 1268 (abstract).

Hartmann, W. (1985). The effect of selection and genetic factors on resistance to disease in fowls. *World's Poultry Science Journal*, **41**, 20–35.

Hutt, F. B. & Cole, R. K. (1947). Genetic control of lymphomatosis in the fowl. *Science*, **106**, 379–84.

Johnson, L. W. & Edgar, S. A. (1986). Ea-B and Ea-C cellular antigen genes in Leghorn lines resistant and susceptible to acute cecal coccidiosis. *Poultry Science*, **65**, 241–52.

Kim, C. D., Lamont, S. J. & Rothschild, M. F. (1987). Genetic associations of body weight and immune response with the major histocompatibility complex in White Leghorn chicks. *Poultry Science*, **66**, 1258–63.

Kinney, T. B. (1969). A summary of reported estimates of heritabilities and of genetic and phenotypic correlations for traits of chickens. *Agriculture Handbook No. 363*. Washington, DC: US Dept of Agriculture.

Lamont, S. J., Hon, Y.-H., Young, B. H. & Nordskog, A. W. (1987). Differences in major histocompatibility complex gene frequencies associated with feed efficiency and laying performance. *Poultry Science*, **66**, 1064–6.

Leclerq, B. (1983). The influence of dietary protein content on the performance of genetically lean or fat growing chickens. *British Poultry Science*, **24**, 581–7.

Leclerq, B., Blum, J. C. & Boyer, J. P. (1980). Selecting broilers for low and high abdominal fat: initial observations. *British Poultry Science*, **21**, 107–13.

Leclercq, B., Simon, J. & Ricard, F. H. (1987). Effects of selection for high and low plasma glucose concentration in chickens. *British Poultry Science*, **28**, 557–65.

Liljedahl, L. E., Gavora, J. S., Fairfull, R. W. & Gowe, R. S. (1984). Age changes in genetic and environmental variation in laying hens. *Theoretical and Applied Genetics*, **67**, 391–401.

Marks, H. L. (1985). Direct and correlated responses to selection for growth. In *Poultry Genetics and Breeding*, ed. W. G. Hill, J. M. Manson and D. Hewitt, pp.47–58. Harlow: British Poultry Science Ltd.

Marks, H. L. (1987). Selection for 8 week body weight in normal and dwarf chickens under different water/feed environments. *Poultry Science*, **66**, 1252–7.

Marks, H. L. & Washburn, K. W. (1983). The relationhip of altered water/feed intake ratios on growth and abdominal fat in commercial broilers. *Poultry Science*, **62**, 263–72.

Mercer, J. T. & Hill, W. G. (1984). Estimation of genetic parameters for skeletal defects in broiler chickens. *Heredity*, **53**, 193–203.

Munro, S. S. (1936). The inheritance of egg production in the domestic fowl. *Science and Agriculture*, **16**, 591–607.

Naito, M., Nirasawa, K., Oishi, T & Komiyama, T. (1989). Selection experiment for increased egg production under 23h and 24h light–dark cycles in the domestic fowl. *British Poultry Science*, **30**, 49–60.

Pym, R. A. E. & Nicholls, P. J. (1979). Selection for food conversion in broilers: body composition of birds selected for increased body-weight gain, food consumption and food conversion ratio. *British Poultry Science*, **20**, 73–86.

Pym, R. A. E. & Thompson, J. M. (1980). A simple caliper technique for the estimation of abdominal fat in live broiler chickens. *British Poultry Science*, **21**, 281–6.

Ricard, F. H. (1974). Etude de la variabilité génétique de quelques caracteristiques de carcasses en vue de selectionner un poulet de qualité. *Proceedings of the First World Congresss on Genetics Applied to Livestock Production, Madrid*, **1**, 931–40.
Ricard, F. H. & Rouvier, R. (1969). Etude de la composition anatomique du poulet. III. Variabilité de la repartition des parties corporelles dans la souche de type Cornish. *Annales de Génétique et Selection Animaux*, **1**, 151–65.
Saeki, Y. (1957). Inheritance of broodiness in Japanese Nagoya fowl with special reference to sex-linkage and notice in breeding practice. *Poultry Science*, **36**, 378–83.
Schierman, L. W. & Collins, W. H. (1987). Influence of major histocompatibility complex on tumor regression and immunity in chickens. *Poultry Science*, **66**, 812–18.
Sheridan, A. K. (1990). Genotype × environment interactions. In *Poultry Breeding and Genetics*, ed. R. D. Crawford, pp. 897–912. Amsterdam: Elsevier.
Siegel, P. B. (1962). Selection for body weight at 8 weeks of age. 1. Short term response and heritabilities. *Poultry Science*, **41**, 954–62.
Siegel, P. B. & Dunnington, E. A. (1985). Reproductive complications associated with selection for broiler growth. In *Poultry Genetics and Breeding*, ed. W. G. Hill, J. M. Manson and D. Hewitt, pp. 59–72. Harlow: British Poultry Science Ltd.
Simonsen, M., Crone, M., Koch, C. & Hála, K. (1982). The MHC haplotypes of the chicken. *Immunogenetics*, **16**, 513–32.
Stansfield, W. D. (1983). Genetics, 2nd edn. New York: McGraw Hill.
Suzuki, D. T., Griffiths, A. J .F., Miller, J. H. & Lewontin, R. C. (1989). *An Introduction to Genetic Analysis*, 4th edn. New York: W.H. Freeman.
Tabor, C. W. & Tabor, H. (1984). Polyamines. *Annual Review of Biochemistry*, **53**, 749–90.
Thomas, C. H., Blow, W. L., Cockerham, C. C. & Glazener, E. W. (1958). The heritability of body weight gain, feed consumption, and feed conversion in broilers. *Poultry Science*, **37**, 862–9.
Twinning, P. V., Thomas, O. P. & Bossard, E. H. (1978). Effect of diet and type of birds on carcass composition of broilers at 28, 49, and 59 days of age. *Poultry Science*, **57**, 492–7.
Washburn, K. W. (1990). Genetic variation in egg composition. In*Poultry Breeding and Genetics*, ed. R. D. Crawford, pp.781–804. Amsterdam: Elsevier.
Waters, N. F. (1945). Breeding for resistance and susceptibility to avian lymphomatosis. *Poultry Science*, **24**, 259–69.
Whitehead, C. C. & Griffin, H. D. (1985). Direct and correlated response to selection for decreased body fat in broilers. In *Poultry Genetics and Breeding*, ed. W. G. Hill, J. M. Manson and D. Hewitt, pp. 113–23. Harlow: British Poultry Science Ltd.
Whiting, T. S. & Pesti, G. M. (1983). Effects of the dwarfing gene (*dw*) on egg weight, chick weight, and chick weight:egg weight ratio in a commercial broiler strain. *Poultry Science*, **62**, 2297–302.
Williams, J. B. & Sharp, P. J. (1978). Control of the preovulatory surge of luteinizing hormone in the hen (*Gallus domesticus*): the role of progesterone and androgens. *Journal of Endocrinology*, **77**, 57–65.
Wilson, S. P. (1969). Genetic aspects of food efficiency in broilers. *Poultry Science*, **48**, 487–95.

Yoo, B. H., Sheldon, B. L. & Podger, R. N. (1988). Genetic parameters for oviposition time and interval in a White Leghorn population of recent commercial origin. *British Poultry Science*, **29**, 627–37.

Zelenka, D. J., Siegel, P. B. & Van Krey, H. P. (1986). Ovum formation and multiple ovulation in lines of White Plymouth Rocks and other crosses. *British Poultry Science*, **27**, 409–14.

Zijpp, A. J. van der (1983). Breeding for Immune responsiveness and disease resistance. *World's Poultry Science Journal*, **39**, 118–31.

## *References for further reading*

Bowman, J. C. (1984). *An Introduction to Animal Breeding*, 2nd edn. London: Arnold.

Crawford, R. G. (ed.) (1990). *Poultry Breeding and Genetics*, Part IV. Amsterdam: Elsevier.

Falconer, D. S. (1989). *Introduction to Quantitative Genetics*, 3rd edn. Edinburgh: Longman.

Nicholas, F. W. (1987). *Veterinary Genetics*, Chapters 12–19. Oxford: Clarendon Press.

Strickberger, M. W. (1985). *Genetics*, 3rd edn, Chapters 14 and 15. London: Macmillan.

Suzuki, D. T., Griffiths, A. J. F., Miller, J. H. & Lewontin, R. C. (1989). *An Introduction to Genetic Analysis*, 4th edn, Chapters 22 and 23. New York: W.H. Freeman.

Van Vleck, L. D., Pollak, E. J. & Oltenacu, E. A. B. (1987). *Genetics for the Animal Sciences*, Chapters 11–15. New York: W. H. Freeman.

# 10

# *Protein evolution and polymorphism*

## 10.1 Introduction

Mutations alter the genotype by changing the nucleotide sequence in DNA, but natural selection operates on the phenotype, which is largely dependent on the particular proteins made by an organism. A study of the structure and sequence of individual proteins can therefore be useful in furthering the understanding of evolution in two important ways. Firstly, by comparing the structures of a specific protein, e.g. cytochrome *c*, that occur in different species, it is possible to establish or confirm phylogenetic relationships amongst organisms, and to build up phylogenetic trees involving phyla, classes, orders, etc. Secondly, a number of proteins are found to exist in more than one closely related form, e.g. ovalbumin A and B; these are known as **polymorphisms**. A study of the distribution of different polymorphic forms within a population of a given species, together with a knowledge of their dominance relationships, can be used to explain their more recent history. These two areas are interrelated, and in this chapter a number of proteins are examined, some of which have been primarily of importance in establishing and confirming phylogenetic relationships, e.g. the haem proteins, and others have been more useful in determining relationships between breeds of the domestic fowl, e.g. egg-white proteins.

The evidence that led Darwin to propose his evolutionary theory, and the evidence that has been acquired since, and has substantiated his theory, has arisen in three chronological phases. Initially, the major supporting evidence for the course of evolution, as proposed by Darwin, was from **palaeontology**. This evidence compares the macroscopic anatomical features of fossils of various ages with that of the present day forms. If the ages of the sedimentary rocks containing the fossils can be accurately dated, then a phylogenetic tree with an appropiate time scale can be built up.

Since about the mid-1960s sufficient information on **protein sequences** has been acquired to be able to compare specific proteins between different species. Thus evolution of particular proteins may be examined on a molecular basis. It is, of course, the genes themselves in which the mutations occur. Since the mid-1970s it has become generally easier to determine **DNA sequences** than protein sequences (see Chapter 12) and thus the number of known gene sequences is increasing rapidly. This enables comparisons now to be made between the genes coding for specific proteins.

There are a number of ways in which a phylogenetic tree can be constructed using protein sequences from a single type of protein obtained from different species. Proteins or genes having a significant number of similarities are said to be **homologous**. Once the sequences of a number of homologous proteins are known (e.g. cytochrome *c* has now been sequenced in over 80 species), the differences in sequences can be compared. If only point mutations have occurred then aligning the sequences is straightforward, but if deletions, additions or inversions have occurred, then alignment may be more difficult, although computer programs are available in which protein sequences can be compared on a matrix to determine the extent of homology (Chaplin, 1983; Lipman & Pearson, 1985).

When considering the differences between two particular species, the ancestral form from which they are believed to have arisen is deduced by assuming the minimum number of mutations. These deductions are generally made after examining the differences in the protein sequences, and then by using the genetic code (Appendix V), to deduce the minimum number of nucleotide changes in DNA this would require. In this way a 'tree' can be built up. In comparing two protein molecules of differing lengths a knowledge of the three dimensional structure is an advantage in determining where the additions/deletions have occurred, since the tertiary structure is generally conserved in order that the protein remains functional. (The tertiary structure refers to the overall folding arrangement of the protein chain.) A phylogenetic tree (Fig. 10.1) is then generated by considering the minimum number of mutational events required to produce the change in sequence (for a description of the methods used, see Parkin, 1979). The results from this approach have generally been in excellent agreement with the fossil evidence. However, when only a small number of differences is found, since random events are involved, some discrepancies have occurred especially where only a single protein has been considered.

A number of significant findings have emerged from these sequence studies, and while it is beyond the scope of this chapter to discuss them in detail (see Creighton, 1984), some of the most important generalisations will be mentioned briefly. If a single type of protein is analysed in this way it is generally found that the rate of amino acid replacement is constant throughout evolution, e.g. in haemoglobin the rate of amino acid replacement is 1% per 5.8 million years. However, if two types of protein are compared, such as cytochrome *c* and fibrinopeptides, they show different replacement rates, 1% per 20 million years and 1% per 1.1 million years, respectively. The reason is that certain proteins have to specified very precisely in order to maintain their function, e.g. haemoglobin has to bind and release oxygen at the physiological partial pressures that exist in the lungs and in the tissues, whereas other proteins such as fibrinopeptides that are cleaved from fibrinogen in the process of blood clotting, can tolerate more changes without loss of function.

When the positions in the protein sequence where the replacements occur are analysed, and the particular amino acids that are replaced are identified, it is found that replacement is non-random (i.e. certain amino acids are replaced more frequently than others), and that most of the replacements are what are termed as **neutral**, since they do not affect appreciably the functioning of the protein. Replacements occur most frequently in the regions of the protein that do not have to be accurately specified. Thus, the rate of amino acid replacement reflects largely neutral replacements in non-essential positions. The rate of mutation is similar for all genes, but the amount of change in the structure of a protein consistent with its function-

**Fig. 10.1** Evolutionary tree based on the amino acid sequences of cytochrome *c* (Dayhoff, Park & McLaughlin, 1972). The numbers are the number of amino acid substitutions per 100 amino acid residues.

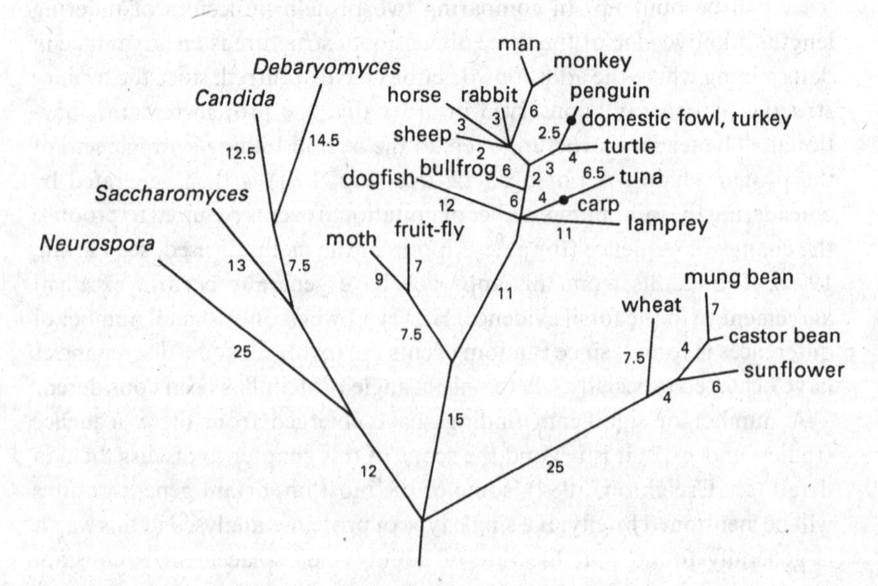

ing efficiently varies amongst proteins, and thus determines the acceptable amino acid replacements of each protein.

Amino acid replacements are also responsible for a second and related phenomenon, namely **protein polymorphism**. Since the advent of electrophoretic methods for the separation of proteins, it has been found that protein polymorphism is extremely widespread. Polymorphism will be detected if two enzymes or other proteins differ in their overall electric charge and hence their mobility in an electric field. A number of polymorphic forms will not be detected by this method, known as electrophoresis, especially if they involve **conservative** mutations (mutations in which like amino acids are replaced with like, e.g. polar with polar, or hydrophobic with hydrophobic). In a survey of 242 species Nevo (1978) found that 26% of all the loci studied were polymorphic, and of the birds included in this survey 15% of loci were detected as polymorphic. Lewontin (1974) pointed out that this is undoubtedly an underestimate because of the undetected polymorphisms, and he estimated that about two thirds of all loci are polymorphic.

When a population of a particular species appears to be well adapted to its environment it might be expected that a single genotype would have been progressively selected over many generations, i.e. the genotype best adapted to the environment. However, this rarely occurs and most well adapted populations show extensive polymorphisms at many loci. The reasons for this are probably varied, dependent on the particular population. A number of proposals have been made to account for polymorphism, but most need substantiation. One suggestion is that in selecting for one particular character the overall 'fitness' is reduced. Genes interact with one another, and the transmission of some genes is closely linked; both of these factors may affect the selection process. Another hypothesis is that heterozygotes have a selective advantage. Sickle cell anaemia in humans is a good example of this. The homozygous HbS/HbS is often fatal because the red blood cells assume a sickle shape and tend to aggregate in the capillaries, and cause damage to tissues through lack of oxygen. The heterozygote HbS/HbA also shows the sickling condition, but in a much less pronounced form, and it confers the advantage over the normal HbA/HbA of resistance to malaria.

Several workers have suggested that many of the mutations that have produced new alleles and hence polymorphisms are neutral and scarcely affect the functioning of the protein. In these cases there is no selective pressure to eliminate them and they become fixed in the population by **random drift**. This process is an important 'plank' of the **neutralist hypothesis** of evolution (see Chapter 1, section 1.2). It should also be noted that

Fisher's Theory (Chapter 1, section 1.2) states, 'The greater the genetic variability upon which selection for fitness may act, the greater the improvement in fitness'. Thus polymorphism may be an important factor in determining a population's ability to adapt to changing environmental conditions.

The two themes, evolution of phylogenetic trees and protein polymorphism, will be considered in the succeeding sections in the cases of individual proteins, and groups of proteins, in the domestic fowl.

## 10.2 Haemoglobin and myoglobin

Haemoglobin and myoglobin are important oxygen carrying proteins in vertebrates. Haemoglobin is a tetramer that comprises four polypeptide chains or globin subunits. Each globin subunit has attached to it an iron-containing prosthetic group called haem. Haemoglobin occurs in the blood and is responsible for the transport of oxygen from the lungs to the tissues. Myoglobin is a single polypeptide chain or monomer and occurs in muscle, particularly skeletal muscle, where it acts as an oxygen reserve; it can release its bound oxygen at low oxygen tensions. An enormous amount of detailed structural information is available for both proteins (see Dickerson & Geiss, 1983).

The evolution of haemoglobin and myoglobin is a classic example in which evolution is believed to have occurred by a process involving successive gene duplications, as shown in Fig. 10.2. This scheme is based largely on the protein sequences of human haemoglobin and myoglobin. The predominant form of haemoglobin in normal adults comprises two α-

**Fig. 10.2** Probable evolutionary relationships of human globins, $\alpha$, $\beta$, $\gamma$, $\delta$, $\epsilon$ and $\zeta$. The numbers indicate millions of years since divergence (Proudfoot, Gill & Maniatis, 1982). A gene duplication is believed to have occurred at each branch point.

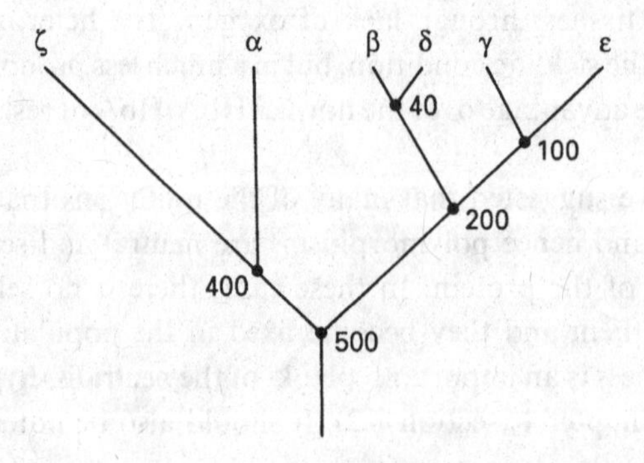

and two $\beta$-globin subunits, but there is also a small amount of the variant $\alpha_2\delta_2$. (The type of haemoglobin is designated using Roman letters, e.g. HbA, and the globin subunits by Greek letters, e.g. $\alpha_2\beta_2$.) Of the other forms of globin, $\gamma$ occurs in the foetus and $\epsilon$ occurs in the very young embryo.

In the species of mammals and birds in which the haemoglobins have so far been examined it is possible to assign the globin chains to one of two basic types, $\alpha$ and $\beta$. The division is based on the extent of homology between the globin in question and that of $\alpha$ and $\beta$ globin, and is generally thought to reflect their evolutionary origin. Thus in humans, the $\beta$-types include $\beta$, $\gamma$, $\delta$ and $\epsilon$, and the $\alpha$-types, $\alpha$ and $\zeta$ (Proudfoot, Gill & Maniatis, 1982). The $\alpha$-types are thought to have diverged about 400 million years ago, i.e. before the divergence of mammals and birds from reptiles, whereas the $\beta$-types diverged about 200 million years ago, i.e. about the same time as the divergence of mammals and birds. The foetal and embryonic haemoglobins differ from the adult haemoglobins in their physiological properties; for example, the haemoglobin tetramer $\alpha_2\gamma_2$ that occurs in the foetus has a higher affinity for oxygen than the normal adult $\alpha_2\beta_2$, and this enables the foetus to cope with lower oxygen tensions than the adult.

In the domestic fowl there are two main forms of adult haemoglobin, haemoglobin A (HbA) and haemoglobin D (HbD), and at least two minor forms, haemoglobin H (HbH) (Moss & Hamilton, 1974) and haemoglobin L (HbL) (Cirotto, Panaca & Arangi, 1987). As in the case of mammals, there are at least three forms of haemoglobin in the embryo. These are present in embryonic erythrocytes (red blood cells) which develop in the yolk sac up until the fifth day of incubation. Between the fifth and the twelfth day of incubation the embryonic erythroid cells gradually become replaced by the definitive erythrocytes, that produce the adult haemoglobin (Keane & Abbott, 1980). Table 10.1 summarises the principal types of haemoglobins of the domestic fowl.

The different globin chains were first distinguished by their electrophoretic properties, and by mapping their tryptic peptides. Moss & Hamilton (1974), using White Leghorns, showed that there were two main types of haemoglobin in the adult domestic fowl, HbA and HbD, and that these two forms were identical with those from the later stages of developing embryos. (These are also referred to as definitive haemoglobins.) They also detected the minor form HbH in late embryos but not in the adult; nevertheless it is classed as a definitive form. More recently additional globin types have been discovered, and some of the globins have been completely sequenced. The two principal $\alpha$ globins, $\alpha^A$ and $\alpha^D$ have been sequenced from their cDNA (Dodgson & Engel, 1983) and are found to be

Table 10.1. *Adult and embryonic haemoglobins from the domestic fowl*

| Primitive haemoglobins | | | Definitive haemoglobins | | |
|---|---|---|---|---|---|
| Haemoglobin | Composition | Proportion | Haemoglobin | Composition | Proportion |
| E | $\alpha_2{}^A\epsilon_2$ | 20% | H | $\alpha_2{}^A\beta_2{}^H$ | trace[a] |
| P | $\pi_2\rho_2$ <br> $\pi_2{}'\rho_2$ | 70% | A | $\alpha_2{}^A\beta_2$ | 70% |
| M | $\alpha_2{}^D\epsilon_2$ | 10% | D | $\alpha_2{}^D\beta_2$ | 30% |
| | | | L | $\alpha_2{}^D\beta_2{}^H$ | trace[a] |

[a] A trace in 7 day embryos but not adults.
Data from Brown & Ingram (1974); Moss & Hamilton (1974); Keane & Abbott (1980); Chapman, Hood & Tobin (1982a & b); Cirotto *et al.* (1987).

widely divergent, differing in a number of amino acid residues. The major embryonic α-type globins are the π globins (referred to as primitive globins). These have been sequenced and found to differ from both $\alpha^A$ and $\alpha^D$ by over 40% (Chapman, Tobin & Hood, 1980). The π and π′ differ by a single amino acid residue at position 124, alanine in the case of π, and glutamic acid in π′. The π-globins resemble other α-type globins from mammals such as ζ-globin in humans (Engel *et al.*, 1983). It is not yet clear whether π and π′ are allelic. There is also uncertainty whether another distinct α-type globin exists in HbE. Chapman, Hood & Tobin (1982a) found 22 differences in the sequences of the α-globin ($\alpha^E$) present in HbE from that of $\alpha^A$, but Knochel *et al.* (1982) have re-examined the sequences of $\alpha^A$ and $\alpha^E$ and found them to be identical.

Although there may be as many as six or seven β-type globin genes the weight of evidence at present favours four, namely β, $\beta^H$, ε and ρ (Dolan *et al.*, 1981). A second type of π-globin, ρ′, has been detected as a cDNA sequence which translates to a ρ-like polypeptide differing from ρ in four positions (Roninson & Ingram, 1981). The globin ε has been sequenced and differs from the β chain in 18 positions (Chapman *et al.*, 1982a) whilst ρ differs from β at 19 positions (Chapman, Tobin & Hood, 1981). The differences in both cases all appear at 'non-critical' positions as far as the functioning of haemoglobin is concerned. Chapman *et al.* (1981) suggest that the splitting of ρ and ε from β occurred as a relatively recent gene duplication.

Both the α- and β-type globin chains are coded for in two **multigene families**. Multigene families are groups of related genes clustered together on particular chromosomes. The genes concerned are assumed to have arisen by successive gene duplications. The gene duplications occur during meiosis as a result of unequal crossing over (Fig. 10.3). A number of multigene families have now been identified (see Hood, Campbell & Elgin, 1975). The $\beta$-type genes from the domestic fowl have been mapped using a DNA library of domestic fowl DNA linked to the vector $\lambda$ Charon 4A (see Chapter 12, section 12.2). Dolan *et al.* (1981) have shown that the arrangement is:

$$\rho\text{---}\beta^{H}\text{---}\beta\text{---}\epsilon$$

This arrangement is unique among vertebrate clusters so far examined in that the adult (definitive) genes are flanked by the embryonic (primitive) genes.

The $\alpha$-globin family has been shown to be arranged as follows:

$$\pi \text{ or } \pi'\text{---}\alpha^{D}\text{---}\alpha^{A}$$

The globin gene families are often regarded as a paradigm for the study of developmentally regulated gene expression. The avian erythrocyte is particularly useful in this respect since it has a number of well-defined stages of development. Since both the $\alpha$- and $\beta$-type globin families have been sequenced it is possible to study the developmental regulation in detail. Figure 10.4 shows the proposed evolution of the gene family. This is based on the differences in amino acid sequence of the globin chains (Chapman *et al.*, 1980).

Gene expression depends initially on transcription, and this requires the binding of RNA polymerase to promoter sites adjacent to the gene. There are also DNA sequences outside the promoter region known as enhancers

**Fig. 10.3** Unequal crossing over occurring during meiosis and leading to gene duplication. A,A′; B,B′ and C,C′ represent homologous pairs of genes and the broken line the position of the chiasma.

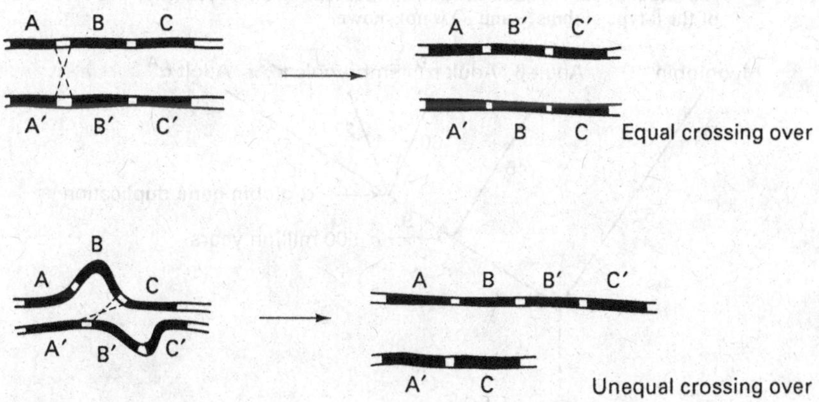

that are able to increase the rate of transcription, often by as much as 100-fold. These enhancers are often of great importance in the developmental regulation of genes, and they may be tissue specific. They are able to act at a distance, i.e. they do not have to be located adjacent to the gene, and may be located on the 5′ or 3′ side of the gene. The enhancers interact with specific proteins and this is believed in some way to make the gene more accessible to the RNA polymerase. Enhancers have been found to be located adjacent to both $\alpha$- and $\beta$-globin families (Nikol & Felsenfeld, 1988; Knezetic & Felsenfeld, 1989), and a protein (Evans, Reitman & Felsenfeld, 1988) known as erythroid-specific factor (Eryf1) has been shown to bind the enhancer (Fig. 10.5). The enhancer region, located between the $\beta$-gene and the $\epsilon$-gene, strongly stimulates transcription in 9–12 day embryonic erythrocytes, but transcriptional activity is much lower in the 5 day embryo. Enhancers of the $\alpha$-gene cluster have been identified which strongly stimulate transcription of the $\alpha^D$- and $\alpha^A$-genes.

The $\alpha$- and $\beta$-type globin multigene families are located on different chromosomes. Hughes *et al.* (1979) used a method of differential centrifugation to fractionate metaphase chromosomes from cultured cells of the domestic fowl. This method only partially separated the chromosomes, but was sufficient to show that the genes for $\beta$, $\rho$ and $\epsilon$ are located on chromosomes 1 or 2, whereas $\alpha^A$ and $\alpha^D$ are located on either a small macrochromosome or a large microchromosome.

Besides considering the evolution of a particular group of proteins within one species, as is done in Fig. 10.4, it is possible to relate these evolutionary stages within one species to the evolution and divergence of species, genera,

**Fig. 10.4** The evolution of globins (myoglobin and haemoglobin) in the domestic fowl. The tree is based on differences in amino acids between the aligned sequences, as proposed by Chapman *et al.* (1980, 1981 & 1982a, b). The number of amino acid substitutions per 100 amino acid residues are shown on the branches. The evolution of the $\beta$-type globins ($\rho$ and $\rho'$) is not shown.

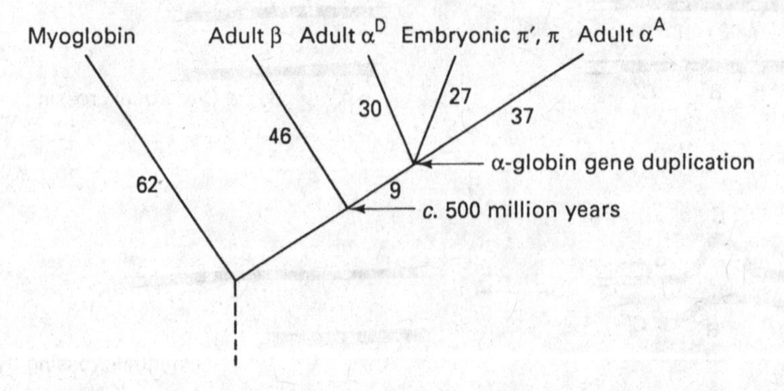

families, etc., from one another. The $\epsilon$-globin chain is believed to have evolved from the $\beta$-globin by gene duplication. Based on the amino acid differences between these globins it appears that the $\epsilon$-globin gene in the domestic fowl has evolved since the origin of the class Aves, and that the mammalian $\epsilon$-type globins have evolved from their $\beta$-type globins independently (Chapman *et al.*, 1982b). The domestic fowl is somewhat unusual in having two $\alpha$-type globins, $\alpha^A$ and $\alpha^D$, each present in significant amounts. The $\alpha^D$ globin is also present in the American rhea, the ostrich, the pheasant, the tree sparrow, the greylag goose and the Andean condor (Schneeganss *et al.*, 1985; Bauer *et al.*, 1985; Oberthur, Godovac-Zimmerman & Braunitzer, 1986), although many species of birds, e.g. penguins, egrets, pigeons, cuckoos and parakeets do not possess the $\alpha^D$ globin (Saha & Ghosh, 1965; Paul *et al.*, 1974) and it is not represented in mammals. It is assumed that those species of birds having no $\alpha^D$ have lost the ability to express this gene, that had been expressed earlier in their ancestry (Chapman *et al.*, 1982b). It is uncertain whether it is essential in the domestic fowl. There is a higher proportion of $\alpha^D$ haemoglobin in the late embryo than in the adult (Moss & Hamilton, 1974) and the presence of $\alpha^D$ haemoglobin increases its oxygen affinity (Vandecasserie, Schnek & Leonis, 1971).

Although myoglobin has been sequenced in a large number of vertebrate species, it has been sequenced only in two species of birds, the domestic fowl and the penguin (*Apterodytes forsteri*); thus at present there is insufficient information from this source to help understand evolution within the class Aves (Romero-Herrera *et al.*, 1978).

It is possible in the case of the haemoglobins to compare not only the

**Fig. 10.5** The arrangement of the $\alpha$ and $\beta$ gene families in the domestic fowl. Exons are shown in solid black. Eryf1, erythroid-specific DNA-binding factor (Nikol & Felsenfeld, 1988; Knezetic & Felsenfeld, 1989).

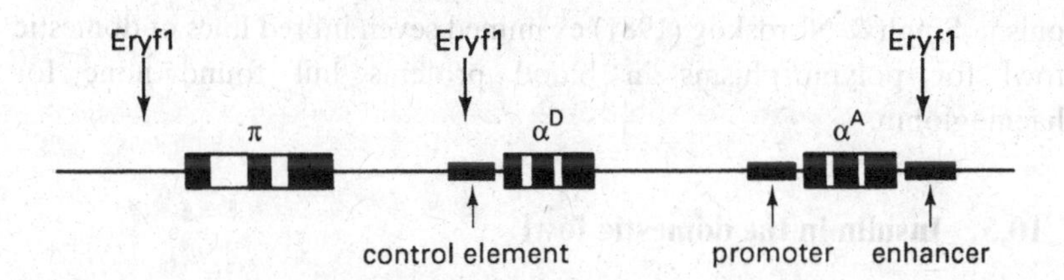


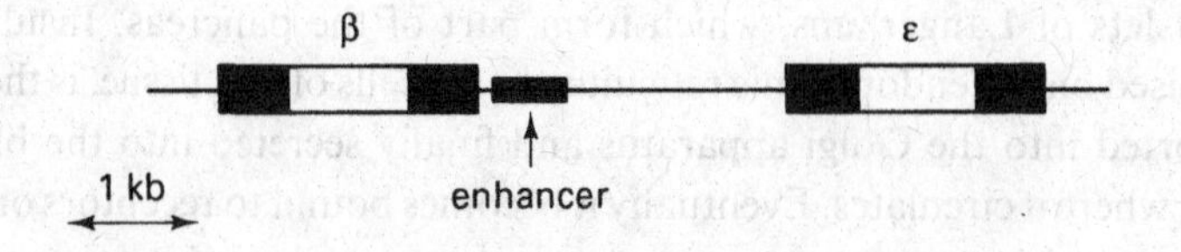

amino acid sequences but also the nucleotide sequences, since sequence data are available for a large number of $\alpha$- and $\beta$-globins and also a number of globin pseudogenes. (**Pseudogenes** are DNA sequences which are closely related to those of known genes, but are not expressed.) It is then possible to compare the number of changes in the nucleotide sequence that result in a change in amino acid sequence with those which result in no change (silent replacements). The latter arise because of redundancy in the genetic code, in which there are 64 triplet sequences coding for 20 amino acids plus three termination signals (see Appendix V). The data show that there are many more changes in nucleotide sequence that result in amino acid sequence replacements than result in silent replacements (Czelusniak *et al.*, 1982). This supports the theory that globin evolution resulted from natural selection acting on mutations in duplicated genes, and is against the neutralist view that random drift has largely been responsible for the changes in amino acid sequences.

Genetic polymorphisms in human haemoglobin are very extensive: over 300 have been found in haemoglobin A. By contrast there appear to be few in the domestic fowl, although they have been much less extensively investigated. Polymorphisms have been found in HbD in the domestic fowl (Washburn, 1968; Callegarini, Baragelles & Conconi, 1969; Keane *et al.*, 1974); in each study they were detected by electrophoresis. Two of the variants behave as single codominant alleles that change the electrophoretic mobility of both HbD and HbM. This is because it is the $\alpha^D$-globin that is changed, and it is common to both haemoglobins. A comparison of the genomic sequence, the cDNA sequence and the amino acid sequence for $\alpha^D$-globin reveals a few inconsistencies (Dodgson & Engel, 1983). Some of these may be attributable to errors in sequencing but others may be due to genetic polymorphisms. Dodgson & Engel (1983) suggested that the differences in codon 107, in particular, are likely to be due to polymorphism. Singh & Nordskog (1981) examined seven inbred lines of domestic fowl for polymorphisms in blood proteins but found none for haemoglobin.

## 10.3 Insulin in the domestic fowl

Insulin is a hormone present in vertebrates which is concerned with the regulation of growth and metabolism. In higher vertebrates it is synthesised in the Islets of Langerhans, which form part of the pancreas. Insulin is synthesised on the endoplasmic reticulum in the cells of this tissue, is thence transported into the Golgi apparatus and finally secreted into the blood stream, where it circulates. Eventually it becomes bound to receptors on the surface of target tissues, is taken up into the target tissues, and, after triggering an appropriate response, becomes degraded.

A particularly interesting aspect of the evolution of insulin is the different constraints imposed on different parts of the molecule which serve different roles of its overall mechanism of action. Insulin is first synthesised as a single polypeptide (Fig. 10.6) precursor called preproinsulin. A section of the molecule (the N-terminus) comprising 24 amino acids acts as a signal peptide (also known as the prepeptide). This prepeptide is needed for insulin to traverse the membrane of the endoplasmic reticulum. Once across the membrane the prepeptide is excised leaving proinsulin. The proinsulin molecule is still a single polypeptide chain, whereas the active hormone comprises two polypeptide chains (known as A and B) joined to one another by disulphide bridges. The linking piece of proinsulin, the C-peptide, serves to bring the A and B chains into juxtaposition in order that the disulphide bridges can form (Fig. 10.6). The C-peptide is then dispensed with and excised, leaving the physiologically active and fully functional

**Fig. 10.6** Biosynthesis and circulation of insulin. (1), Preproinsulin, synthesised on the rough endoplasmic reticulum of' the islets of Langerhans, (2), proinsulin, transported to the Golgi apparatus, and (3), insulin, stored in granules prior to secretion from the pancreas. –SS–, disulphide bridge.

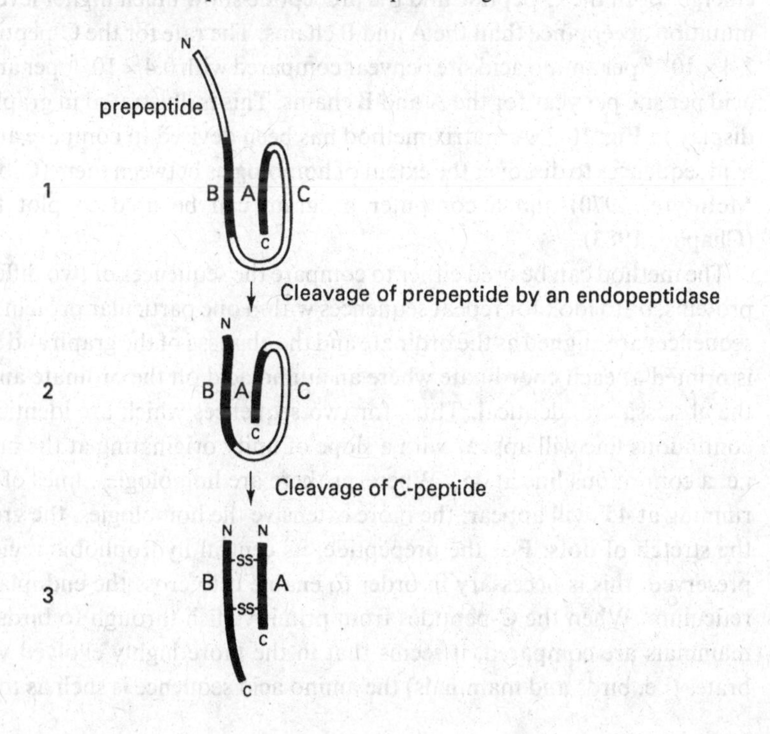

insulin molecule. Insulin, once circulating in the blood stream, must be able to recognise specifically, and bind strongly to, the receptors on the target tissues. The evolutionary constraints on the A and B chain, the C-peptide and the prepeptide differ considerably, and thus it is perhaps not surprising that they have evolved at different rates. Their evolution, particularly in relation to the domestic fowl, is discussed in the following section.

The amino acid sequences for insulin have been determined for over 25 vertebrate species, and the preproinsulin gene sequences and/or cDNA sequences are known for over a dozen vertebrates (Steiner *et al.*, 1985). From the avian standpoint, the amino acid sequences of the domestic fowl, turkey and domestic duck insulins have been determined (see Croft, 1980), the C-peptide sequence for duck is known (Markussen & Sundby, 1973), and the complete preproinsulin gene sequence has been determined for domestic fowl (Perler *et al.*, 1980). The A and B chains from the domestic fowl and from the turkey have identical sequences, but there are five differences between those of the domestic fowl and the duck. The duck and the domestic fowl C-peptides are very similar except for two additional amino acids at the C-terminus in the domestic fowl.

When comparisons are made of the amino acid sequences of the various regions of preproinsulin in a number of vertebrates, the following points emerge. Both the C-peptide and the prepeptide show much higher levels of mutation acceptance than the A and B chains. The rate for the C-peptide is $2.4 \times 10^{-9}$ per amino acid site per year compared with $0.4 \times 10^{-9}$ per amino acid per site per year for the A and B chains. This is illustrated in graphical display in Fig. 10.7. A matrix method has been devised to compare amino acid sequences to discover the extent of homologies between them (Gibbs & McIntyre, 1970) and a computer program can be used to plot them (Chaplin, 1983).

The method can be used either to compare the sequences of two different proteins, or to look for repeat sequences within one particular protein. The sequences are aligned as the ordinate and the abscissa of the graph and a dot is printed at each coordinate where an amino acid on the ordinate and on the abscissa are identical. Thus, for two sequences which are identical, a continuous line will appear with a slope of unity originating at the origin, i.e. a continuous line at 45°. Wherever there are homologies, lines of dots running at 45° will appear; the more extensive the homologies, the greater the stretch of dots. For the prepeptide, its central hydrophobic region is preserved; this is necessary in order to enable it to cross the endoplasmic reticulum. When the C-peptides from primitive fish through to birds and mammals are compared, it seems that in the more highly evolved vertebrates (i.e. birds and mammals) the amino acid sequence is such as to fold readily into a $\beta$-sheet. This is a type of protein structure that enables the proinsulin molecule to fold and bring the A and B chains into juxtaposition more quickly, and hence to assemble the definitive insulin.

The typical complete vertebrate insulin gene is shown on Fig. 10.8. It contains two intervening sequences, I-1 and I-2; the length of I-1 is quite similar for all those so far determined, but the I-2 of the domestic fowl is much longer than I-2 sequences from other species (Steiner *et al.*, 1985).

Although in certain species there are two non-allelic insulin genes, e.g. rats, mice, tuna, bonito and toadfish, there is no evidence for more than one insulin gene, or pseudogene, in the domestic fowl (Perler *et al.*, 1980). When the potencies of different insulins are tested, both the domestic fowl and the turkey insulins are active at lower concentrations than those of mammals. In the domestic fowl this appears to be entirely due to a higher affinity of the insulin for the insulin receptor (Simon, Freychet & Rosselin, 1974). When potency experiments were carried out using insulin and insulin receptors from various species it became clear that the differences were due to changes in the insulin structure rather than that of the insulin receptor. Muggeo *et*

**Fig. 10.7** A comparison of the 103 amino acid sequence of preproinsulin from the domestic fowl with that from the rat, using the dot matrix plot described by Gibbs & McIntyre (1970). Note the greater homology between the A- and B-peptides, than between the C-peptide and the signal sequence.

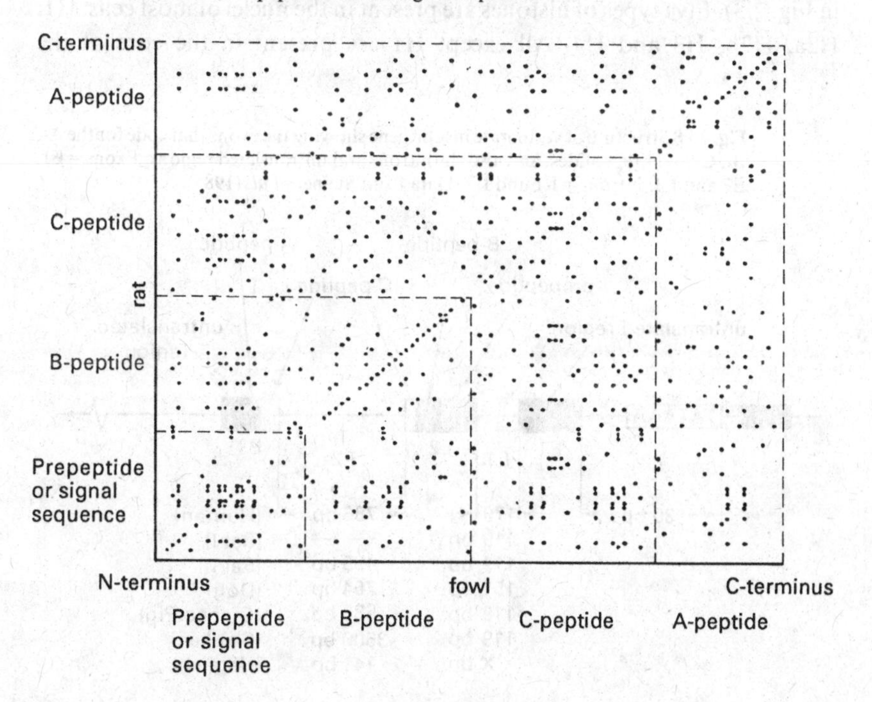

*al.* (1979) concluded that the receptor is functionally more conserved during evolution than insulin itself.

Insulin from the domestic fowl and the turkey contains the amino acid histidine at position 8 in the A chain, while in mammals threonine or alanine is usually present in this position. Pullen *et al.* (1976) have shown, in the three dimensional structure of insulin, that this position is just on the periphery of the putative receptor binding region and thus the histidine substitution could account for the stronger binding.

It is now clear from sequence studies that there are several other peptide hormones, e.g. insulin-like growth factors that are related to insulin and have evolved from a common ancestor making an insulin superfamily; however, although these have been studied during chick embryo development (Ralphs, Wylie & Hill, 1990), their molecular genetics have not been studied.

## 10.4 Histones in the domestic fowl

The histones are a class of basic proteins present in the nuclei of all eukaryotes where they are bound to DNA forming a compact structural arrangement. This arrangement which forms the basic unit of the chromosomes is the nucleosome (previously described in Chapter 2, section 2.1 and in Fig. 2.3). Five types of histones are present in the nuclei of most cells: H1, H2a, H2b, H3 and H4. All except H1 are present in the core of the

**Fig. 10.8** Structure of vertebrate insulin gene showing the exons that code for the A-, B-, C- and prepeptides, and also the introns and untranslated regions. Exons = E1, E2 and E3, introns = I-1 and I-2. Data from Steiner *et al.* (1985).

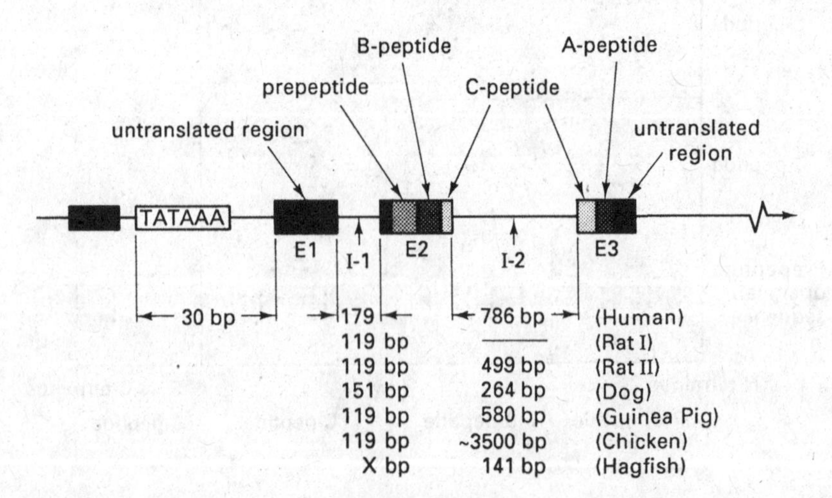

nucleosome, whereas H1 plays a role in the association of adjacent nucleosomes and is located on the outside of the nucleosome. Histones have been well characterised and sequenced in many organisms (Von Holt *et al.*, 1979) and they have been found to be a highly conserved group of proteins, reflecting the importance of precise specification of the nucleosome unit for all eukaryotic organisms. The hydrophobic regions within their structure are the most highly conserved and this correlates with those areas that are involved in the histone–histone contacts.

The rates of evolution of each of the histone classes differ, being fastest in the H1 and slowest in the H4, as shown below.

| Histone | Accepted Point Mutations/100 residues /$10^8$ years |
|---|---|
| H4 | 0.10 |
| H3 | 0.15 |
| H2b | 0.39 |
| H2a | 0.54 |
| H1 | 2.40 |

(Hunt & Dayhoff, 1982)

An additional class of histones, designated H5, is found in the nucleated erythrocytes of birds, amphibians, reptiles and fish. It is thought that this class is synthesised in place of part of the H1 in the mature erythrocyte, since the sum of the amounts of H1 + H5 in erythrocytes is approximately equal to the amount of H1 in other tissues (Garel *et al.*, 1975). The H5 from the erythrocytes of the domestic fowl shows a strong sequence homology with the H1 from mammalian tissues such as rabbit thymus (Sautiere *et al.*, 1975), and it seems that the H1 and H5 histones have evolved from a common ancestor through gene duplication (Yaguchi, Roy & Seligy, 1979). The complete sequences of H5 from domestic goose and domestic fowl have been determined (Yaguchi *et al.*, 1979; Briand *et al.*, 1980). They have 193 and 189 amino acid residues respectively and there are 30 differences between them of which 14 are non-neutral and 4 are deletions.

Histone H4 is the most highly conserved histone and the sequences from the domestic fowl and calf thymus are identical. Also, no polymorphisms of H4 have been found in the domestic fowl. In contrast, non-allelic polymorphic forms have been found in the erythrocytes of the domestic fowl for H2A, H2B and H3 (Urban, Franklin & Zweidler, 1979). Two forms have been found for each of the three classes. They involve a small number of conservative substitutions. The two forms of H3 both have identical structures with the corresponding forms in mammals, but the major forms of H2A and H2B have four differences with the counterparts from calf thymus. It seems that these polymorphic forms have been preserved in parallel throughout the evolution of the vertebrates.

## 10.5 Egg-white proteins

The proteins from egg white provide the most extensive group of protein polymorphisms that have been established in the domestic fowl. Egg proteins, being readily available, have been more thoroughly studied than many other proteins. Those from egg white are more easily purified than those from egg yolk since there is no lipid to interfere with the purification process. There are 10 major groups of proteins in egg white, although there may be over 50 different proteins. Polymorphisms have been found in ovalbumin, ovomucoid, ovotransferrin, lysozyme, and ovoglobulins. They may exist in other proteins, but have not yet been detected. Ovomucoid is a good example in which to study protein evolution, since it comprises three domains, that may have originated by gene duplication, one of which has been sequenced in over 100 species of birds (Laskowski *et al.*, 1987).

This section therefore considers the major groups of proteins from egg white as examples in which protein polymorphism and protein evolution have been studied. Polymorphisms in domestic fowl egg white and serum have been reviewed by Baker *et al.* (1970) and only a few additional polymorphisms have been identified since then. On the other hand, many more sequences of ovomucoid from avian species (Laskowski *et al.*, 1987) are now known, and this has helped in understanding the molecular basis of evolution. The major proteins of egg white are listed in Table 10.2.

### *10.5a Ovalbumins*

There are two well-established genetic variants of ovalbumin, *OvA* and *OvB*. They are distinguished by electrophoresis; ovalbumin A migrates more slowly than ovalbumin B on starch gels run at pH 5.4. In addition to these genetic variants there may be up to three bands of ovalbumin A ($A_1$, $A_2$ and $A_3$) and of ovalbumin B ($B_1$, $B_2$ and $B_3$). These three forms differ from each other by having two, one or no phosphate groups respectively, on two serine residues (see Baker *et al.*, 1970). Many flocks of domestic fowl are monomorphic and contain only either ovalbumin A or B. Ovalbumin A differs from ovalbumin B in having an asparagine residue in place of aspartic acid (Wiseman, Fothergill & Fothergill, 1972). This was first shown by amino acid sequencing but was confirmed when the cDNA was sequenced. It was caused by a point mutation in residue 998 of the nucleotide sequence (McReynolds *et al.*, 1978). Ovalbumin is a glycopro-

Table 10.2. *Principal egg-white proteins in the domestic fowl*

| Protein | Percentage in egg white | $M_r$ | Polymorphic forms |
|---|---|---|---|
| Ovalbumin | 54% | 46000 | A, B |
| Ovotransferrin | 12% | 76000 | A, B, C, BW |
| Ovomucoid | 11% | 28000 | see section 11.5c |
| Lysozyme | 3.4% | 14300 | F and S |
| Ovomucin | 2.9% | 8300000 | |
| Ovoinhibitor | 1.5% | 49000 | |
| Ovoglycoprotein | 1.0% | 24000 | |
| Flavoprotein | 0.8% | 29200 | |
| Ovomacroglobulin | 0.5% | 900000 | |
| Avidin | 0.05% | 68300 | |
| $G_2$Globulin | 1.0% | 47000 | A, B |
| $G_3$Globulin | | ? | A, $A^S$, $A^F$, B, $B^S$, $B^F$, J-$B^1$, J-$B^2$, J, M |

Data from Baker (1968a); Feeney & Allison (1969); Osuga & Feeney (1977); Hamazume, Mega & Ikenaka (1984); Kato *et al.* (1985); Stevens & Duncan (1988).

tein having oligosaccharide chains attached to the polypeptide chain through asparagine residues (Huang, Mayer & Montgomery, 1970). There are indications of additional variants, differing from one another in their oligosaccharide chains (Iwase, Kato & Hotta, 1984).

## *10.5b Ovotransferrins*

The second most abundant protein in egg white is ovotransferrin (also known as conalbumin). It is a glycoprotein and its sequence has been determined both directly as the amino acid sequence (Williams *et al.*, 1982) and as a complete mRNA sequence (Jeltsch & Chambon, 1982). The molecule has a bilobular structure in which the two domains are the N-terminus and the C-terminus of the sequence and it has many disulphide bridges, six in the N-domain and nine in the C-domain. This is of particular interest since disulphide bridges, once evolved in a protein, are rarely lost (Williams, 1982). Also, the positions of the disulphides in ovotransferrin highlight the evolutionary relationships between the two domains. Transferrins occur in both egg white and egg yolk, and also in serum. All three proteins have the same amino acid sequences and differ only in the way they are glycosylated (Williams, 1962, 1969). In the hen, serum and egg-yolk transferrin are synthesised in the liver, and egg-white transferrin in the oviduct, but it is clear that they are the products of identical genes, since any genetic variation found in serum transferrin is also found in that of egg white (see Baker *et al.*, 1970).

On the basis of sequences determined so far, Williams (1982) proposed that transferrins evolved from an ancestral protein of $M_r$ 40 000 by gene duplication occurring about 500 million years ago which gave rise to the N-domain and the C-domain. Approximately 40% of the residues are homologous between the N- and the C-domains in hen ovotransferrin, and about 50% are homologous between the corresponding domains in the hen and in human serum transferrin. There have been suggestions that gene quadruplication rather than duplication may have occurred during their evolution. However, when a matrix plot (Gibbs & McIntyre, 1970: see section 10.3)) is carried out, the duplication shows up clearly but quadruplication is not convincing (Williams *et al.*, 1982).

The role of transferrin in serum is clear, namely the transport of iron from the intestine to other tissues of the body. There are specific receptors on the membranes of cells enabling them to bind the iron-charged transferrin. In egg white the function is less clear although the ovotransferrin has the ability to bind iron strongly and thus reduces available iron that may prevent the growth of bacteria. Any polymorphisms found in the ovotransferrin are paralleled by variants in serum transferrins.

Four polymorphic forms of ovotransferrins have been demonstrated by starch gel electrophoresis and are believed to be alleles at a single locus *Tf* (see Baker,1968b, and Baker *et al.*, 1970). They are designated $Tf^A$, $Tf^B$, $Tf^C$ and $Tf^{BW}$. The commonest is $Tf^B$, and many flocks are monomorphic for this allele. $Tf^A$ occurs together with $Tf^B$ in Light Sussex and Old English Game. $Tf^C$ occurs together with $Tf^B$ in Marans (Baker, 1968b). $Tf^{BW}$ has not yet been found in the domestic fowl, but it occurs in *Gallus sonnerati* (Baker *et al.*, 1970). The differences in the structures of the proteins coded by these alleles have not been determined, only the $Tf^B$ variant has been sequenced (Williams *et al.*, 1982). It is found that a more complex pattern arises if the transferrins are not saturated with $Fe^{2+}$ ions, since the apotransferrins separate from those with $Fe^{2+}$ bound. Another method of separating the transferrins is by isoelectric focusing on polyacrylamide gels, which separates them on the basis of different isoelectric points. Three variants of the main serum transferrin have been found by this method, *Tf-0*, *Tf-1* and *Tf-2*; they differ in having 0, 1 and 2 sialic acid residues. These are the result of post-translational modification. Kimura (1983) has shown that in the embryo *Tf-0* and *Tf-1* predominate, but on approaching maturity the proportion of *Tf-2* increases at the expense of the other two.

### 10.5c Ovomucoid

Ovomucoid makes up 11% of the egg-white protein in the domestic fowl. It is a glycoprotein having a $M_r$ of 28 000. It has been well characterised and has been sequenced (Kato *et al.*, 1987). One of its most interesting properties is that it inhibits certain proteinases. Ovomucoids from different species of birds each have a characteristic pattern of proteinase inhibition. These differences are useful in studying the rate of evolution of different regions of the protein and in examining evolutionary relationships between species. Figure 10.9 shows the complete sequence of ovomucoid from the domestic fowl; the molecule has nine disulphide bridges and divides readily into three domains. These domains are believed to have arisen by a process of gene duplication.

In Fig. 10.10 the ovomucoid sequence of the domestic fowl is compared with itself using the matrix plot devised by Gibbs and McIntyre (see section 10.3). Apart from the continuous line running from the origin other lines of dots can be seen corresponding with the homologies between domains I and II, I and III, and II and III. From the patterns of continuous dots with slopes of 45° it can be seen that the homologies are greater between I and II than between II and III, and I and III. From this it has been deduced that the ovomucoids evolved by two successive gene duplications; the first of these involved the splitting of domain III from the ancestral form of I and II, and the second the splitting of I from II (Kato, Kohr & Laskowski, 1978).

Proteinase inhibitors other than ovomucoids have been isolated from widely differing species (Laskowski & Kato, 1980) and many show similarities with ovomucoids. For example, the dog submandibular inhibitor has been sequenced and consists of two domains, I and II. Domain II shows close homology with ovomucoid domain III. This, together with other evidence, suggests that the first gene duplication occurred before mammals and birds evolved from reptilian stock, but the second occurred after the divergence of mammals (about 200 million years ago). Ovomucoids have been isolated from a wide range of orders of birds, and all have three domains showing homologues that suggest that the second duplication occurred before the separate orders evolved. The positions of the introns in the gene sequence of ovomucoid have been determined (Stein *et al.*, 1980) and found to be excised from equivalent positions (Fig. 10.9, showing the positions in the amino acid sequence) in each of the three domains, which further strongly supports the gene duplication hypothesis.

When comparisons were made between different species of birds, the proteinase inhibitory properties of ovomucoids fell into distinct groups

(Feeney & Allison, 1969). The ovomucoids were tested as potential inhibitors of trypsin and chymotrypsin, and the stoichiometry of inhibition was measured. The results showed that the ovomucoids could be described as single-headed, double-headed or triple-headed, depending on the number of molecules of trypsin and/or chymotrypsin they inhibited (Fig. 10.11 and Table 10.3).

**Fig. 10.9** Complete amino acid sequence of ovomucoid from the domestic fowl, arranged to show the three domains, and also the positions from which the introns (Stein *et al.*, 1980) are excised during processing and before translation (see Chapter 2, section 2.2). The amino acids are shown using the single letter code (see Appendix IV). C–C, disulphide bridge.

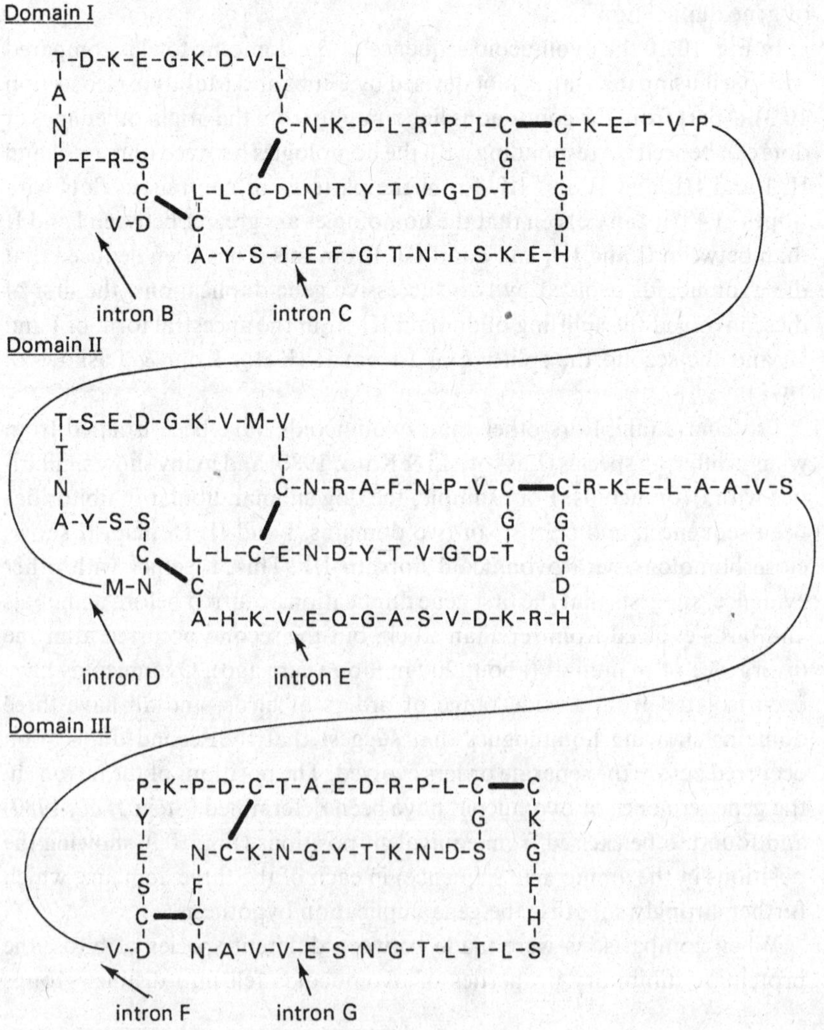

Domain I

T-D-K-E-G-K-D-V-L
A V
N C-N-K-D-L-R-P-I-C—C-K-E-T-V-P
P-F-R-S G E
C L-L-C-D-N-T-Y-T-V-G-D-T G
V-D C D
A-Y-S-I-E-F-G-T-N-I-S-K-E-H

intron B intron C

Domain II

T-S-E-D-G-K-V-M-V
T L
N C-N-R-A-F-N-P-V-C—C-R-K-E-L-A-A-V-S
A-Y-S-S G G
C L-L-C-E-N-D-Y-T-V-G-D-T G
M-N C D
A-H-K-V-E-Q-G-A-S-V-D-K-R-H

intron D intron E

Domain III

P-K-P-D-C-T-A-E-D-R-P-L-C—C
Y G K
E N-C-K-N-G-Y-T-K-N-D-S G
S F F
C—C H
V-D N-A-V-V-E-S-N-G-T-L-T-L-S

intron F intron G

A surprising finding was that ovomucoids from closely related species have different specificities, e.g. golden pheasant inhibits chymotrypsin, but domestic fowl inhibits trypsin and turkey inhibits both. The differences in the inhibitory properties of the ovomucoid depend on the amino acid sequence at the binding site, i.e. the region where the ovomucoid binds the proteinase. After binding to the proteinase, a peptide bond is cleaved in the ovomucoid inhibitor, and the modified inhibitor remains bound to the active site of the proteinase, thereby inhibiting its action. It is the amino acid on the carboxyl side of the bond which is cleaved that largely determines the specificity. For example, the presence of lysine in this position often leads to trypsin inhibitory activity, whereas if methionine or leucine is in this position chymotrypsin is inhibited (Kato *et al.*, 1978). However, other amino acid residues adjacent also affect its inhibitory activity. Since all ovomucoids have three domains, whether a given ovomucoid is single-, double- or triple-headed depends on whether one, two or three of the domains are inhibitory (Fig. 10.11).

It is possible, by using specific proteinases such as *Staphylococcus* V8 proteinase, to cleave between the domains in ovomucoids and then separate the three domains. Using this method the third domain has been separated and sequenced by Laskowski *et al.* (1987) from 101 species of birds. By comparing these domains they have shown that the seven positions in the

**Fig. 10.10** A matrix plot (Gibbs & McIntyre, 1970) to show homologies between different domains of ovomucoid. There is greater homology between domains I and II than between either domains I and III or between II and III.

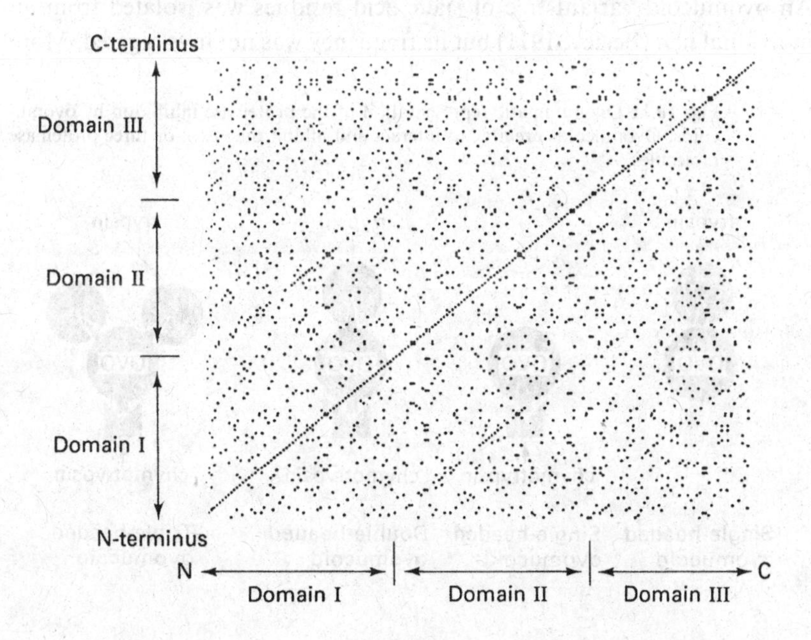

sequence that show greatest variability are all among the eleven positions that are in contact with the proteinase in the enzyme inhibitor complexes. These are referred to as the hypervariable amino acid residues because of their large variation between closely related species. This finding at first seemed rather surprising and contrary to expectation. Normally positions in a protein which are important in determining function are the most highly conserved. However, the function of ovomucoid is not yet clear (Laskowski *et al.*, 1987). Some possibilities are: (i) a storage protein, serving as a source of protein for the developing embryo, (ii) a protein inhibitor which regulates proteinase activities at particular stages of development, and (iii) a proteinase inhibitor which protects the development of the embryo from bacterial proteinases. In the first of these possibilities it would not be expected that the protein would have to be highly conserved if it were simply a source of amino acids. For the second, the inhibitor active site would be expected to be conserved if the specificities of the developmental proteinase were conserved between species. On the other hand, a high rate of evolution could be a means of adapting to a different spectrum of bacteria which invade different species. At present there is no clear explanation.

Ovomucoids differ from most of the other proteins present in egg white in that they are not resolved into sharp bands when separated either by starch gel or polyacrylamide gel electrophoresis; in both they usually give rather diffuse bands. Thus, none of the early separations of egg white indicated any polymorphism of ovomucoid (Lush, 1961, and see Baker *et al.*, 1970). An ovomucoid variant free of sialic acid residues was isolated from an individual hen (Beeley, 1971) but its frequency was not investigated. More

**Fig. 10.11** Diagrammatic representation of the proteinase inhibition by ovomucoids. Ovomucoids are able to interact and inhibit one, two, or three proteinase molecules.

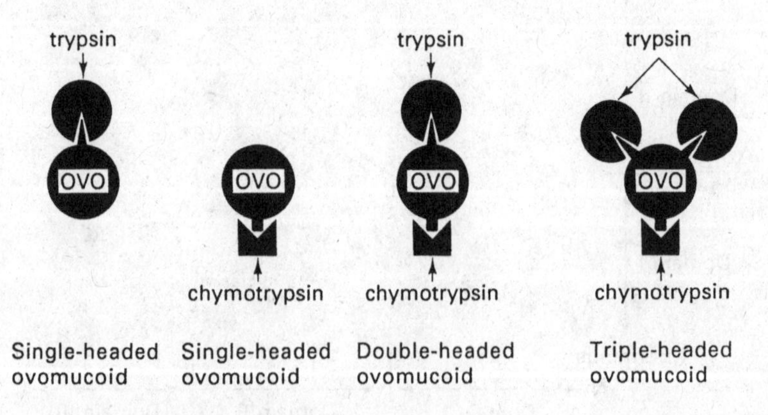

Table 10.3. *Inhibitory patterns of ovomucoids*

| Ovomucoid (species) | Moles of trypsin inhibited by ovomucoid | Moles of chymotrypsin inhibited by ovomucoid |
|---|---|---|
| *Single-headed* | | |
| Domestic fowl | 1 | — |
| Ostrich | 1 | — |
| Cassowary | 1 | — |
| Pheasant | — | 1 |
| *Double-headed* | | |
| Turkey | 1 | 1 |
| Penguin | 1 | 1 |
| Tinamou | 1 | 1 |
| Rhea | 1 | 1 |
| Guinea fowl | 1 | 1 |
| Goose | 2 | — |
| Japanese quail | 2 | — |
| *Triple-headed* | | |
| Duck | 2 | 1 |
| Emu | 2 | 1 |

From Kato *et al.* (1976).

recently in their survey of third domain sequences from 101 species of birds Laskowski *et al.* (1987) found some to be polymorphic. These were from the plumed whistling duck (*Dendrocygna eytoni*), Montezuma quail (*Cyrtonyx montezumae*), the scaled quail (*Callipepia squamata*), the Japanese quail (*Coturnix coturnix japonica*) and the grey jungle fowl (*Gallus sonnerati*). In each case, the polymorphisms were consistent with two allelic genes, involving a single nucleotide replacement in the third domain. Of these only that in the Japanese quail has been extensively studied (Bogard, Kato & Laskowski, 1980). It was found to be the product of two allelic autosomal genes at a single locus, and was identified in all populations of Japanese quail sampled both in the USA and in Japan. The third domain of the domestic fowl was found not to be polymorphic in its amino acid sequence in eggs obtained from a variety of sources (Laskowski *et al.*, 1987).

Looking for polymorphisms or genetic variants by determining the amino acid sequences is time consuming and requires specialised equipment, but it has the advantage of indicating precisely how the polymorphic forms differ. A much simpler method for carrying out a survey is to run Cleveland gels (Cleveland *et al.*, 1977). In this technique the protein in question is subjected to proteolysis by a specific proteinase. The digest is separated by polyacrylamide gel electrophoresis in the presence of sodium dodecyl sulphate (SDS). This separates the fragments on a size basis. If there is an amino acid difference in a position that affects the way in which it becomes degraded by the proteinase, then a different pattern of fragments is obtained on the gels. In a preliminary study Stevens, Scott & Duncan (1986) found that chymotrypsin and thermolysin were the best proteinases for showing differences in the ovomucoids from different breeds of domestic fowl.

Kato *et al.* (1987) determined the complete amino acid sequence of ovomucoid from the domestic fowl. They found that 25% of the ovomucoid molecules contained a dipeptide deletion (valine–serine) between domain II and III. This could be explained by observations of Stein *et al.* (1980), who carried out DNA sequencing of the intron–exon junctions corresponding to this region and found a repeated triplet (GTG).

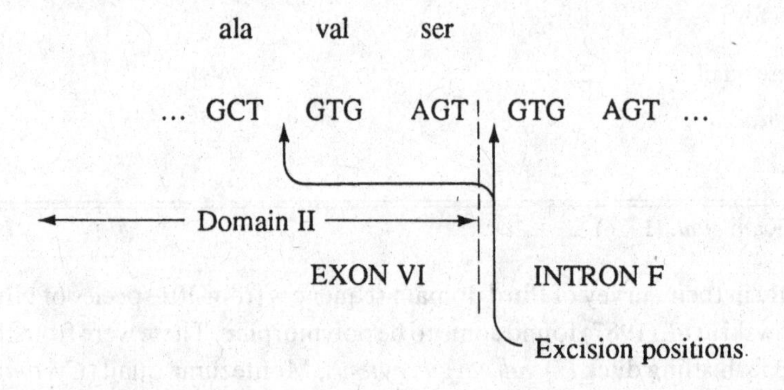


They suggest that when the pre-mRNA is processed to excise the introns, cleavage may occur at either of the GTG triplets. If this is the case it would require only a single gene to account for the normal and 'deleted' forms of ovomucoid, since it is an 'error' in processing the transcript.

### *10.5d Lysozyme*

One of the most extensively studied enzymes from the domestic fowl is egg-white lysozyme. Its three dimensional structure was determined by Philips (1967) and much detail is known about its mechanism of catalysis (Imoto *et al.*, 1972). However, from the genetic standpoint, only two polymorphic forms have been detected (Baker, 1968b). Longsworth, Cannon & MacInnes (1940) demonstrated the presence of three globulins in egg white, $G_1$, $G_2$ and $G_3$, the first of which was later shown to be lysozyme. It is still referred to in some papers as globulin $G_1$. The two forms are $G_1^F$ and $G_1^S$; the former is more common and the latter has been reported only in a single flock of Polish bantams (Baker, 1968b). The two were distinguished by their electrophoretic mobility, but nothing is known of their structural differences.

Lysozymes have been isolated and sequenced from the egg white of a number of different species of birds and have been found to fall into two distinct types: **c**-type (**c**hicken) and **g**-type (**g**oose). They differ in $M_r$, the c-type being about 14000, and the g-type about 21000. Their amino acid sequences are significantly different and they are antigenically distinct. Although the c-type lysozyme has been sequenced from a number of species, it appears that the g-type has a wider distribution. The c-type has been found in the egg whites of several Galliformes, some Anseriformes and, most recently in one of the Columbiformes (Menendez-Arias, Gavilanes & Rodriguez, 1985). Prager, Wilson & Arnheim (1974) detected the g-type immunologically in nine different orders of birds: Anseriformes, Struthioniformes, Rheiformes, Apterygiformes, Tinamiformes, Podicipediformes, Sphenisciformes, Casuariiformes and Charadriiformes. From the sequences known it is clear that the c-type lysozymes from different species are sufficiently similar to have evolved from a single ancestral form (Weaver *et al.*, 1985). A comparison of the c-type sequences from four species of pheasant with those of the turkey and the domestic fowl suggests that the turkey is more closely related to the domestic fowl than are the pheasants (Jolles *et al.*, 1979). This has already been discussed (Chapter 1, section 1.3).

Three g-type lysozymes have been sequenced: those from the Embden goose (Simpson & Morgan, 1983), the ostrich (Schoentgen, Jolles & Jolles, 1982) and the black swan, *Cygnus atratus* (Simpson *et al.*, 1980). They are clearly very similar to one another, but show only a slight sequence similarity with the c-type lysozymes (Schoentgen *et al.*, 1982). Thus it is debatable whether the similarities that exist between the c-type and g-type lysozymes are the result of convergent or divergent evolution.

The three dimensional structure of goose lysozyme was determined by Grutter, Weaver & Matthews (1983), and can be compared with that of the c-type lysozyme from domestic fowl determined several years previously. Ninety of the $\alpha$-carbon atoms in the two structures are spatially equivalent. Two amino acid residues essential for catalytic activity in the lysozyme of domestic fowl, namely $glu_{35}$ and $asp_{52}$, have their equivalents or near equivalents in goose lysozyme in $glu_{73}$ and $asp_{86}$. When the overall secondary structures are compared, both are found to have $\alpha$-helical structures on both sides of a stretch of $\beta$-sheet structure, although the goose lysozyme has rather more $\alpha$-helical structure in total. This is accounted for by the additional 56 amino acid residues in the goose lysozyme which occur largely at the N-terminus (Fig. 10.12). Weaver *et al.* (1985) conclude that it

**Fig. 10.12** Comparison of *A*, amino acid sequences of hen and duck, and *B*, the amino acid sequences of hen and goose; *C*, the three-dimensional structures hen and goose egg-white lysozymes. The sequences are compared using a dot matrix plot (Gibbs & McIntyre, 1970; see section 10.3). Duck and hen egg-white lysozymes have very similar sequences, whereas that of the goose is quite different. However, similarities in their three dimensional structures can be seen when they are superimposed. The dark areas are those common to both hen and goose, and the light areas are only present in the goose lysozyme (Weaver *et al.*, 1985).

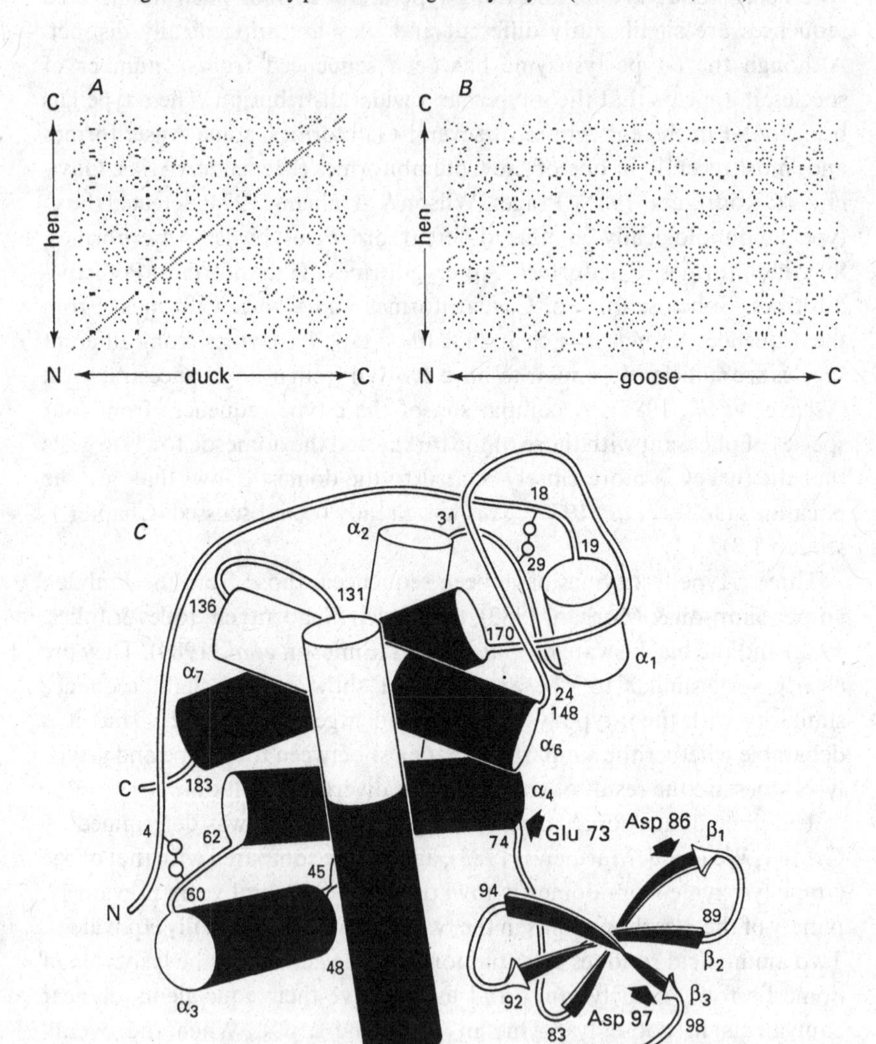

is most likely that both the c-type and g-type have evolved from a more distant ancestral form by divergent evolution.

Although most egg whites so far examined have only a single type of lysozyme, those from the black swan and the Canada goose have both c-type and g-type. Hindenburg, Spitznagel & Arnheim (1974) examined tissues of the domestic fowl and the Embden goose for evidence of both types of lysozyme, using an immunological method. They found that the polymorphonuclear leucocytes of domestic fowl and the bone marrow of the Embden goose contained both c-type and g-type lysozyme antigenic material. Thus it seems probable that, although most egg whites contain only one type of lysozyme, the genes for both types are present in the genomes of many birds, but only one of these is commonly expressed in the oviduct. The g-type and c-type genes appear to have distinct loci that are regulated independently.

### *10.5e Ovoglobulins*

The ovoglobulins were first separated by Longsworth *et al.* (1940), using zone electrophoresis. This was long before starch gel and polyacrylamide gel electrophoretic methods were available. In the older schemes for protein classification, globulins were proteins that are heat denaturable, soluble in dilute salt solutions, but insoluble in water, and those isolated from egg white were designated ovoglobulin $G_2$ and $G_3$. By comparison with most egg-white proteins already discussed, the ovoglobulins show greater polymorphism, but as a group they are less well defined and little is known of their structure. Both have been purified recently (Stevens & Duncan, 1988; S. E. Wilson and L. Stevens, unpublished data).

The two principal polymorphic forms of $G_2$ are $G_2A$ and $G_2B$ (in early papers designated $G_2C$ and $G_2D$, respectively). Baker (1968b) found in a survey of 37 breeds of domestic fowl that the $G_2B$ form was the most abundant, and that $G_2A$ generally did not occur in breeds of Northern European and Mediterranean origin. The finding of a higher proportion of $G_2B$ was also supported by more recent work (Singh & Nordskog, 1981) in which the distribution in six lines of Leghorns, one Fayoumi and one Spanish breed was examined. It was found that the frequency of $G_2B$ was between 75 and 100%. Baker & Manwell (1972) found that $G_2$ was absent in Sonnerat's jungle fowl (*Gallus sonnerati*) but was present in the red jungle fowl (*Gallus gallus*). This is one piece of evidence in favour of the monophyletic origin of the domestic fowl (see Chapter 1, section 1.4). Stevens & Duncan (1988) have shown by peptide mapping using Cleveland gels that ovoglobulin $G_2A$ and $G_2B$ differ in their sensitivities to chymotrypsin and V8 protease. Ovoglobulin $G_2A$ has an additional site sensitive to both chymotrypsin and V8 proteinase. The two forms also differ in their isoelectric points, $G_2A$ being more acidic than $G_2B$.

Other apparently rare variants are $G_2B'$, $G_2B''$ and $G_2L$ found in a red jungle fowl, a Sebright bantam and *Gallus lafayettei* respectively (Baker *et al.*, 1970). The extent of their distribution in the domestic fowl is not known. Ten variants of ovoglobulin $G_3$ have been distinguished by electrophoresis, using a variety of different buffers to separate them (Baker *et al.*, 1970). $G_3A$ and $G_3B$ are the common alleles and show codominance (Baker, 1968b). $G_3A$ is the more abundant, $G_3B$ being absent or only occurring in low frequency in breeds of English and American origin (Baker *et al.*, 1970). The variant $G_3J$ is the only one found in the red jungle fowl that has not been found so far in any of the domesticated breeds. Ovoglobulins $G_3A$ and $G_3B$ have been examined by peptide mapping but have not yet shown clear differences (S. E. Wilson and L. Stevens, unpublished data).

### *10.5f Other egg-white proteins*

In addition to the proteins mentioned above there are other minor ones in which no polymorphisms have been discovered so far, and which in some cases have not yet been fully characterised. Some of these will be mentioned briefly here, but those showing enzyme activity will be described in the next section.

There are a number of proteinase inhibitors besides ovomucoid but they occur in much lower concentrations. **Ovoinhibitor** resembles ovomucoid in having multiple domains with a characteristic pattern of disulphide bridges (Laskowski & Kato, 1980). It has six domains compared with three in ovomucoid, and thus further gene duplications are thought to have occurred in its evolution. From studies of the nucleotide sequences, and of the positions of introns and exons in the genes for ovomucoid and ovoinhibitor, Scott *et al.* (1987) have proposed a scheme whereby ovomucoid and ovoinhibitor evolved from a primordial single-domain inhibitor. An identical proteinase inhibitor to ovoinhibitor, $\alpha_2$-proteinase inhibitor, has been found in the plasma of the domestic fowl (Barrett, 1974).

**Cystatin** is a proteinase inhibitor which inhibits cysteine proteinases and has been isolated and characterised by Anastasi *et al.* (1983). It is present in low concentrations in egg white (80 $\mu$g/ml) and also lower concentrations in serum (1 $\mu$g/ml). It occurs in two forms, but it is not known whether these are the products of two different genes. The cystatin gene has been cloned from the oviduct of the domestic fowl (Colella *et al.*, 1989). The mRNA for cystatin is expressed in many adult tissues, the highest steady state levels being in lung, gizzard, brain and heart; it is also expressed in embryonic tissues. A much larger proteinase inhibitor, **ovostatin**, has been isolated and found to have a $M_r$ of 380000, but no variants of it have been discovered (Nagase *et al.*, 1983). It seems likely that it is identical with what was formerly known as ovomacroglobulin.

Egg white also contains proteins that are able to bind B vitamins, and may be responsible for transporting the vitamins into the egg. **Avidin**, which binds biotin, was one of the first proteins from egg white to be sequenced (see Green, 1975). The sequence shows a slight resemblance to that of lysozyme, although it is doubtful whether they are related. No variants have been discovered. A **thiamin-binding protein** has been purified but, again, no variants have been discovered (Muniyappa & Adiga, 1979). The **riboflavin-binding protein** (listed as flavoprotein in Table 10.2) is of interest since it is lacking in a hereditary defect discovered in a strain of White Leghorns. Winter *et al.* (1967) showed that the homozygous progeny (*rd*/*rd*) died of riboflavin deficiency on or near the 13th day of incubation, and that they could be rescued by injecting riboflavin. Homozygous *rd*/*rd* hens produced riboflavin deficient eggs even when the diet was supplemented with riboflavin. The riboflavin was excreted causing riboflavinuria. The genes show incomplete dominance; the heterozygote *Rd*/*rd* produced about half the concentration of the binding protein compared with the normal homozygote *Rd*/*Rd*, and the homozygous *rd*/*rd* lacks the binding protein.

The riboflavin-binding protein is unusual among egg proteins in that it occurs in egg white and egg yolk, and in the serum. That present in the serum and egg yolk is synthesised in the liver, whereas that present in the egg white is synthesised in the oviduct (White & Merrill, 1988). The riboflavin-binding protein has been sequenced by conventional amino acid sequencing ((Norioka *et al.*, 1985) and by sequencing the cDNA (Zheng *et al.*, 1988). There is a difference between riboflavin-binding protein from the yolk and the white in that the former lacks 11 to 13 of the C-terminal amino acid residues that are present in the latter, and there are also differences in glycosylation (White & Merrill, 1988). These differences are assumed to be due to post-translational modification. Otherwise the amino acid sequences are identical, and they are products of identical genes. There is a particular interest in the expression of the gene for the riboflavin-binding protein in the domestic fowl, since it is under different hormonal control in the liver and oviduct (DurgaKumari & Adiga, 1986). Riboflavin-binding protein isolated from all three sources (yolk, white and serum) shows the same polymorphism at residue 14, in having either asparagine or lysine. This is because of a point mutation and would require only a single base change (codons AA(U or C) for asparagine to AA(A or G) for lysine: see Appendix

V). So far the distribution of these two polymorphic forms in different breeds and varieties has not been studied.

### *10.5g Enzyme polymorphism*

Relatively few enzyme polymorphisms have been reported in the domestic fowl, and most of those which have are in blood or egg white. The enzymes studied so far have all been hydrolytic enzymes, with the exception of catalase, and this is undoubtedly a reflection of the fact that these are amongst the most easily detected on gels after electrophoresis. **Catalase** activity has been studied in erythrocytes. Ueda & Hachinohe (1984) detected two catalase variants on starch gels, $Ct^A$ and $Ct^B$. They are autosomic alleles showing codominance, both being expressed in the heterozygote. In a survey of different breeds of domestic fowl $Ct^B$ only was present in White Leghorns, New Hampshires and White Plymouth Rocks, whereas both alleles were present in the Australorp.

Two forms of **alkaline phosphatase** have been detected in serum using naphthyl phosphate as substrate (Law & Munro, 1965; Wilcox, 1966). The faster migrating form is completely dominant (*Akp*) over the slower migrating form (*akp*). When crossing birds homozygous for the two different forms, the expected 3:1 ratio is found in the progeny. In surveys of different strains and breeds of domestic fowl *akp* was found to be the most frequent, with the exception of one strain of Leghorns (Wilcox, 1966; Singh & Nordskog, 1981). Alkaline phosphatase was also one of the enzymes used by Okada *et al.* (1984) in a study on the phylogenetic relationships of Japanese breeds (see Chapter 1, section 1.5).

**Esterases** have been widely studied because they can easily be detected on gels using a range of naphthyl esters. Grunder (1968) found three staining zones on starch gels: I and II correspond to aliesterases and III to cholinesterases. In zone I five bands could be detected, corresponding to two phenotypes. Phenotype A showed bands 1, 3 and 4, and phenotype B showed bands 2, 4 and 5. The two alleles ($Es^A$ and $Es^B$) show codominance and so the phenotype AB has bands 1, 2, 3, 4, and 5. $Es^B$ occurs most frequently (Grunder, 1968; Singh & Nordskog, 1981; Okada *et al.*, 1984). Other alleles which have not been detected in the domesticated fowl (Okada *et al.*, 1984) have been found in the green jungle fowl (*G. varius*).

Lundin & Wilhelmson (1989) have detected polymorphic forms of a peptidase and pyrophosphatase in the liver, spleen and kidney of White Leghorns and Rhode Island Reds. Three alleles of the peptidase and two of the pyrophosphatase have been found, each at a single locus. Yardley *et al.* (1988) have detected two polymorphic forms of amylase in pancreas, intestine and serum. Glutamyl peptidase has been detected in two different forms, the faster migrating in *G. sonnerati* and the slower in *G. gallus* (Baker & Manwell, 1972).

For a comprehensive list of enzyme polymorphisms in the domestic fowl, see Grunder (1990).

## References

Anastasi, A., Brown, M. A., Kembhavi, A. A., Nicklin, M. J. H., Sayers, C. A., Sunter, D. C. & Barrett, A. J. (1983). Cystatin, a proteinase inhibitor of cysteine proteinases. *Biochemical Journal*, **211**, 129–38.

Baker, C. M. A. (1968a). The proteins of egg white. In *Egg Quality*, ed. T. C. Carter, pp. 67–108. Edinburgh: Oliver & Boyd.

Baker, C. M. A. (1968b). Molecular genetics of avian proteins. IX. Interspecific and intraspecific variation of egg white proteins of the genus *Gallus*. *Genetics*, **58**, 211–26.

Baker, C. M. A., Crozier, G., Stratil, A. & Manwell, C. (1970). Identity and nomenclature of some protein polymorphisms of chicken eggs and sera. *Advances in Genetics*, **15**, 147–74.

Baker, C. M. A. & Manwell, C. (1972). Molecular genetics of avian proteins. XI. Egg proteins of *Gallus gallus*, *G. sonnerati* and hybrids. *Animal Blood Groups and biochemical Genetics*, **3**, 101–7.

Barrett, A. J. (1974). Chicken $\alpha_2$-proteinase inhibitor: a serum protein homologous with ovoinhibitor of egg white. *Biochimica et Biophysica Acta*, **371**, 52–62.

Bauer, H., Braunitzer, G., Oberthur, W. Kosters, J. & Girm, F. (1985). The Hemoglobin of the Andean Condor (*Vultur gryphus*). The amino-acid sequence of the Major (HbA) and minor component (HbD). *Biologische Chemie Hoppe-Seyler*, **366**, 1141–8.

Beeley, J. G. (1971). The isolation of ovomucoid variants differing in carbohydrate composition. *Biochemical Journal*, **123**, 399–405.

Bogard, W. C., Kato, I. & Laskowski, M. (1980). A $ser^{162}/gly^{162}$ polymorphism in Japanese quail ovomucoid. *Journal of Biological Chemistry*, **255**, 6569–74.

Briand, G., Kmiecik, D., Sautiere, P., Wonters, D., Borie-Loy, O., Biserte, G., Mazen, A. & Champagne, M. (1980). Chicken erythrocyte histone H5. IV. Sequence of carboxy-terminal half of the molecule (96 residues) and complete sequence. *Federation of European Biochemical Societies Letters*, **112**, 147–51.

Brown, J. L. & Ingram, V. M. (1974). Structural studies on chick embryo hemoglobins. *Journal of Biological Chemistry*, **249**, 3960–72.

Callegarini, C., Baragelles, A. & Conconi, F. (1969). A $\gamma$-globin-chain mutation in the hemoglobin 1 (HB1) of the domestic chicken. *Experientia*, **25**, 537–8.

Chaplin, M. F. (1983). The use of a microcomputer for comparing protein and nucleic acid sequences. *Biochemical Education*, **11**, 151.

Chapman, B. S., Tobin, A. J. & Hood, L. E. (1980). Complete amino acid sequence of the major early embryonic $\alpha$-like globins of the chicken. *Journal of Biological Chemistry*, **255**, 9051–9.

Chapman, B. S., Tobin, A. J. & Hood, L. E. (1981). Complete amino-acid sequence of major early embryonic β-like globin in chickens. *Journal of Biological Chemistry*, **256**, 5524–31.

Chapman, B. S., Hood, L. E. & Tobin, A. J. (1982a). Amino acid sequences of the ε and $\alpha^E$ globins of HbE, a minor early embryonic hemoglobin of the chicken. *Journal of Biological Chemistry*, **257**, 643–50.

Chapman, B. S., Hood, L. E. & Tobin, A. J. (1982b). Minor early embryonic chick hemoglobin H: Amino acid sequences of ε and $\alpha^D$ chains. *Journal of Biological Chemistry*, **257**, 651–8.

Cirotto, C., Panaca, F. & Arangi, I. (1987) The minor haemoglobins of primitive and definitive erythrocytes of the chicken embryo. Evidence for haemoglobin L. *Development*, **101**, 805–13.

Cleveland, D. W., Fischer, S. G., Kirschner, M. W. & Laemli, I. (1977). Peptide mapping by limited proteolysis in sodium dodecyl sulfate and analysis by gel electrophoresis. *Journal of Biological Chemistry*, **252**, 1102–6.

Colella, R., Sakaguchi, Y., Nagase, H. & Bird, J. W. C. (1989). Chicken egg white cystatin. *Journal of Biological Chemistry*, **264**, 17164–9.

Creighton, T. E. (1984). *Proteins: Structure and Molecular Properties*. New York: W. H. Freeman.

Croft, L. R. (1980). *Handbook of Protein Sequence Analysis*, 2nd edn. Chichester: John Wiley.

Czelusniak, J., Goodman, M., Hewett-Emmett, D., Weiss, M. L., Venta, P. J. & Tashian, R. E. (1982). Phylogenetic origins and adaptive evolution of avian and mammalian haemoglobin genes. *Nature*, **298**, 297–300.

Dayhoff, M. O., Park, C. M. & McLaughlin, P. J. (1972). Building a phylogenetic tree: cytochrome *c*. In *Atlas of Protein Sequence and Structure* Vol. 5, pp.7–16. ed. M. O. Dayhoff. Washington DC: National Biomedical Research Council.

Dickerson, R. E. & Geiss, I. (1983). *Haemoglobin: Structure, function, evolution and pathology*. California: Benjamin/Cummings Publishing Co.

Dodgson, J. B. & Engel, J. D. (1983). The nucleotide sequence of the adult chicken α-globin genes. *Journal of Biological Chemistry*, **258**, 4623–9.

Dolan, M., Sugarman, B. J., Dodgson, J. B. & Engel, J. D. (1981). Chromosomal arrangement of the chicken β-type globin genes. *Cell*, **24**, 669–77.

DurgaKumari, B. & Adiga, P. R. (1986). Hormonal induction of riboflavin carrier protein in the chicken oviduct and liver: a comparison of kinetics and modulation. *Molecular and Cellular Endocrinology*, **44**, 285–92.

Engel, J. D., Rusling, D. J., McCune, K. C. & Dodgson, J. B. (1983). Unusual structure of the chicken embryonic α-globin gene π′. *Proceedings of the National Academy Sciences of USA*, **80**, 1392–6.

Evans, T., Reitman, M. & Felsenfeld, G. (1988). An erythrocyte-specific DNA-binding factor recognizes a regulatory sequence common to all chicken globin genes. *Proceedings of the National Academy of Sciences of USA*, **85**, 5976–80.

Feeney, R. E. & Allison, R. G. (1969). *Evolutionary Biochemistry of Proteins*. New York: John Wiley.

Garel, A., Mazen, A., Champagne, M., Sautiere, P., Kmiecik, D., Loy, O. & Biserte, G. (1975). Chicken erythrocyte Histone H5: I. Amino terminal sequence (70 residues). *Federation of European Biochemical Societies Letters*, **50**, 195–9.

Gibbs, A. J. & McIntyre, G. A. (1970). The diagram, a method for computing sequences: Its use with amino acid and nucleotide sequences. *European Journal of Biochemistry*, **16**, 1–11.

Green, N. M. (1975). Avidin. *Advances in Protein Chemistry*, **29**, 85–133.

Grunder, A. A. (1968). Inheritance of electrophoretic variants of serum esterases in domestic fowl. *Canadian Journal of Genetics and Cytology*, **10**, 961–7.

Grunder, A. A. (1990). Genetics of biochemical variants in chickens. In *Poultry Breeding and Genetics*, ed. R. D. Crawford, pp. 239–55. Amsterdam: Elsevier.

Grutter, M. G., Weaver, L. H. & Matthews, B. W. (1983). Goose lysozyme structure: an evolutionary link between hen and bacteriophage lysozymes? *Nature*, **303**, 828–31.

Hamazume, Y., Mega, T. & Ikenaka, T. (1984). Characterization of hen egg white- and yolk-riboflavin binding proteins and the amino acid sequence of egg white–riboflavin binding protein. *Journal of Biochemistry*, **95**, 1633–44.

Hindenburg, A., Spitznagel, J. & Arnheim, N. (1974). Isozymes of lysozyme in leukosis and egg white: Evidence for the species-specific control of egg-white lysozyme synthesis. *Proceedings of the National Academy of Sciences of USA*, **71**, 1653–7.

Hood, L. E., Campbell, J. & Elgin, S. (1975). The organization, expression and evolution of antibody genes and other multigene families. *Annual Review of Genetics*, **9**, 305–53.

Huang, C.-C., Mayer, H. E. & Montgomery, R. (1970). Microheterogeneity and paucidispersity of glycoproteins. Part 1. The carbohydrate of chicken ovalbumin. *Carbohydrate Research*, **13**, 127–37.

Hughes, S. H., Stubblefield, E., Payvar, F., Engel, J. D., Dodgson, J. B., Spector, D., Cordell, B., Schimke, R. T. & Varmus, H. E. (1979). Gene localization by chromosome fractionation: globin genes are on at least two chromosomes and three estrogen inducible genes are on three chromosomes. *Proceedings of the National Academy of Sciences of USA*, **76**, 1348–52.

Hunt, L. T. & Dayhoff, M. O. (1982). Evolution of chromosomal proteins. In *Macromolecular Sequences in Systematic and Evolutionary Biology*, ed. M. Goodman, pp. 193–239. New York: Plenum.

Imoto, T., Johnson, L. N., North, A. C. T., Phillips, D. C. & Rupley, J. A. (1972). Vertebrate lysozymes. In *The Enzymes*, 3rd edn, Volume 7, ed. P. D. Boyer, pp. 666–868. New York: Academic Press.

Iwase, H., Kato, Y. & Hotta, K. (1984). Comparative study of ovalbumins from various avian species by Con A/Sepharose chromatography. *Comparative Biochemistry and Physiology*, **77B**, 743–7.

Jeltsch, J.-M. & Chambon, P. (1982). The complete nucleotide sequence of the chicken ovotransferrin mRNA. *European Journal of Biochemistry*, **122**, 291–5.

Jolles, J., Ibrahimi, I. M., Prager, E. M., Schoentgen, F., Jolles, P. & Wilson, A. C. (1979). Amino acid sequence of pheasant lysozyme. Evolutionary change affecting processing of prelysozyme. *Biochemistry*, **18**, 2744–52.

Kato, A., Oda, S., Yanaka, Y., Matsudomi, N. & Kobayashi, K. (1985). Functional and structural properties of ovomucin. *Agricultural and Biological Chemistry*, **49**, 3501–4.

Kato, I., Kohr, W. J. & Laskowski, M. (1978). Evolution of avian ovomucoids. In *Proceedings of the 11th Federation of European Biochemical Societies*, **47**, 197–206. Oxford: Pergamon Press.

Kato, I., Schrode, J., Kohr, W. J. & Laskowski, M. (1987). Chicken Ovomucoid: Determination of its amino acid sequence, determination of the trypsin reactive site, and preparation of all three of its domains. *Biochemistry*, **26**, 193–201.

Kato, I., Schrode, J., Wilson, K. A. & Laskowski, M. (1976). Evolution of proteinase inhibitors. *Protides in Biologial Fluids*, **23**, 235–43.
Keane, R. W. & Abbott, U. K. (1980). Erythropoiesis in normal and mutant chick embryos. *Developmental Biology*, **75**, 442–53.
Keane, R., Abbott, U., Brown, J. & Ingram, V. (1974). Ontogeny of haemoglobins: Evidence of hemoglobin M. *Developmental Biology*, **38**, 229–36.
Kimura, I. (1983). Developmental change in microheterogeneity of serum transferrin of chickens. *Development, Growth and Differentiation*, **25**, 531–5.
Knezetic, J. A. & Felsenfeld, G. (1989). Identification and characterization of a chicken α-globin enhancer. *Molecular and Cellular Biology*, **9**, 893–901.
Knochel, W., Wittig, B., Wittig, S., John, M. E., Grundmann, U., Oberthur, W., Godovac, J. & Braunitzer, G. (1982). No evidence for 'stress' α-globin genes in chicken. *Nature*, **295**, 710–12.
Laskowski, M. & Kato, I. (1980). Protein inhibitors of proteinases. *Annual Review of Biochemistry*, **49**, 593–626.
Laskowski, M., Kato, I., Ardelt, W., Cook, J., Denton, A., Empie, M. W., Kohr, W. J., Park, S. J., Parks, K., Schatzley, B. L., Schoernberger, O. L., Tashiro, M., Vichot, G., Whatley, H. E., Wieczcorek, A. & Wieczorek, M. (1987). Ovomucoid third domain from 100 avian species: Isolation, sequencing and hypervariability of enzyme–inhibitor contact residues. *Biochemistry*, **26**, 202–21.
Law, G. R. J. & Munro, S. S. (1965). Inheritance of two alkaline phosphatase variants in fowl plasma. *Science*, **149**, 1518.
Lewontin, R. C. (1974). *The Genetic Basis of Evolutionary Change*. New York: Columbia University Press.
Lipman, D. J. & Pearson, W. R. (1985). Rapid and sensitive protein similarity searches. *Science*, **227**, 1435–40.
Longsworth, L. G., Cannon, R. K. & MacInnes, D. A. (1940). Electrophoretic study of the proteins of egg white. *Journal of the American Chemical Society*, **62**, 2580–90.
Lundin, L.-G. & Wilhelmson, M. (1989). Genetic variation of peptidase and pyrophosphatase in the chicken. *Poultry Science*, **68**, 1313–18.
Lush, I. E. (1961). Genetic polymorphisms in the egg albumin proteins of the domestic fowl. *Nature*, **189**, 981–4.
Maizel, J. V. & Lenk, R. P. (1981). Enhanced graphic matrix analysis of nucleic acid and protein sequences. *Proceedings of the National Academy of Sciences of USA*, **78**, 7665–9.
Markussen, J. & Sundby, F. (1973). Isolation and amino acid sequence of the C-peptide of duck proinsulin. *European Journal of Biochemistry*, **34**, 401–8.
McReynolds, L., O'Malley, B. W., Nisbet, A. D., Fothergill, J. E., Givol, D., Fields, S., Robertson, M. & Brownlee, G. G. (1978). Sequence of chicken ovalbumin mRNA. *Nature*, **273**, 723–8.
Menendez-Arias, L., Gavilanes, J. G. & Rodriguez, R. (1985). Amino acid sequence around the cysteine residues of pigeon egg-white lysozyme: comparative study with other type *c* lysozymes. *Comparative Biochemistry and Physiology*, **82B**, 639–42.
Moss, G. & Hamilton, E. (1974). Chicken definitive erythrocyte haemoglobins. *Biochimica et Biophysica Acta*, **371**, 379–91.
Muggeo, M., Ginsburg, B. H., Roth, J., Neville, D. M., DeMeyts, P. & Kahn, C. R.

(1979). The insulin receptor in vertebrates is functionally more conserved during evolution than insulin itself. *Endocrinology*, **104**, 1393–402.

Muniyappa, K. & Adiga, P. R. (1979). Isolation and characterization of thiamin-binding protein from chicken egg white. *Biochemical Journal*, **177**, 887–94.

Nagase, H., Harris, E. D., Woessner, J. F. & Brew, K. (1983). Ovostatin: A novel proteinase inhibitor from chicken egg white. *Journal of Biological Chemistry*, **258**, 7481–9.

Nevo, E. (1978). Genetic variation in natural populations: Patterns and theory. *Theoretical Population Biology*, **13**, 121–77.

Nikol, J.M. & Felsenfeld, G. (1988). Bidirectional control of the chicken $\beta$- and $\epsilon$-globin genes by a shared enhancer. *Proceedings of the National Academy of Science of USA*, **85**, 2548–52.

Norioka, N., Okada, T., Hamazume, Y., Mega, T. & Ikenaka, T. (1985). Comparison of amino acid sequences of hen plasma-, yolk-, and white- riboflavin-binding proteins. *Journal of Biochemistry*, **97**, 19–28.

Oberthur, W., Godovac-Zimmerman, J. & Braunitzer, G. (1986). The expression of $\alpha^D$-chains in the hemoglobin of adult ostrich (*Struthio camelus*) and American Rhea (*Rhea americana*). The different evolution of adult bird $\alpha^A$-, $\alpha^D$- and $\beta$-chains. *Biologische Chemie Hoppe-Seyler*, **367**, 507–14.

Okada, I., Yamamoto, Y., Hashiguchi, T. & Ito, S. (1984). Phylogenetic studies on the Japanese native breeds of chickens. *Japanese Poultry Science*, **21**, 318–29.

Osuga, D. T. & Feeney, R. E. (1977). Egg proteins. In *Food Proteins*, ed. J. R. Whitacher & S. R. Tannenbaum, pp.209–67. Westport, Ct.: Avi Publishing.

Parkin, D. T. (1979). *An Introduction to Evolutionary Genetics*. London: Arnold.

Paul, C., Vandecasserie, C., Schnek, A. & Leonis, J. (1974). N-terminal amino acid sequences of the $\alpha$ and $\beta$ chains of the two chicken haemoglobin components. *Biochimica et Biophysica Acta*, **371**, 155–8.

Perler, F., Efstratiadas, A., Lomedico, P., Gilbert, W., Kolodner, R. & Dodgson, J. (1980). The evolution of genes: the chicken preproinsulin gene. *Cell*, **20**, 555–66.

Petrovic, S. & Vitale, L. (1990). Purification and properties of glutamyl aminopeptidase from chicken egg-white. *Comparative Biochemistry and Physiology*, **95B**, 589–95.

Philips, D. C. (1967). The hen egg-white lysozyme molecule. *Proceedings of the National Academy of Sciences of USA*, **57**, 484–95.

Prager, E. M., Wilson, A. C. & Arnheim, N. (1974). Widespread distribution of lysozyme g in egg white of birds. *Journal of Biological Chemistry*, **249**, 7295–7.

Proudfoot, N. J., Gill, A. & Maniatis, T. (1982). The structure of the human zeta-globin gene and a closely linked, nearly identical pseudogene. *Cell*, **31**, 553–63.

Pullen, R. A., Lindsay, D. G., Wood, S. P., Tickle, I. J., Blundell, T. L., Wollmer, A., Krail, G., Brandenburg, D., Zahn, H., Glieman, J & Gammeltoft, S. (1976). Receptor binding region of of insulin. *Nature*, **259**, 369–73.

Ralphs, J. R., Wylie, L. & Hill, D. J. (1990). Distribution of insulin-like growth factors peptides in the developing chick embryo. *Development*, **109**, 51–8.

Romero-Herrara, A. E., Lehman, H., Joysey, K. A. & Friday, A. E. (1978). On the evolution of myoglobin. *Philosophical Transactions of the Royal Society B*, **283**, 61–163.

Roninson, I. & Ingram, V. M. (1981). cDNA sequence of a new chicken embryonic $\rho$-globin. *Proceedings of the National Academy of Sciences of USA*, **78**, 4782–5.

Saha, A. & Ghosh, J. (1965). Comparative studies on avian haemoglobins. *Comparative Biochemistry and Physiology*, **15**, 217–35.

Sautiere, P., Kmiecik, D., Loy, O., Briand, G., Biserte, G., Garel, A. & Champagne, M. (1975). Chicken erythrocyte histone H5; II Amino acid sequence adjacent to phenylalanine residue. *Federation of European Biochemical Societies Letters*, **50**, 200–3.

Schneeganss, D., Braunitzer, G., Oberthur, W., Kosters, J. & Grim, F. (1985). The Hemoglobin of the tree sparrow (*Passer montanus*). The amino-acid sequence of the major (HbA) and minor (HbD) component. *Biologische Chemie Hoppe-Seyler*, **366**, 893–9.

Schoentgen, F., Jolles, J. & Jolles, P. (1982). Complete amino acid sequence of ostrich (*Struthio camelus*) egg-white lysozyme, a goose-type lysozyme. *European Journal of Biochemistry*, **123**, 489–97.

Scott, M. J., Huckaby, C. S., Kato, I., Kohr, W. J., Laskowski, M., Tsai, M.-J. & O'Malley, B. W. (1987). Ovoinhibitor introns specify functional domains as in the related and linked ovomucoid gene. *Journal of Biological Chemistry*, **262**, 5899–907.

Simpson, R. J. & Morgan, F. J. (1983). Complete amino acid sequence of Embden Goose (Anser anser) egg-white lysozyme. *Biochimica et Biophysica Acta*, **744**, 349–51.

Simpson, R. J., Begg, G. S., Dorow, D. S. & Morgan, F. J. (1980). Complete amino acid sequence of goose-type lysozyme from the egg white of the Black Swan. *Biochemistry*, **19**, 1814–19.

Simon, J., Freychet, P. & Rosselin, G. (1974). Chicken insulin: radioimmunological character and enhanced activity in rat fat cells and liver plasma membranes. *Endocrinology*, **95**, 1439–49.

Singh, H. & Nordskog, A. W. (1981). Biochemical polymorphic systems in inbred lines of chickens: a survey. *Biochemical Genetics*, **19**, 1031–5.

Stein, J. P., Catterall, J. F., Kristo, P., Means, A. R. & O'Malley, B. W. (1980). Ovomucoid intervening sequences specify functional domains and generate protein polymorphism. *Cell*, **21**, 681–7.

Steiner, D. F., Chan, S. J., Welsh, J. M. & Kwok, S. C. M. (1985). Structure and evolution of the insulin gene. *Annual Review of Genetics*, **19**, 463–84.

Stevens, L. & Duncan, D. (1988). Peptide mapping of ovoglobulins G2A and G2B in the domestic fowl. *British Poultry Science*, **29**, 665–9.

Stevens, L., Scott, R. & Duncan, D. (1986). A comparison of ovomucoids isolated from different breeds of chickens. *British Poultry Science*, **27**, 160 (abstract).

Ueda, J. & Hachinohe, Y. (1984). A new method for separation of chicken red cell catalase isozymes. *Animal Blood Groups and Biochemical Genetics*, **15**, 323–5.

Urban, M. K., Franklin, S. G. & Zweidler, A. (1979). Isolation and characterization of the histone variants of chicken erythrocytes. *Biochemistry*, **18**, 3952–60.

Vandecasserie, C., Schnek, A. & Leonis, J. (1971). Oxygen affinity of avian haemoglobins: chicken and pigeon. *European Journal of Biochemistry*, **24**, 284–7.

Von Holt, C., Strickland, W. N., Brandt, W. F. & Strickland, M. S. (1979). More histone structures. *Federation of European Biochemical Societies Letters*, **100**, 201–18.

Washburn, K. (1968). Inheritance of an abnormal hemoglobin in a randombred population of domestic fowl. *Poultry Science*, **47**, 561–4.

Weaver, L. H., Grutter, M.G ., Remington, S. J., Gray, T. M., Isaacs, N. W. & Matthews, B.W. (1985). Comparison of goose-type, chicken-type, and phage-type lysozymes illustrates the changes that occur in both amino acid sequence and three-dimensional structure during evolution. *Journal of Molecular Evolution*, **21**, 97–111.

White, H. B. & Merrill, A. H. (1988). Riboflavin-binding protein. *Annual Review of Nutrition*, **8**, 279–99.

Wilcox, F. H. (1966). A recessively inherited electrophoretic variant of alkaline phosphatase in chicken serum. *Genetics*, **53**, 799–805.

Williams, J. (1962). A comparison of conalbumin and transferrin in the domestic fowl. *Biochemical Journal*, **83**, 355–64.

Williams, J. (1969). A comparison of glycopeptides from ovotransferrin and serum transferrin of hen. *Biochemical Journal*, **108**, 57–67.

Williams, J. (1982). Evolution of Transferrins. *Trends in Biochemical Sciences*, **7**, 394–7.

Williams, J., Elleman, T. C., Kingston, I. B., Wilkins, A. G. & Kuhn, K. A. (1982). The primary structure of hen ovotransferrin. *European Journal of Biochemistry*, **122**, 297–303.

Winter, W. P., Buss, E. G., Clagett, C. O. & Boucher, R. V. (1967). The nature of the biochemical lesion in avian renal riboflavinuria. II. The inherited change of a riboflavin-binding protein from blood and eggs. *Comparative Biochemistry and Physiology*, **22**, 897–906.

Wiseman, R. L., Fothergill, J. E. & Fothergill, L. A. (1972). Replacement of asparagine by aspartic acid in hen ovalbumin and a difference in immunochemical reactivity. *Biochemical Journal*, **127**, 775–80.

Yaguchi, M., Roy, C. & Seligy, V. L. (1979). Complete amino acid sequence of goose erythrocyte H5 histone and homology between H1 and H5 histones. *Biochemical and Biophysical Research Communications*, **90**, 1400–6.

Yardley, D. G., Gapusan, R. A., Jones, J. E. & Hughes, B. L. (1988). The amylase gene–enzyme system of chickens. I. Allozymic and activity variation. *Biochemical Genetics*, **26**, 747–55.

Zheng, D. B., Lim, H. M., Pene, J. J. & White, H. B. (1988) Chicken riboflavin-binding protein cDNA sequence. *Federation of American Societies for Experimental Biology*, **2**, A354 (abstract).

# 11

# *Immunogenetics of the domestic fowl*

## 11.1 Introduction

**Antibodies** that make up part of the major defence system of the body, differ from the proteins discussed in the previous chapter in that the particular set of antibodies possessed by an individual is unique, and depends in part on the **antigens** with which the individual has been challenged during its lifetime. A process of selection occurs during the lifetime of the organism, but that does not imply that immunoglobulins (the antibodies) as a class have not evolved as with other proteins. The various types of immunoglobulins and their mechanisms of action have evolved over millions of years and collectively have the potential to combat a large range of antigens.

Before discussing the evolution and genetics of the immune system it will be necessary to outline the salient features of the system in general, and that of the fowl in particular. Immunogenetics is a subject that has developed greatly following the recent technical advances in the molecular biology of nucleic acids (see Chapter 12). Thus it has been possible to test the theories concerning the origin of antibody diversity proposed in the 1950s, since a number of gene sequences for immunoglobulins have become known (Porter, 1973).

Three separate areas are discussed in this chapter: (i) the genetics of antibody formation and the immune response, (ii) the Major Histocompatibility Complex (MHC), and (iii) the blood group antigens. The Major Histocompatibility Complex is a group of cell surface antigens. They are highly allelic, many alleles occurring at one locus. This has given rise to a wide variety of major histocompatibility antigen combinations, individual organisms each having an almost unique combination. If mismatching the MHC occurs in tissue transplantation this usually leads to graft rejection. Blood group antigens are like the MHC in being cell surface antigens but, instead of having a wide tissue distribution, they are generally located only on the surface of erythrocytes. There are several distinct groups of these and, for a similar reason, it is also necessary to match donor and recipient before blood transfusion.

The next sections (11.2–11.4) summarise the important features of the immunological system in vertebrates in order to provide the background in which to explain the immunogenetics of the domestic fowl.

## 11.2 The nature of the antibody response

The antibodies form a major part of the defence system in higher vertebrates. When a higher organism becomes infected with foreign agents, usually bacteria or viruses, or when it is injected with certain foreign macromolecules, the organism reacts by generating antibodies that bind specifically to that foreign agent. This leads to the eventual clearance of the antigen:antibody complexes from the body and hence elimination of the infecting organism. Once an antibody has become bound to infected cells bearing a foreign antigen, the infected cells are targetted for destruction by macrophages. There are two types of response, known as the **humoral response** and the **cellular response**.

In the humoral response certain cells called **lymphocytes** secrete antibodies which then circulate in the blood and lymphatic system of vertebrates, and bind specifically to foreign antigens. The term 'humor' refers to a fluid, and humoral immunity is immunity brought out by molecules in solution, both the antibodies and the antigens. The second system (cellular response) works in a similar way except that whole cells are responsible for binding the antigens. The binding sites occur on the cell membrane and are not released into solution, thus the interaction in this case is often between cells, that is those of the host and the foreign agent. The cellular response is important in combatting tumour cells, and also viral infections. Although there are several differences in detail between the humoral and the cellular response, the underlying genetics are similar, and since the humoral system has been studied more thoroughly in the domestic fowl, in this résumé the focus will be on it.

The antibodies that combine with foreign agents are highly specific in their interactions. For example, if a protein that is not foreign to an organism is modified by addition of a specific chemical group, it will then become antigenic, and antibodies can be raised in the host that will combine with it. If the same protein is modified with a very similar group, but which is an isomer of the first compound, then the antibody first raised can distinguish the two modified forms of the protein. This high degree of specificity has been demonstrated many times, e.g. antibodies raised against related bacteria. However, closely related bacteria may stimulate the production of cross-reacting antibodies. Understanding this high degree of specificity of antibodies posed an interesting problem: how, in molecular terms, is specificity achieved? A second and equally difficult problem was: if the antibodies are so highly specific, how can an organism make such an enormous repertoire of antibodies? For example, if a protein that has been chemically modified so that it is unlike anything that the organism could have previously encountered, how is that organism nevertheless able to generate a set of highly specific antibodies against it? So remarkable is the immunological system that a higher vertebrate with a well developed system has the capacity to make hundreds of millions of different antibodies.

To understand these processes it was first necessary to isolate specific antibodies and determine their structure in detail in order to see precisely how this specificity could occur. Secondly, it was necessary to determine the mechanism by which these highly specific antibodies were synthesised in response to a signal from a foreign antigen. The structure of antibodies is described in section 11.3, followed in section 11.4 by a discussion on how the diversity of antibodies is generated. In both sections only a brief résumé is provided, since immunology is a major discipline in its own right. The three organisms in which most immunological studies have been carried out are the mouse, rabbit and man, and so the general descriptions given below apply to them, but probably apply to related mammals, and with some modifications to birds also. For further background reading in this area, see Roitt, Brostoff & Male (1985).

## 11.3 The structure of immunoglobulins

The antibodies that circulate in the blood of vertebrates belong to the class of proteins known as **immunoglobulins**. These make up 20–25% of the proteins circulating in the plasma. There are estimated to be more than $10^9$ different immunoglobulins circulating in the blood. There are five different major classes of immunoglobulins (IgG, IgM, IgA, IgD & IgE), that are all built up of similar basic units. There are two types of polypeptide chain called ‘heavy’ and ‘light’, the heavy consisting of around 450 amino acid residues and the light around 220. The basic structure of immunoglobulins comprises two light and two heavy chains, joined by disulphide bridges (Fig. 11.1). This is the form which occurs in the simplest immunoglobulins known as immunoglobulins G (IgG), D (IgD) and E (IgE). Immunoglobulin A (IgA) exists both in a monomeric and a polymeric form, whereas immunoglobulin M (IgM) exists as a pentamer with five of the four-chain units joined together.

All five classes of immunoglobulins have the common property of being able to bind in a highly specific manner to an antigen. They differ in their particular roles in the overall defence system. **IgMs** are the earliest immunoglobulins to appear after the host's first encounter with a particular antigen. The binding of the IgM molecules to the antigens activates a set of proteins known as **complement proteins** that are able to lyse the bacteria to which the antibodies are bound. The IgMs also cause the activation of a group of cells present in blood called **macrophages**, that are able to digest the foreign cells. The **IgG** immunoglobulins are the principal circulating antibodies and are produced in large quantities after a second encounter with an antigen. The **IgAs** are found in secretions such as saliva, tracheo-bronchial secretions, colostrum, milk and genito-urinary secretions, and are thus particularly important in protecting the body from foreign agents. **IgEs** cause the stimulation of **mast cells** which causes them to release histamine and this is associated with, for example, sensitivity to asthma and hay fever. It is not very clear what their beneficial role is, if any. The role of **IgD** is also not clear.

Each immunoglobulin molecule has a region that binds specifically to an

**Fig. 11.1** Diagrammatic representation of a typical immunoglobulin G (IgG) structure. The variable domains are indicated as $V_L$ and $V_H$ for the light and heavy chains respectively, and the constant domains as $C_L$ and $C_\gamma1$, $C_\gamma2$ and $C_\gamma3$ for the light and heavy chains respectively. –S–S–, disulphide bridge.

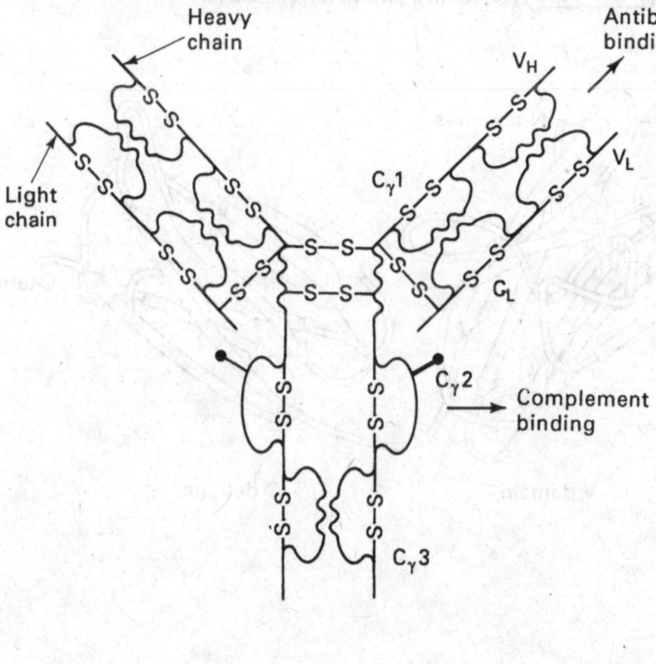

antigen, and a region that triggers processes such as complement activation (Fig. 11.1). If one considers the immunoglobulin class IgG, for example, in a typical vertebrate there may be as many as $10^9$ different IgGs. When the structure of a number of IgGs, differing in their antigenic specificities, are compared, they are found to be made up of polypeptide chains comprising **constant regions** (C region) and **variable regions** (V region). The former differ little in sequence from one IgG to another, but the latter structures show considerable sequence variation, particularly in three regions known as the **hypervariable regions**. Figure 11.2 shows the three dimensional structure of a single light chain, and it can be seen that the hypervariable regions are on the exposed surface where they can make contact with the antigen. A light chain of IgG comprises one V domain joined to a C domain, and a heavy chain comprises one V domain joined to three C domains (Fig. 11.1). The heavy chains of the immunoglobulins IgM and IgE differ in having four C domains joined to a single V domain.

The immunoglobulins have been less well characterised to date in the domestic fowl than in a number of mammals such as the mouse, rabbit and man. However, at least three classes of immunoglobulin have been identified that are the counterparts of the IgG, IgM and IgA. The sizes of the light and heavy chains are not identical with those from mammals (see Table

**Fig. 11.2** Folding pattern of a light chain of an immunoglobulin. Note the two separate C and V domains and the hypervariable regions. The arrows (⟹) represent $\beta$-pleated sheet structure.

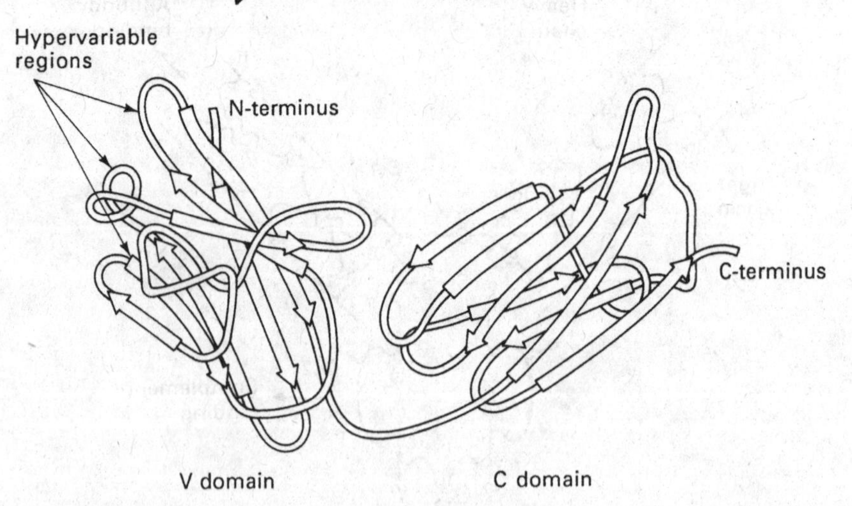

Table 11.1. *Immunoglobulins in human and domestic fowl*

| | Human | | | Domestic fowl | | |
|---|---|---|---|---|---|---|
| | IgG | IgM | IgA | IgY (IgG) | IgM | IgA |
| Total $M_r$ ($\times 10^{-3}$) | 146–170 | 970 | 160–385 | 173 | 860 | 170–>200 |
| Heavy chain $M_r$ ($\times 10^{-3}$) | 51–60 | 65 | 52–56 | 71 | 63–70 | |
| Heavy chain designation | $\gamma$ | $\mu$ | $\alpha$ | $\nu$ | $\mu$ | $\alpha$ |
| Light chain $M_r$ ($\times 10^{-3}$) | 22–25 | 22–25 | 22–25 | 21–25 | 22–24 | 19 |
| Light chain designation | $\lambda$ & $\kappa$ | $\lambda$ & $\kappa$ | $\lambda$ & $\kappa$ | $\lambda$ | $\lambda$ | $\lambda$ |
| Structure $(H_2L_2)_n$ | $n=1$ | $n=5$ | $n=1$–2 | $n=1$ | $n=5$ | $n=1$–2 |
| Average concentration in serum (mg/ml) | 13.5 | 1.5 | 3.5 | 6 | 1.75 | 0.45 |

References: Benedict (1979); Butler (1983); Roitt *et al.* (1985).

11.1) and for this reason the IgG has been referred to as IgY (Butler, 1983). There is homology in the amino acid sequence between IgG from mammalian sources and the IgY from domestic fowl, and also functional similarity in that IgY is the principal immunoglobulin responsible for the humoral response. For these reasons this immunoglobulin from the domestic fowl is often referred to as IgG. Recently it has been shown by Parvari *et al.* (1988) that the heavy chain of this immunoglobulin in the domestic fowl contains four C regions and one V region, in contrast to that from the heavy chain of mammalian IgG, and they have proposed that the name IgY be retained in order to emphasise its distinctiveness.

Each of the immunoglobulin classes has different heavy chains; those from mammalian sources are designated $\gamma$ for IgG, $\mu$ for IgM and $\alpha$ for IgA. It has been proposed that the heavy chain from domestic fowl IgY be designated $\nu$ (Parvari *et al.*, 1988). However, there is an additional subdivision in that two different types of light chain, differing in amino acid sequence, are possible, $\lambda$ and $\kappa$. Thus for each of the major classes IgG, IgM and IgA there are both $\lambda$ and $\kappa$ types. The proportion of $\kappa$ to $\lambda$ types is constant within a species, but there is variation between species. The pattern of distribution for selected animal species is shown in Table 11.2.

Although there are still only limited data, avian species appear to have

Table 11.2. *Distribution of light chain immunoglobulins in various animals*

| Animal | $\lambda$ | $\kappa$ |
|---|---|---|
| Mouse | 5% | 95% |
| Horse | >95% | <5% |
| Human | 50% | 50% |
| Domestic fowl | >95% | |
| Turkey | >95% | |
| Shark | | 100% |

Reference: see Grant *et al.* (1971).

predominantly $\lambda$ type immunoglobulin whereas fishes appear to have predominantly $\kappa$ type (Suran & Papermaster, 1967). Mammals, on the other hand, have both $\lambda$ and $\kappa$ in varying proportions. There are homologies between the heavy and light chains, suggesting their common origin. Thus, Grant, Sanders & Hood (1971) suggested a phylogenetic tree (Fig. 11.3) in which the $\kappa$ and $\lambda$ divergence occurred about 250 million years ago about the time of divergence of fish from the ancestors of present day mammals, and that the evolution of separate light and heavy chains preceded this.

Within each class of immunoglobulin there exist different **allotypes**. An

**Fig. 11.3** Possible evolution of immunoglobulin chains (Grant, Sanders & Hood, 1971).

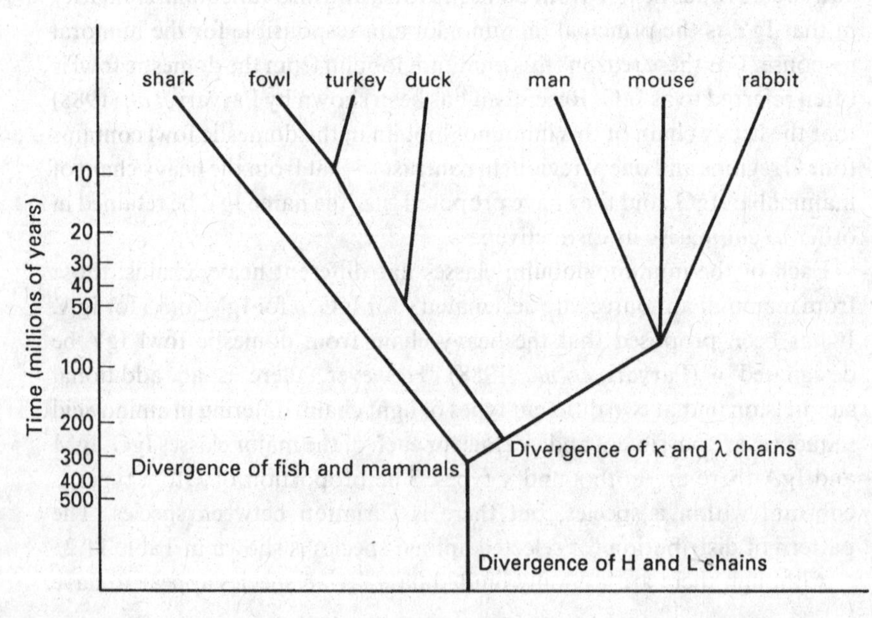

allotype is defined as the product of an allele which may be detectable as an antigen by another member of the same species. Genes at three loci are known to encode for the immunoglobulin allotypes of the domestic fowl (Benedict, 1979). The gene locus, *IgG-1*, encodes the constant (C) region of the heavy (H) chain (Foppoli *et al.*, 1979), and so far nine alleles are known, designated alphabetically from $G\text{-}1^a$ to $G\text{-}1^i$. The gene locus, *IgM-1* encodes the $C_H$ region of IgM (Foppoli & Benedict, 1980) and four alleles are known, $M\text{-}1^a$ to $M\text{-}1^d$. The third locus encodes the light chain and is designated *IgL-1* and two alleles are known, $L\text{-}1^a$ and $L\text{-}1^b$ (Foppoli & Benedict, 1980). Recently Bacon *et al.* (1986) have tested inbred lines of domestic fowl to see whether different alleles at the *IgG-1* locus showed differences in resistance to Marek's disease and to Rous sarcomas. Comparing homozygous $G\text{-}1^a/G\text{-}1^a$ and $G\text{-}1^e/G\text{-}1^e$ and heterozygous $G\text{-}1^a/G\text{-}1^e$ they found that all three showed similar resistance to Marek's disease, but a smaller proportion of $G\text{-}1^a/G\text{-}1^a$ developed Rous sarcoma than with $G\text{-}1^a/G\text{-}1^e$ and $G\text{-}1^e/G\text{-}1^e$.

## 11.4 Generation of antibody diversity

From the studies of the amino acid sequences of immunoglobulins it is apparent that the differences in antibody specificities are due to differences in the amino acid sequences, particularly in the hypervariable regions of the immunoglobulins (Fig. 11.2). Although these are only limited regions of the protein (10–12 amino acids in each polypeptide chain), a very large number of different antibodies are theoretically possible. If, in any position in the immunoglobulin chain one amino acid could be replaced by any of the other 19 amino acids, over a sequence of 10 amino acid residues that would permit $20^{10}$ ($=10^{12}$) different antibodies. But if one assumes that each antibody is coded for by a separate gene, then this would entail an enormous number of genes, corresponding to an amount of DNA greatly exceeding the total DNA present in a typical cell. How is this diversity generated so that an organism can cope with such a large number of antigens?

Well before any of the immunoglobulins had been sequenced, Burnet (1959) proposed his **clone selection theory**. He proposed that undifferentiated **stem cells** from which the mature antibody secreting cells arose were initially pluripotent, i.e. each had the potential to produce a wide variety of antibodies. During the divisions which lead to their maturation they become unipotent, i.e. each cell produced only one type of antibody. If a circulating antigen interacted with one of these cells, the cell would be stimulated to divide and secrete the particular antibody that would be able to bind the antigen (Fig. 11.4). A wide variety of cells is formed but only those 'selected' by their ability to combine with an antigen proliferate and survive. Evidence since Burnet's proposal has shown the theory to be essentially correct.

If each mature cell (**plasma cell** or **lymphocyte**) produces only a single type of antibody and yet each develops from a much smaller number of pluripotent stem cells, how are the genes organised in the stem cells? There is far too little DNA in each cell to include a separate gene for each antibody that the cell has the potential of making. A number of mechanisms have been suggested and there is sufficient evidence from the immunoglobulin gene sequences in certain species to confirm that two main mechanisms operate, namely **somatic mutation** and **somatic recombination**. The relative importance of the two mechanisms varies between species.

Somatic mutation is the process in which the DNA of the stem cells is rapidly mutating whilst cells are proliferating by mitotic division. Somatic recombination suggests that there are several separate regions (DNA sequences) on the chromosomes that code for parts of the genes; these regions can join together in a large number of different combinations

**Fig. 11.4** Assembly of an antibody light chain. $V_1$, $V_2$, . . . $V_n$ are variable domains, $J_1$, $J_2$, $J_3$ and $J_4$ are joining domains, and C is the constant domain.

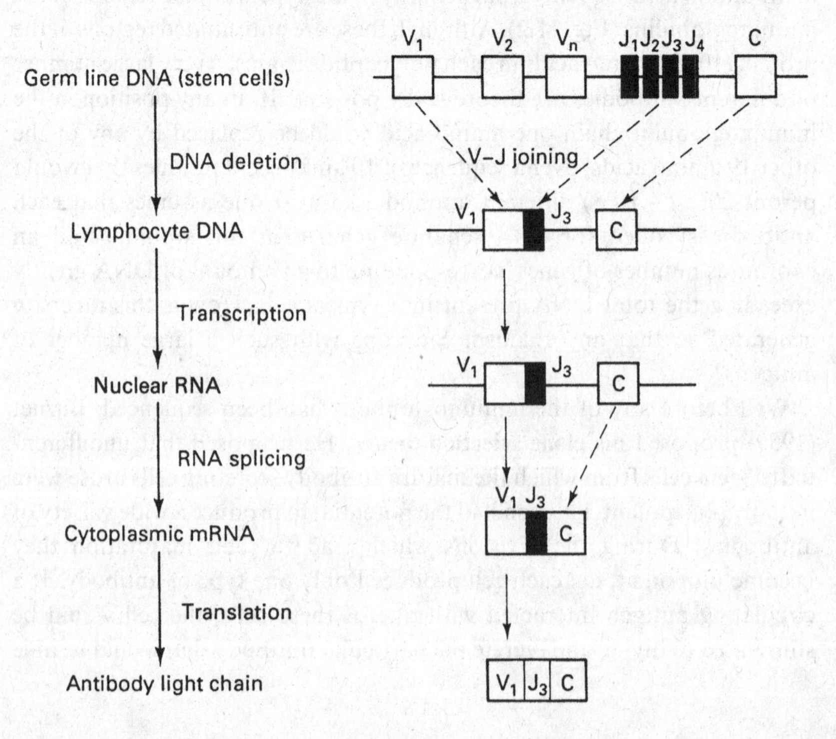

thereby creating a large variety of genes. The details of such recombinations have been worked out in a number of different vertebrate species. The genes for the light chains comprise three types of region: **C**, **V** and **J**. The **C** codes for **constant regions**, **V** for **variable regions** and **J** for **joining regions**. In the case of the heavy chains there is also another type of region, the **diversity region (D)** which increases still further the number of recombinations possible. The way in which the large variety of light chains is generated is illustrated in Fig. 11.4, and a similar mechanism operates for the heavy chain.

To obtain an estimate of the number of antibodies that is possible the situation in the mouse is examined, since that is the species for which most detail exists at present. The estimated number of different V, D and J genetic elements in the mouse (see Williamson & Turner, 1987) is given below:

| | |
|---|---|
| light chain $\kappa$ | $\simeq 250\ V_\kappa$ |
| | $4\ J_\kappa$ |
| light chain $\lambda$ | $3\ V_\lambda$ |
| | $3\ J_\lambda$ |
| heavy chain | $\simeq 250\ V_H$ |
| | $\simeq 10\ D_H$ |
| | $4\ J_H$ |

There is a single constant region for the light and heavy chain. Thus with the $\kappa$ light chain the number of recombinations possible is $250 \times 4 \times 1 = 1000$, and with the heavy chain $250 \times 10 \times 4 \times 1 = 10000$. Considering the possible combinations of $\kappa$ light chains with heavy chains gives $1000 \times 10000 = 10^7$.

Another factor should also be considered in somatic recombination. The precise positions in the DNA sequence at which the breaks in the genetic elements V, J and D occur, show some variation; this is known as junctional diversity. It is estimated to increase each pair of units fivefold, giving 5000 for the light chain and 250000 for the heavy chain, and therefore $1.25 \times 10^9$ in total. In addition to the diversity produced by somatic recombination, there is also that produced by somatic mutation, as described above, but this is more difficult to quantify. However, the combination of these mechanisms has the potential to generate as many as $10^{11}$ different cell types, each capable of generating a different antibody (see Darnell, Lodish & Baltimore, 1986). The details of the mechanism as it occurs in the domestic fowl is described in section 11.6.

## 11.5 The development of B cells

Both the circulating antibodies and the cellular immune system originate from the lymphocytes. In mammals the **T lymphocytes** differentiate in the thymus gland, and the **B lymphocytes** in foetal liver and spleen and in adult bone marrow. However, the B lymphocytes were so designated because they were first characterised in the domestic fowl in a gland known as the **Bursa of Fabricius**. The function of this gland, first described by Hieronymus Fabricius in the seventeenth century, was unknown until Glick (1956; for a review see Glick, 1983) showed that it had an immunological function. It is unique to the Class Aves; its counterpart, if any, in reptiles, amphibians and mammals is unknown.

The lymph tissues are divided into 'primary' and 'secondary' organs. The preliminary phase of establishing surface receptors on stem cells occurs in the primary organs, such as the thymus and foetal liver, the bone marrow or the Bursa of Fabricius. The stem cells themselves originate in the mesenchyme tissue whence they 'seed' the primary organs (Fig. 11.5). From the primary organs they migrate to the secondary sites such as the spleen, lymph nodes and caecal tonsils. That the secondary sites are dependent on the primary organs for their source of lymph cells can be demonstrated by surgical removal of the primary organ at the appropriate stage of development. Because of the unique role of the bursa in birds and its ease of removal, the domestic fowl is well suited to studies on the functions of B cells. It is much more difficult to eliminate B cells selectively from mammalian species.

The B and T lymphocytes cannot be distinguished on the basis of their size or morphology, but they can be distinguished by chemical or immuno-

**Fig. 11.5** Schematic representation of bursal B cell morphogenesis in the domestic fowl (Weill & Reynaud, 1987).

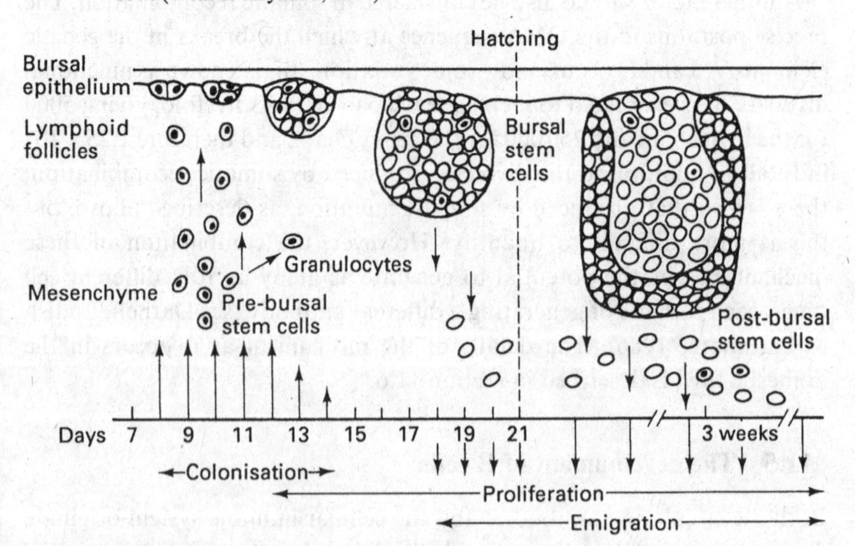

logical methods. The B cells are responsible for the humoral response and thus produce all the soluble immunoglobulins. The T cells are more heterogeneous in function and can be subdivided into three groups. The **cytotoxic** or **killer T cells ($T_c$)** have surface receptors similar in structure to the soluble antibodies. These surface receptors can bind foreign cells and thence set in train the steps leading to their destruction. They are therefore responsible for **cellular immunity**. The second type of T cells, the $T_H$ or **helper cells**, assist the B cells and certain other T cells by stimulating them to produce antibodies. The third type, $T_s$ or **suppressor T cells**, suppress B cell activity and $T_c$ activity. They are particularly important in suppressing responses to self antigens. This ensures that antibodies are only produced against foreign antigens. Defective or deficient $T_s$ cells lead to **autoimmune diseases**, in which the immune system destroys part of its own tissues. Such diseases are known in humans and have also been found in the domestic fowl. An example is the **Obese strain (OS)** domestic fowl, in which thyroid autoantibodies occur spontaneously and the thyroid undergoes progressive destruction associated with a chronic inflammatory lesion (see Wick *et al.*, 1989).

For studying the early development of the B cell lineage, the bursa has the advantage over the nearest mammalian counterpart, i.e. foetal liver, of not being responsible for a number of other processes, e.g. haemopoiesis, that would complicate the study. The bursa is an oval sac-like structure in the dorsal region of the cloaca at its junction with the colon. It is relatively small, less than 0.5% of body weight. The bursa rudiment develops as a diverticulum of the cloaca and can be seen in the chick embryo by 5 days' incubation. LeDouarin *et al.* (1975) and Houssaint, Belo & LeDouarin (1976) showed that between the 8th and 15th day the bursa becomes 'seeded' by undifferentiated cells (Fig. 11.5). A similar seeding of the thymus begins about the 7th day. In both cases the undifferentiated cells arise in blood islands of the yolk sac (Cooper & Burrows, 1989). Unlike the case of mammals, the embryonic bird liver is never a major site of haemopoietic activity. The stem cells divide, with a doubling time of about 10 h, and developing lymphocytes that have IgM molecules on their surface membranes can be detected by the 12th day (Lydyard, Grossi & Cooper, 1976). By the 20th day of incubation 90% of the bursal cells can be detected having IgM immunoglobulins. During the last few days of embryonic life the lymphocytes migrate from the bursa to the secondary lymphoid tissues (Fig. 11.5). This process continues after hatching so that after about 5 weeks most of the stem cells have left the bursa for the periphery where they continue to proliferate B cells for the rest of the fowl's life.

The bursa reaches its maximum weight of about 3 g between 4 and 12 weeks, when it contains $10^4$–$10^5$ discrete follicles and then undergoes regression at the onset of sexual maturity, its role having been completed (Payne, 1971). The IgM class of antibodies is the first to appear in the bursa (16–18 days) and the IgG (IgY) class is detectable by hatching. The IgA class is the predominant immunoglobulin class to be synthesised by the intestinal mucosa in the mature fowl. Cells synthesising IgA arise in the bursa and follow the IgG (IgY) synthesising cells in the order of migration. This has been demonstrated by carrying out bursectomies at different stages of development and observing the effects on the production of certain classes of immunoglobulins. The switchover in production in the bursa is believed to be: IgM→IgG (IgY)→IgA (Kincade & Cooper, 1973; Cooper & Burrows, 1989).

## 11.6 Generation of antibody diversity in the domestic fowl

The first complete nucleotide sequence of the gene for a $\lambda$ light chain immunoglobulin from the domestic fowl was reported by Reynaud, Dahan & Weill (1983). Its C region is 61% homologous with human $C_\lambda$ and mouse $C_{\lambda 1}$. The $V_\lambda$ domain shows a similar degree of homology with human $V_\lambda$ sequences, but only 42% homology with mouse $V_{\lambda 1}$. Since 1983 the nucleotide sequences of the complete genetic elements (V, J and C regions) for the diversity of light chains have been determined (see Reynaud *et al.*, 1987a) and a somewhat unexpected pattern has been discovered. Unlike the mammalian systems, that of the domestic fowl consists of a very compact organisation, the complete locus for the light chain is contained within less than 30 kilobase pairs of DNA. This includes a single $V_\lambda$ locus and a single J locus, in contrast to the multiplicity in the mouse (see section 11.5), but it has, in addition 25 V pseudogenes. How do these genetic elements generate the diversity of antibodies?

Jalkanen *et al.* (1984) examined the light chains of immunoglobulins produced in bursectomised domestic fowl. The bursectomies were carried out very early, at 60 h (before the bursal primordium appears) and the immunoglobulins produced by these chickens were examined at 10 weeks after hatching and compared with controls. The bursectomised chickens produced circulating IgM and IgG (IgY), although the IgG (IgY) concentrations were about one tenth normal. The light and heavy chains had normal $M_r$ values. However, when the light and heavy chains were subjected to proteolysis and separated on isoelectric focusing gels, they showed far fewer bands than controls, suggesting that they were far less heterogeneous than the controls. Also, the chickens developed immunity when challenged with a range of antigens. They concluded that although bursectomised chickens are capable of synthesising immunoglobulins, they cannot generate such a diversity of antibodies.

The bursa is thought to provide the microenvironment in which the diversity of lymphocytes is generated. It comprises a number of follicles, usually in the region of $10^4$, and it is possible experimentally to separate out individual follicles. Weill *et al.* (1986) examined the individual follicles to see whether each follicle is committed to synthesising a single type of immunoglobulin, and is therefore monoclonal. They isolated DNA both from the complete bursa, and from individual follicles. The DNA samples were then fragmented using restriction endonucleases (see Chapter 12, section 12.2a) and separated by gel electrophoresis. The DNA isolated from individual follicles showed far fewer bands than that from the total DNA, suggesting that each follicle produces a very limited set of rearrangements. The data are consistent with approximately two rearrangements per follicle. The authors thus propose that the follicles of the embryonic bursa become colonised by a very few stem cells and these are committed to a particular rearrangement of the IgG gene at the beginning of development.

As a result of cloning and sequencing the DNA from the entire light chain region, Reynaud *et al.* (1987a) have proposed a molecular mechanism for generation of diversity. Their work has been complemented by that of Thompson & Neiman (1987), who have used restriction endonuclease mapping to assess the diversity of DNA sequences generated. Diversification occurs in two stages, as illustrated in Fig. 11.6.

The first stage involves a rearrangement which entails joining the $V_\lambda$ and J sequences; the second involves gene conversion (Wysocki & Gefter, 1989). In the former it can be seen from Fig. 11.6 that this involves deletion of a 2 kilobase sequence between $V_\lambda$ and the J element. In the latter step a wide

**Fig. 11.6** Organisation and generation of diversity by reorganisation of the light chain (Reynaud *et al.*, 1987b).

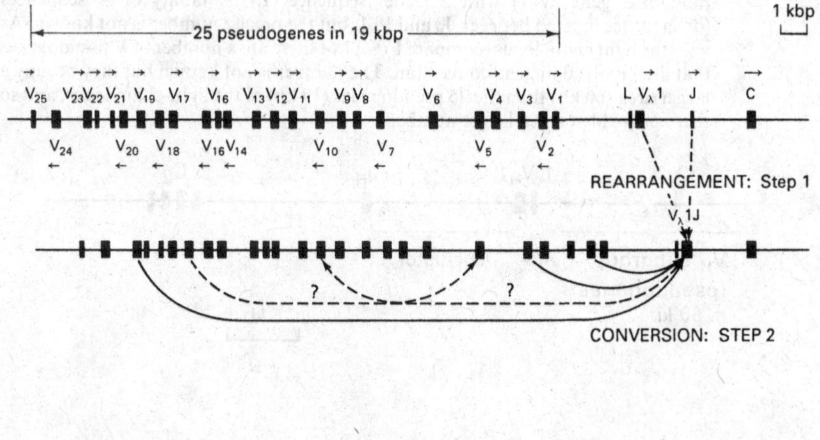

range of $V_\lambda$ sequences is generated by repeated recombination with combinations of the pseudogenes. Fragments from the pseudogenes ($V_1$ to $V_{25}$) become incorporated into the $V_\lambda$ causing gene conversion and hence diversity. This may occur on several occasions, and may also occur between pseudogenes, as well as between a pseudogene and $V_\lambda$. Reynaud, Dahan & Weill (1987b) point out that the term 'pseudogene' is misleading in this context and that 'partial gene' might be appropriate. These V elements are truncated in some part of the V domain and have no leader sequence or transcription signals. They are therefore not transcribed until they have become incorporated into the $V_\lambda$ gene, which has both a leader sequence and transcription signals. Although less work has been carried out on the heavy chain gene locus, it seems likely that analogous mechanisms operate (Reynaud *et al.*, 1989). There are known to be three types of heavy chain in the domestic fowl, $\mu$, $\nu$ and $\alpha$, but there is very little of the $\alpha$ (Kincade & Cooper, 1973; Parvari *et al.*, 1988). As has been mentioned previously, the heavy chain differs from that of mammalian IgG in having four C regions. The work of Parvari *et al.* (1988) suggests that in the domestic fowl there is a single $V_H$ region and at least two D genes, and that the $J_H$ region has a high degree of homology with the $V_H$ region. They propose that the diversity of H chains arises through the multiple D genes joining to the V and J genes. The probable arrangement of the heavy chain locus is given in Fig. 11.7.

## 11.7 The histocompatibility antigens in the domestic fowl

The earlier sections of this chapter have been concerned with the proteins which the body makes in order to combat bacteria, viruses and other foreign material. This section is concerned with the antigenic proteins,

**Fig. 11.7** Probable arrangement of the heavy-chain locus in the domestic fowl (Reynaud *et al.*, 1989). The locus comprises a $C_\mu$ gene (shown here as 4 exons, although the precise number is not yet known), a single $J_H$ element, and a unique functional gene ($V_H1$) with a leader sequence (L). A family of D sequences (diversity) is located between $J_H$ and $V_H1$, but the precise number is not known. As with the light chain locus (compare Fig. 11.6) there are a number of V pseudogenes that are involved in gene conversion. The number is not known but they occupy a larger area (60 kb) than the 25 pseudogenes (19 kb) of the light-chain locus, and so there probably exist a larger number.

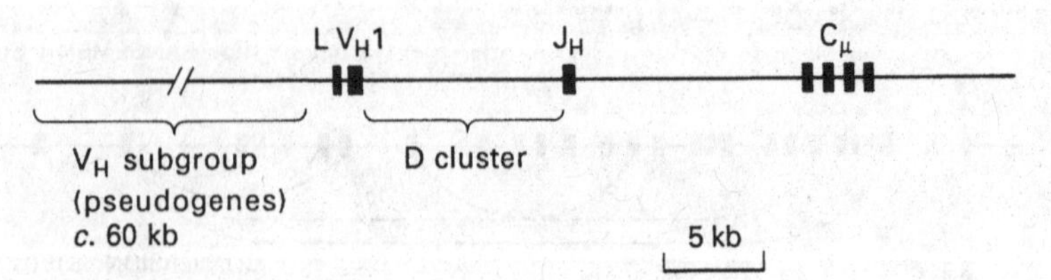

Table 11.3. *Histocompatibility antigens in the domestic fowl*

| Type | B Region of genome | Cells in which expressed | Polypeptide structure & $M_r$[a] |
|---|---|---|---|
| Class I | *B-F* | All nucleated cells | $\alpha_1$-$\alpha_2$-$\alpha_3$ 45 000 & $\beta_2$ 12 000 |
| Class II | *B-L* | Macrophages, B and T lymphocytes | $\alpha_1$-$\alpha_2$ 32–34 000 & $\beta_1$-$\beta_2$ 27–29 000 |
| Class III | Genes not identified | — | — |
| Class IV | *B-G* | Erythrocytes | 2 × 46–48 000 linked by disulphide bridge |

[a] Relative molecular mass ($M_r$) is a dimensionless number and is the ratio of the molecular mass of a molecule to 1/12 the mass of one atom of $^{12}C$.

mainly produced on the surface of cells, which account for immunological reactions between B and T cells, different types of T cells, or between T cells and foreign cells introduced by tissue transplants. These antigens, known as histocompatibility antigens, were originally discovered as a result of transplantation studies. Unless two individuals are very closely related genetically, tissue transplanted from one to the other is rejected. In a sense the ABO blood group system in humans, discovered by Landsteiner in 1900, was the first group of histocompatibility antigens to be discovered; it is, however, specific for one type of tissue, namely erythrocytes.

Many of the early studies on tissue transplantation were carried out using inbred strains of mouse. The genetic differences between individual mice were reduced by successive generations of inbreeding, and it was eventually possible to pinpoint the region on a particular chromosome which was responsible for transplant rejection. In the mouse this is a significant part of chromosome 17 and became known as the **Major Histocompatibility Complex (MHC)**. This region contains not only the genes for antigenic determinants on cells, but also a number of other genes related to the immunological system. An analogous region is found to reside on human chromosome 6. The region of the chromosome responsible for the MHC is known as the **H-2** region in mice, and **HLA** in man. Similar systems have been found in many vertebrates, although those in mice and man have been by far the most extensively studied. The blood group systems are alloantigens that reside specifically on erythrocyte cell membranes. An **alloantigen** is an antigen present in an individual that is recognised as foreign by another individual of the same species.

The histocompatibility antigens have been divided into three classes (Table 11.3) largely as the result of work on mammals, particularly mice and man, but an additional class exists in the domestic fowl. Each antigenic determinant is due to cell surface proteins that are often glycosylated. The class I antigens are present throughout the tissues of the body, although the specific type present in an individual is usually determined using lymphocytes. Class II antigens have a much more restricted distribution, being present mainly on B lymphocytes, with few if any present on T lymphocytes. Class III antigens are concerned with certain components of the complement system (C2, C4 and factor B) and are poorly understood in the domestic fowl. An additional class in the domestic fowl, class IV, is restricted to erythrocytes. The structures of classes I, II and IV are shown in Fig. 11.8. Class I type consists of two polypeptide chains, a glycosylated protein of $M_r$ 45000 constituting the heavy chain, and a protein of $M_r$ of 12000 known as $\beta_2$-microglobulin, the second chain. The two glycosylated proteins of the class II are more uniform in size, having $M_r$ values of approximately 34000 and 28000. The families of proteins that include classes I and II of the histocompatibility complex are generally thought to have evolved from a single ancestral immunoglobulin-like domain (Robertson, 1982). Class III antigens, which are soluble proteins present in the plasma, appear to be distinct in both structure and origin. The class IV antigens comprise two polypeptide chains ($M_r = 2 \times 48000$) which are linked by disulphide bonds (Kline *et al.*, 1988).

**Fig. 11.8** Structure of classes I (coded at the B-F locus), II (coded at the B-L locus) and IV (coded at the B-G locus) histocompatibility antigens in the domestic fowl (Guillemot *et al.*, 1989).

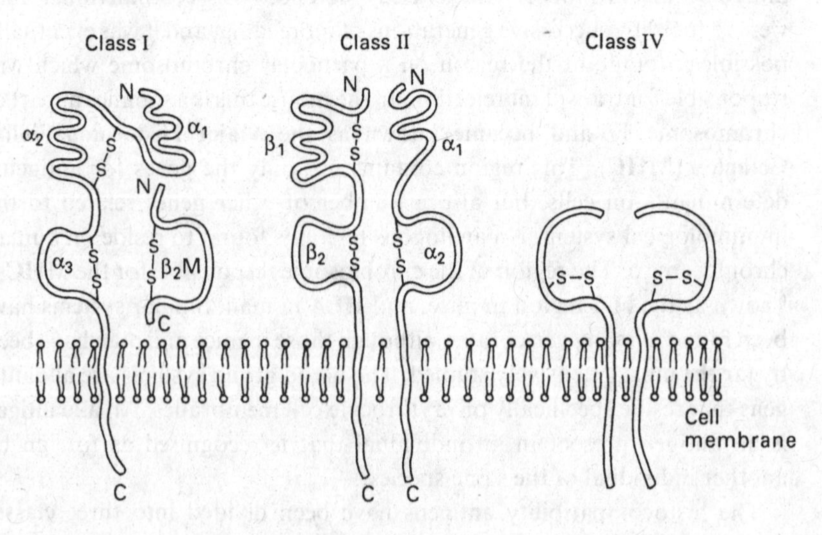

Work on the domestic fowl histocompatibility complex (see Lamont, 1989; Lamont & Dietert, 1990) is not as far advanced as that in mice and man, although it is being studied quite extensively as it is very important in the development of commercial breeds to produce maximum resistance to disease (Nordskog, 1983), as will be seen in section 11.8.

The first blood groups to be identified in the domestic fowl, during 1950–1 (see Briles, 1984), were designated A, B, C and D. Schierman & Nordskog (1961) showed that blood group B was rather distinct from the other groups. When they carried out skin homografts (grafts between different individuals of the same species) some were accepted, others rejected. They found that rejection occurred only when the particular blood group B present in the donor was not present in the host. It thus became apparant that the locus of the blood group B was also strongly linked to or identical with the histocompatibility locus. The B locus is the equivalent of the H-2 locus in mice, and the HLA locus in man. Originally it was thought that the B locus was a single locus having multiple alleles, i.e. a large number of alternative genes could be present at the one locus, and these could account for the differences between individuals; however, Pink *et al.* (1977) showed that the B locus was complex, comprising at least three distinct sub-regions. These sub-regions of the B locus are known as *B-F*, *B-L* and *B-G* and correspond with the classes of histocompatibility antigens given in Table 11.3 (Hála, 1977; Zaleski *et al.*, 1983; Nordskog, 1983).

When mapping experiments were carried out the B locus did not show any linkage to any markers on any of the macrochromosomes. However, by an ingenious set of experiments that involved producing trisomic chickens (see Chapter 2, section 2.5), Bloom & Bacon (1985 and also Bloom *et al.*, 1987) were able to show that the B locus is on a microchromosome adjacent to the **nucleolar organiser region** (see Chapter 2, section 2.4), as shown on Fig. 11.9.

A trisomic domestic fowl having 2n + 1 chromosomes may be trisomic for any of the 39 chromosomes. The nucleolar organiser region is responsible for organising the synthesis of ribosomal RNA and it can easily be identified histologically by its dense staining. A normal diploid cell will have two nucleolar organiser regions, whereas one which is trisomic for the particular chromosome carrying the nucleolar organiser region will have three such regions. In this case the trisomies associated with three nucleolar organiser regions were found to be associated with an additional microchromosome. In addition, trisomic chickens, that had been bred to be heterozygous at the B locus, were able to show three different B antigens. Trisomic chickens have also been found to be useful in studying the effects of gene dosage on the level of expression of histocompatibility antigens

(Delany, Dietart & Bloom, 1988), where it has been found that there is an increase in the amount of glycoprotein on the surface of the erythrocytes (Delany *et al.*, 1987).

The close linkage between the B locus and the nucleolar organiser region has also been demonstrated in experiments in which a DNA fragment has been cloned and shown to contain both loci (Guillemot *et al.*, 1988). The nucleolar organiser region comprises about 150 tandem repeats of ribosomal RNA genes and these occupy about 6000 kb of the telomeric long arms of microchromosome 17 (17th in size order). The complete microchromosome is about 8000 kb and so the B locus must reside within the 2000 kb which includes the centromere (Guillemot *et al.*, 1989 and Fig. 11.9).

Individual domestic fowl differing at the B locus are referred to as different haplotypes. A **haplotype** is defined as a set or group of genetic determinants on one chromosome, in this particular case the determinants are those for the sub-regions of the B locus, namely the *B-F*, *B-L* and *B-G* loci. These loci of the major histocompatibility complex (B locus) are so closely linked that they are usually inherited as a single unit, i.e. the haplotype. An individual will inherit a haplotype from both parents, and these may be either identical (homozygous) or non-identical (heterozy-

**Fig. 11.9** Microchromosome 17 of the domestic fowl, showing the loci of the nucleolar organiser region and the B complex. The chromosome comprises about 8000 kb DNA of which the nucleolar organiser region occupies about 6000 kb of the long arm. The latter comprises about 150 tandemly repeated transcriptional units coding for ribosomal RNA. The distance from the nucleolar organiser region to the proximal end of the B complex is about 10 kb. (Bloom & Bacon, 1985; Guillemot *et al.*, 1989; Kroemer *et al.*, 1990).

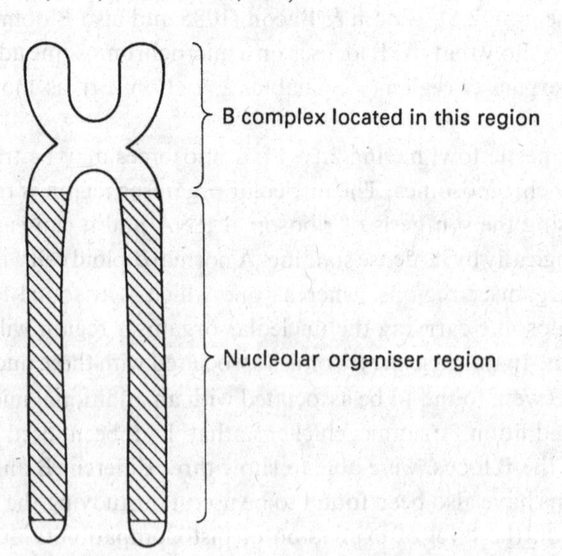

gous). Briles & Briles (1982) have identified 27 different haplotypes at the B locus. Each can be distinguished by a unique pattern of reactions with 15 different antisera, or by using monoclonal antibodies (Lamont, Warner & Nordskog, 1987). There may well be more haplotypes to be discovered in other populations of domestic fowl and it may become easier in future to use the pattern of restriction fragmentation of DNA (see Chapter 12, section 12.2a) to carry out the typing. Miller, Abplanalp & Goto (1988a) have shown that the B-G sub-regions of 17 different haplotypes of the major histocompatibility complex each show a distinct fragmentation pattern. Some of the haplotypes are discussed further in section 11.8.

All three sub-regions of the B locus, i.e. *B-F*, *B-L* and *B-G*, are highly polymorphic, and are closely linked. Recombinants between the B-F and B-G are less than 0.04% (they are therefore less than 0.04 map units apart) and no recombinants were discovered between the B-F and B-L loci (Hála, Boyd & Wick, 1981a; Hála *et al.*, 1988). By two dimensional gel electrophoresis of the B-G antigens, Miller, Goto & Briles (1988b) showed that recombination can occur within the B-G region. The role of the B-G antigen is not understood although it may cooperate with the B-F antigen, since the co-presence of both B-F and B-G antigens on the surface of erythrocytes affects the production of antisera to B-F (Hála, Plachy & Schulmanova, 1981b). Clones containing the DNA for the B-G region and for the whole of the major histocompatibility complex have been obtained and these have revealed more details of the genetic organisation (Goto *et al.*, 1988; Guillemot *et al.*, 1988).

The B locus of the domestic fowl has a much more compact arrangement than the corresponding loci in humans and mice; a molecular map is given in Fig. 11.10. The total size of the B locus is not yet known but the minimum size for the B-F/B-L region is about 250 kb compared with 2400 kb for the corresponding HLA class I and class II regions in mammals (Guillemot *et al.*, 1989). The B-G region is at least as complex as that of B-F/B-L, but less information is at present available. The compact nature of the B-F/B-L loci, in which the introns are only about one tenth the size of the corresponding mammalian ones, accounts for the very low rates of recombination between B-F and B-L genes.

Complement, which comprises at least 11 serum proteins in humans, has not been well characterised in the domestic fowl (for review see Linscott, 1986). Koch (1986) studied the complement component, factor B, in the fowl. By electrophoretic separation he showed that the protein is polymorphic, corresponding to three phenotypes, and that the segregation of the gene for factor B is not linked to MHC. It thus appears to differ from the mammalian systems in not being part of major histocompatibility complex.

## 11.8 Properties associated with different B haplotypes in domestic fowl

As already discussed the B locus is multiallelic, and to date 27 haplotypes have been distinguished by serological methods (Briles & Briles, 1982). Studies in which the DNA containing the B locus was fragmented using restriction endonucleases and the restriction fragment length polymorphisms (RFLP) were examined, suggested that there may be significantly more (Chaussé *et al.*, 1989; Tilanus *et al.*, 1989). The principle underlying this method is as follows (see also Chapter 5, section 5.6 on DNA fingerprinting). Restriction endonucleases are enzymes that catalyse the cleavage of DNA at specific positions, depending on the base sequence of the DNA and on the specificity of the particular restriction endonuclease used (see Fig. 12.1, later). The DNA fragments so obtained by this method can be separated on a size basis by agarose gel electrophoresis. The pattern of fragments on the gel, indicating fragments of different sizes, is characteristic for each DNA with its own base sequence. If two corresponding regions of DNA from, for example, two different haplotypes, differ by a

**Fig. 11.10** Molecular map of the major histocompatibility locus (B locus) in the domestic fowl. The arrangement of the genes in the B complex was determined using three non-overlapping cosmid clusters (see Chapter 12, section 12.2b) and therefore the distance separating each of the three regions shown (labelled X, Y & Z) is not yet known. It can be seen that although the B-G region is clearly separated from the B-L and B-F regions, there is overlap between the B-L and B-F regions. The nucleolar organiser region which is responsible for ribosomal RNA synthesis is closest to the B-L region. (Guillemot *et al.*, 1989; Kroemer *et al.*, 1990).

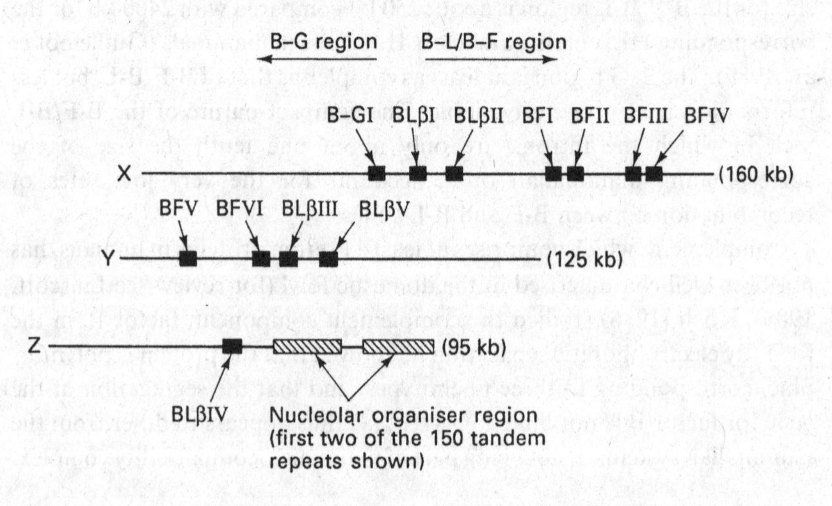

small number of bases, this may alter their susceptibility to restriction endonucleases and hence the fragmentation pattern.

The haplotypes at the B locus, that have been distinguished by serotyping, have been designated $B^1$ to $B^{29}$, $B^{16}$ having been lost, and $B^{20}$ having been found to be identical with $B^{21}$. Since the B locus is concerned with antigens, there has been much interest to see whether certain haplotypes show greater resistance to disease than others, with the possible advantage of being able to breed in the best haplotypes into commercial flocks (see Nordskog, 1983 and Zijpp, 1983). The susceptibility to a herpes-induced malignant lymphoma, Marek's disease, varies considerably among haplotypes: $B^{21}$ imparts strong resistance, whereas $B^{19}$ renders the strain susceptible (Briles, Stone & Cole, 1977; Schierman & Collins, 1987; Steadham *et al.*, 1987). $B^2$, $B^6$ and $B^{14}$ also confer resistance whereas $B^1$ and $B^{13}$ confer susceptibility. For a complete list of haplotypes and their susceptibility to Marek's disease see Calnek *et al.* (1989). By making a recombinant involving the B locus, Briles *et al.* (1983) were able to show that a gene or genes within the B-F sub-region conferred the resistance.

The effect of different B haplotypes on the susceptibility to Rous Sarcoma Virus has been studied in detail (see Collins & Zsigray, 1984). The B haplotype possessed by an individual is a major determinant of whether Rous sarcomas regress or continue to grow and ultimately kill the host. Various B haplotypes have been tested, and also the $F_1$ generation resulting from crossing two fowl homozygous at the B locus. The possession of $B^2$ or $B^6$ caused regression in a large proportion of fowl, whereas little regression was observed with $B^5$ or $B^{13}$. In some cases heterozygous fowl showed intermediate character, e.g. $B^2/B^5$, whereas $B^{23}/B^{26}$ showed a greater tendency for tumour regression than the parent homozygotes. The latter was interpreted as evidence of genetic complementation (Brown *et al.*, 1982). Evidence from recombinants suggests that the gene influencing regression is located in the B-F region (Collins & Briles, 1980).

In addition to measuring tumour regression, the reverse trend, i.e. incidences of metastases, has been measured. Strains homozygous for $B^5$ showed significantly higher instances of metastases than those homozygous for $B^{24}$. There is evidence that other loci besides those of MHC are involved in determining resistance to Rous Sarcoma Virus (Schierman & Collins, 1987). Although $B^2/B^2$ fowl are very susceptible and $B^5/B^5$ resistant, there is considerable variation from among individuals within these haplotypes, suggesting other factors are involved. Gilmour *et al.* (1986) have identified two additional genetic loci that are involved. These autosomal loci are designated Ly-4 and Bu-1. Their results show that higher degrees of regression are associated with interaction between the Ly-4 and Bu-1 loci.

The Ly-4 locus determines surface alloantigens of T lymphocytes, whereas the Bu-1 locus determines alloantigens of the B lymphocytes.

The B complex has also been shown to be involved in the immune response to bacteria (Pevzner, Nordskog & Kaeberle, 1975) and also to antigenic synthetic polypeptides (Gunther *et al.*, 1974; Benedict, Pollard & Maurer, 1977). Autoimmune thyroiditis is also associated with the B complex (Bacon, Kite & Rose, 1974). The disease was mentioned earlier in connection with the role of B and T cells in section 11.5, and a hypothesis for its genetic regulation is described by Hála (1988). As a result of examining its occurrence in relation to B haplotypes, it has been proposed that $B^3$ determines a high response to the antigenic determinant thyroglobulin, $B^1$ a moderate response, and $B^2$ and $B^4$ low responses (Rose, Bacon & Sundick, 1976). Gyles *et al.* (1986) have tested a number of different haplotypes (including $B^1$, $B^3$ and $B^4$) for resistance to Newcastle disease vaccine, infectious bronchitis vaccine, infectious bursal disease viral agent and *Salmonella pullorum*. They found significant differences in susceptibility between different haplotypes, but no single type showed maximum resistance best when challenged with each of the four agents.

Because of the close proximity of the B-F and B-L sub-regions, which means that very few recombinants occur between these loci, it is difficult to demonstrate directly which of the B-F or B-L antigens is implicated with a certain property, e.g. resistance to autoimmune disease (Hála *et al.*, 1989). However, this could be resolved by restriction fragment length polymorphism (RFLP), and this method is being used increasingly as a method for molecular genotyping (Hála *et al.*, 1989; Chaussé *et al.*, 1989; Tilanus *et al.*, 1989; Pitcovski *et al.*, 1989; Warner *et al.*, 1989).

## 11.9 Blood groups in the domestic fowl

Blood groupings arise from the different patterns of cell surface antigens on red blood cells. If red cells from a donor are given to a recipient of the same species, and are recognised as foreign by the latter because of the particular surface antigens, then reaction with the recipient's antibodies causes agglutination of the red blood cells. In the previous section the blood group B was described. It differs from the other blood groups in also being the locus of the major histocompatibility complex. Thirteen blood groups (Table 11.4) have so far been identified, and their history is described by Briles (1984), who identified the first locus in 1950. Of the 13 only four (*Ea-B*, *Ea-H*, *Ea-J* and *Ea-P*) have been precisely mapped, although linkage relations have been established for others. The blood groups are identified by their expression on erythrocytes, but the B and C group antigens, at

Table 11.4. *Blood groups in the domestic fowl*

| Group | Number of alleles | Chromosome[a] | Comments |
|---|---|---|---|
| *EaA* | 7 | 1, pu | Also expressed in lymphocytes, close to *EaE* |
| *EaB* | 30 | Microchromosome | Also MHC, adjacent to nucleolar organiser region |
| *EaC* | 8 | 1, pu | Present on lymphocytes |
| *EaD* | 5 | ? | Associated with egg weight |
| *EaE* | 11 | 1, pu | Close linkage with *EaA* |
| *EaH* | 3 | 1 | |
| *EaI* | 8 | ? | |
| *EaJ* | 2 | 1, | |
| *EaK* | 4 | ? | |
| *EaL* | 2 | ? | |
| *EaN* | 2 | ? | |
| *EaP* | 10 | 1 | |
| *EaR* | 2 | ? | |

[a] All those shown as chromosome 1 were found on linkage group III. This is equated with chromosome 1 on the basis of translocation markers (see Chapter 5, section 5.3). pu, position on the chromosome unknown.
References: Briles & Gilmour (1979); Somes (1988).

least, are known to be expressed also on lymphocytes. There are a number of alloantigens of the A group, some of which have been distinguished using monoclonal antibodies (Fulton, Briles & Lamont, 1990). These antigens can be detected as cell surface antigens on erythrocytes in 3 day embryos (Fulton, Hall & Lamont, 1990) and their expression increases with age until maturation. The alleles at the A locus are codominant and these alleles appear to be related to erythrocyte maturation. The functions of most of the blood group antigens are not known, although the B group, as already mentioned, is the major histocompatibility complex, and the C group is linked to a minor histocompatibility complex (Lamont & Dietert, 1990). The salient feaures of the different groups are summarised in Table 11.4.

## References

Bacon, L. D., Kite, J. H. & Rose, N. R. (1974). Relationship between the major histocompatibility (B) locus and autoimmune disease in obese chickens. *Science*, **186**, 274–5.

Bacon, L. D., Ch'ng, L. K., Spencer, J., Benedict, A. A., Fadly, A. M., Witter, R. L. & Crittenden, L. B. (1986). Tests of association of immunoglobulin allotype genes and viral oncogenesis in chickens. *Immunogenetics*, **23**, 213–20.

Benedict, A. A. (1979). Immunoglobulin allotypes: Chicken. In *Inbred and Genetically Defined Strains of Laboratory Animals*, Part 2, ed. P. L. Altmann and D. D. Katz, pp. 661–4. Bethesda: Federation of American Societies for Experimental Biology.

Benedict, A. A., Pollard, A. L. W. & Maurer, P. H. (1977). Genetic control of immune responses in chicken II. Responses to methylated bovine serum albumin poly($glu^{60}$ $ala^{30}$ $tyr^{10}$) aggregates. *Immunogenetics*, **4**, 199–204.

Bloom, S. E. & Bacon, L. D. (1985). Linkage of the major histocompatibility (*B*) complex and the nucleolar organizer in the chicken: Assignment to a microchromosome. *Journal of Heredity*, **76**, 146–54.

Bloom, S. E., Briles, W. E., Briles, R. W., Delany, M. E. & Dietert, R. R. (1987). Chromosomal localization of the major histocompatibility (*B*) complex (MHC) and its expression in chickens aneuploid for the major histocompatibility complex/ribosomal deoxyribonucleic acid microchromosome. *Poultry Science*, **66**, 782–9.

Briles, W. E. (1984). Early chicken blood group investigations. *Immunogenetics*, **20**, 217–26.

Briles, W. E. & Briles, R. W. (1982). Identification of haplotypes of the chicken major histocompatability complex *B*. *Immunogenetics*, **15**, 449–59.

Briles, W. E., Briles, R. W., Taffs, R. E. & Stone, H. A. (1983). Resistance to a malignant lymphoma in chickens is mapped to subregion of major histocompatibility (*B*) complex. *Science*, **219**, 977–9.

Briles, W. E. & Gilmour, D. G. (1979). Blood group systems: Chicken. In *Inbred and Genetically Defined Strains of Laboratory Animals*, Part 2, ed. P. L. Altman and D. D. Katz, pp. 650–2. Bethesda: Federation of American Societies for Experimental Biology.

Briles, W. E., Stone, H. A. & Cole, R. K. (1977). Marek's disease: Effects of *B* histocompatibility alloalleles in resistant and susceptible chicken lines. *Science*, **195**, 193–5.

Brown, D. W., Collins, W. M., Ward, P. H. & Briles, W. E. (1982). Complementation of major histocompatability haplotypes on regression of Rous sarcoma virus-induced tumours in noninbred chicken. *Poultry Science*, **61**, 409–13.

Burnet, F. M. (1959). *The Clonal Selection Theory of Acquired Immunity*. Cambridge: Cambridge University Press.

Butler, E.J. (1983) Plasma proteins. In *Physiology and Biochemistry of the Domestic Fowl*, Volume 4, ed. B. M. Freeman, pp. 321–38. London: Academic Press.

Calnek, B. W., Adene, D. F., Schat, K. A. & Abplanalp, H. (1989). Immune response *versus* susceptibility to Marek's disease. *Poultry Science*, **68**, 17–26.

Chaussé, A.-M., Coudert, F., Dambrine, G., Guillemot, F., Miller, M. M. & Auffray, C. (1989). Molecular genotyping of four chicken *B*-complex haplotypes with B-$L_{\beta}$, B-F, and B-G probes. *Immunogenetics*, **29**, 127–30.

Collins, W. M. & Briles, W. E. (1980). Response of two *B* (MHC) recombinants to Rous sarcoma virus-induced tumours. *Animal Blood Groups and Biochemical Genetics*, **2** (Suppl. 1), 38.

Collins, W. M. & Zsigray, R. M. (1984). Genetics of the response to Rous sarcoma virus-induced tumours in chickens. *Animal Blood Groups and Biochemical Genetics*, **15**, 159–71.

Cooper, M. D. & Burrows, P. D. (1989). B-cell differentiation. In *Immunoglobulin Genes*, ed. T. Honjo, pp. 1–21. New York: Academic Press.

Darnell, J., Lodish, H. & Baltimore, D. (1986). *Molecular Cell Biology*, Chapter 24. New York: Scientific American Books.

Delany, M. E., Briles, W. E., Briles, R. W., Dietert, R. R., Willand, E. M. & Bloom, S. E. (1987). Cellular expression of MHC glycoproteins on erythrocytes from normal and aneuploid chickens. *Developmental and Comparative Immunology*, **11**, 613–24.

Delany, M. E., Dietert, R. R. & Bloom, S. E. (1988). MHC-chromosome dosage effects: evidence for increased expression of Ia glycoprotein and alteration of B cell subpopulations in neonatal aneuploid chickens. *Immunogenetics*, **27**, 24–30.

Foppoli, J. M. & Benedict, A. A. (1980). Localization of two chicken IgM allotypes to H chains. *Molecular Immunology*, **17**, 439–44.

Foppoli, J. M., Ch'ng, L. K., Benedict, A. A., Ivanyi, J., Derka, J. & Wakeland, F. K. (1979). Genetic nomenclature for chicken immunoglobulin allotypes: An extensive survey of inbred lines and antisera. *Immunogenetics*, **8**, 385–404.

Fulton, J. E., Briles, R. W. & Lamont, S. J. (1990). Monoclonal antibody differentiates chicken A system alloantigens. *Animal Genetics*, **21**, 39–45.

Fulton, J. E., Hall, V. J. & Lamont, S. J. (1990) Ontogeny and expression of chicken A blood group. *Animal Genetics*, **21**, 47–57.

Gilmour, D. G., Collins, W. M., Fredricksen, T. L., Urban, W. E., Ward, P.F. & DiFronzo, N.L. (1986). Genetic interaction between non-MHC T- and B-cell alloantigens in response to Rous sarcomas in chickens. *Immunogenetics*, **23**, 1–6.

Glick, B. (1956). Normal growth of the bursa of Fabricius in chickens. *Poultry Science*, **35**, 843–51.

Glick, B. (1983). The Bursa of Fabricius. In *Avian Biology*, Volume VII, ed. D. S. Farner, J. R. King and K. C. Parkes, pp. 443–500. New York: Academic Press.

Goto, R., Miyada, C. G., Young, S., Wallace, R. B., Abplanalp, H., Bloom, S. E., Briles, W. E. & Miller, M. M. (1988). Isolation of a cDNA clone from the *B-G* subregion of the chicken histocompatibility (*B*) complex. *Immunogenetics*, **27**, 102–9.

Grant, J. A., Sanders, B. & Hood, L. (1971). Partial amino acid sequences of chicken and turkey immunoglobulin light chains. Homology with mammalian $\gamma$ chains. *Biochemistry*, **10**, 3123–32.

Guillemot, F., Billault, A., Pourquie, O., Behar, G., Chausse, A.-M., Zoorob, R., Kreibich, G. & Auffray, C. (1988). A molecular map of the chicken major histocompatibility complex: the class II $\beta$ genes are closely linked to the class I genes and the nucleolar organiser. *European Molecular Biology Organization Journal*, **7**, 2775–85.

Guillemot, F., Kaufman, J. F., Skjoedt, K. & Auffray, C. (1989). The major histocompatibility complex in the chicken. *Trends in Genetics*, **5**, 300–4.

Gunther, E., Balcarova, J., Hála, K., Rude, E. & Hraba, T. (1974). Evidence for an association between immune responsiveness of chicken to (T,G)-A-L and the major histocompatibility system. *European Journal of Immunology*, **4**, 548–53.

Gyles, N. R., Fallah-Moghaddan, H., Patterson, L. T., Skeeles, J. K., Whitfill, C. E. & Johnson, L. W. (1986). Genetic aspects of antibody responses in chickens to different classes of antigens. *Poultry Science*, **65**, 223–32.

Hála, K. (1977). The major histocompatibility system of the chicken. In *The Major Histocompatibility System in Man and Animals*, ed. G. Götze, pp.291–312. Berlin: Springer-Verlag.

Hála, K. (1988). Hypothesis: Immunogenetic analysis of spontaneous autoimmune thyroiditis in Obese strain (OS) chickens: A two-gene family model. *Immunobiology*, **177**, 354–73.

Hála, K., Boyd, R. & Wick, G. (1981a). Chicken major histocompatibility complex and disease. *Scandinavian Journal of Immunology*, **14**, 607–17.

Hála, K., Plachy, J. & Schulmanova, J. (1981b). The role of the B-G region antigen in the humoral immune response to the B-F region antigen of chicken MHC. *Immunogenetics*, **14**, 393–401.

Hála, K., Chaussé, A.-M., Bourlet, Y., Lassila, O., Hasler, V. & Auffray, C. (1988). Attempt to detect recombination between *B-F* and *B-L* genes within the chicken *B* complex by serological typing, *in vitro* MLR, and RFLP analyses. *Immunogenetics*, **28**, 433–8.

Hála, K., Sgonc, R., Auffray, C. & Wick, G. (1989). Typing of MHC haplotypes in OS chicken by means of RFLP analysis. In *Recent Advances in Avian Immunology Research*, pp. 177–86. New York: Alan R. Liss.

Houssaint, E., Belo, M. & LeDouarin, N. M. (1976). Investigations on cell lineage and tissue interactions in the developing bursa of Fabricius through interspecific chimeras. *Developmental Biology*, **53**, 250–64.

Jalkanen, S., Jalkanen, N., Granfors, K. & Toivanen, P. (1984). Defect in the generation of light-chain diversity in bursectomized chickens. *Nature*, **311**, 69–71.

Kincade, P. W. & Cooper, M. D. (1973). Immunoglobulin. Site and sequence of expression in developing chicks. *Science*, **179**, 398–400.

Kline, K., Briles, W. E., Bacon, L. & Sanders, B. G. (1988). Characterization of different B-F (MHC Class I) molecules in the chicken. *Journal of Heredity*, **79**, 249–56.

Koch, C. (1986). A genetic polymorphism of the complement component factor B in chickens not linked to the major histocompatibility complex (MHC). *Immunogenetics*, **23**, 364–7.

Kroemer, G., Bernot, A., Behar, G., Chaussé, A.-M., Gastinel, L.-N., Guillemot, F., Park, I., Thoraval, P., Zoorob, R. & Auffray, C. (1990). Molecular genetics of the chicken MHC: Current status and evolutionary aspects. *Immunological Reviews*, **113**, 119–145.

Lamont, S. J. (1989). The chicken major histocompatibility complex in disease resistance and poultry breeding. *Journal of Dairy Science*, **72**, 1328–33.

Lamont, S. J. & Dietert, R. R. (1990). Immunogenetics. In *Poultry Breeding and Genetics*, ed. R. D. Crawford, pp. 497–541. Amsterdam: Elsevier.

Lamont, S. J., Warner, C. M. & Nordskog, A. W. (1987). Molecular analysis of the chicken major histocompatibility complex gene and gene products. *Poultry Science*, **66**, 819–24.

LeDouarin, N. M., Houssaint, E., Jotereau, F. V. & Belo, M. (1975). Origin of hemopoietic stem cells in embryonic bursa of Fabricius and bone marrow studied through interspecific chimeras. *Proceedings of the National Academy of Sciences of USA*, **72**, 2701–5.

Linscott, W. D. (1986). Biochemistry and biology of the complement system in domestic animals. *Progress in Veterinary Microbiology and Immunology*, **2**, 54–77.

Lydyard, P., Grossi, C. E. & Cooper, M. D. (1976). Ontogeny of B cells in the chicken. I. Sequential development of clonal diversity in the Bursa. *Journal of Experimental Medicine*, **144**, 79–97.

Miller, M. M., Abplanalp, H. & Goto, R. (1988a). Genotyping chickens for the *B-G* subregion of the major histocompatibility complex using restriction fragment length polymorphisms. *Immunogenetics*, **28**, 374–9.

Miller, M. M., Goto, R. & Briles, W. E. (1988b). Biochemical confirmation of recombination within the *B-G* region of the chicken major histocompatibility complex. *Immunogenetics*, **27**, 127–32.

Nordskog, A. W. (1983). Immunogenetics as an aid to selection for disease resistance in the fowl. *World's Poultry Science Journal*, **39**, 199–209.

Parvari, R., Avivi, A., Lentner, F., Ziv, E., Tel-Or, S., Burstein, Y. & Schechter, I. (1988). Chicken immunoglobulin $\gamma$-heavy chains: limited VH gene repetoire, combinatorial diversification by D gene segments and evolution of the heavy chain locus. *European Molecular Biology Organization Journal*, **7**, 739–44.

Payne, L. N. (1971). The lymphoid system. In *Physiology and Biochemistry of the Domestic Fowl*, Volume 2, ed. D. J. Bell and B. M. Freeman, pp. 985–1038. London: Academic Press.

Pevzner, I., Nordskog, A. W. & Kaeberle, M. L. (1975). Immune response and the B blood group locus in chickens. *Genetics*, **80**, 753–9.

Pink, J. R. L., Droege, W., Hála, K., Miggiano, V. C. & Ziegler, A. (1977). A three locus model for the major histocompatibility complex. *Immunogenetics*, **5**, 203–16.

Pitcovski, J., Lamont, S. J., Nordskog, A. W. & Warner, C. M. (1989). Analysis of *B-G* and immune response genes in the Iowa State University S1 chicken line by hybridization of sperm deoxyribonucleic acid with a major histocompatibility complex class II probe. *Poultry Science*, **68**, 94–9.

Porter, R. R. (1973). Structural studies of immunoglobulins. *Science*, **180**, 713–6.

Reynaud, C.-A., Anquez, V., Dahan, A. & Weill, J.-C. (1985). A single rearrangement event generates most of the chicken immunoglobulin light chain diversity. *Cell*, **40**, 283–91.

Reynaud, C.-A., Anquez, V., Grimal, H. & Weill, J.-C. (1987a). A hyperconversion mechanism generates the chicken light chain preimmune repertoire. *Cell*, **48**, 379–88.

Reynaud, C.-A., Dahan, A., Anquez, V. & Weill, J.-C. (1989). Development of the chicken antibody repertoire. In *Immunoglobulin Genes*, ed. T. Honjo, pp. 151–62. New York: Academic Press.

Reynaud, C.-A., Dahan, A. & Weill, J.-C. (1983). Complete sequence of a chicken $\lambda$ light chain immunoglobulin derived from the nucleotide sequence of its mRNA. *Proceedings of the National Academy of Sciences of USA*, **80**, 4099–103.

Reynaud, C.-A., Dahan, A. & Weill, J.-C. (1987b). A gene conversion program during the ontogenesis of chicken B cells. *Trends in Genetics*, **3**, 248–51.

Robertson, M. (1982). The evolutionary past of the major histocompatibility complex and the future of cellular immunology. *Nature*, **297**, 629–32.

Roitt, I., Brostoff, J. & Male, D. (1985). *Immunology*, Chapters 1–9. Edinburgh: Churchill Livingstone.

Rose, N. R., Bacon, L. D. & Sundick, R. S. (1976). Genetic determinants of thyroiditis in OS chickens. *Transplantation Reviews*, **31**, 264–85.

Schierman, L. W. & Collins, W. M. (1987). Influence of the major histocompatibility complex on tumor regression and immunity in chickens. *Poultry Science*, **66**, 812–8.

Schierman, L. W. & Nordskog, A. W. (1961). Relationship of blood type to histocompatibility in chickens. *Science*, **134**, 1008–9.

Steadham, E. M., Lamont, S. J., Kujdych, I. & Nordskog, A. W. (1987). Association of Marek's disease with Ea-B and immune response genes in subline and $F_2$ populations of the Iowa State S1 Leghorn line. *Poultry Science*, **66**, 571–5.

Somes, R. G. (1988) *International Registry of Genetic Stocks*, Bulletin 476. Storrs, Ct.: University of Connecticut.

Suran, A. A. & Papermaster, B. (1967). N-terminal sequences of heavy and light chains of leopard shark immunoglobulins: evolutionary implications. *Proceedings of the National Academy of Sciences of USA*, **58**, 1619–23.

Thompson, C. B. & Neiman, P. E. (1987). Somatic diversification of the chicken immunoglobulin light chain gene is limited to the rearranged variable gene segment. *Cell*, **48**, 369–78.

Tilanus, M. G. J., Ginkel, R. V., Engelen, I., Rietveld, F. W., Hepkema, B. G., Van der Zijpp, A. J. Egberts, E. & Blankert, H. (1989). The chicken B-complex in commercial pure lines: comparison of polymorphisms defined by serotyping and DNA analysis. *Recent Advances in Avian Immunology Research*, pp. 187–96. New York: Alan R. Liss.

Warner, C., Gerndt, B., Xu, Y., Bourlet, Y., Auffray, C., Lamont, S. & Nordskog, A. (1989). Restriction fragment length polymorphism analysis of major histocompatibility complex class II genes from inbred chicken lines. *Animal Genetics*, **20**, 225–31.

Weill, J.-C, Reynaud, C.-A., Lassila, O. & Pink, J. R. L. (1986). Rearrangement of chicken immunoglobulin genes is not an ongoing process in the embryonic bursa of Fabricius. *Proceedings of the National Academy of Sciences of USA*, **83**, 3336–40.

Weill, J.-C. & Reynaud, C.-A. (1987). The chicken B cell compartment. *Science*, **238**, 1094–8.

Wick, G., Brezinschek, H. P., Hála, K., Dietrich, H., Wolf, H. & Kroemer, G. (1989). The obese strain of chickens: an animal model with spontaneous autoimmune thyroiditis. *Advances in Immunology*, **47**, 433–500.

Williamson, A. R. & Turner, M. W. (1987) *Essential Immunogenetics*. Oxford: Blackwell Scientific Publications.

Wysocki, L. J. & Gefter, M. L. (1989). Gene conversion and the generation of antibody diversity. *Annual Review of Biochemistry*, **58**, 509–31.

Zaleski, M. B., Dubiski, S., Niles, E. G. & Cunningham, R. K. (1983). *Immunogenetics*. Boston: Pitman.

Ziegler, A. & Pink, R. (1976). Chemical properties of two antigens controlled by the major histocompatibility complex of the chicken. *Journal of Biological Chemistry*, **251**, 5391–6.

Zijpp, A. J. (1983). Breeding for immune responsiveness and disease resistance. *World's Poultry Science Journal*, **39**, 118–31.

# 12

# *Gene cloning, sequencing and transfer in the domestic fowl*

## 12.1 Introduction

Since the early 1970s it has become possible to isolate DNA and fragment it in a highly specific manner into pieces comparable in size to that of a gene, and then by joining such pieces to a suitable vector, which is normally viral DNA or a plasmid, to insert the chimaeras so formed into another species or organism. Over the last decade this technology, generally referred to as **genetic engineering**, has become refined and many of the initial technical difficulties overcome. This technology triggered what is generally regarded as the third rapid advance in the understanding of genetics (the first being the rediscovery of Mendel's work in 1901, and the second the advent of molecular genetics which began about 1940). It is now potentially possible to isolate a gene and transfer it either into another individual of the same species or into a different species. Further, by transferring it to a host with a short generation time, e.g. a bacterium such as *E. coli* with a generation time as short as 20 min, multiple copies or **clones** of the single gene may be produced. The much greater number of gene copies then available for isolation means that detailed structural studies can be carried out on the gene, including the determination of the sequence of its nucleotides. It is also possible, knowing the DNA sequence to carry out **site-directed mutagenesis**. This means that a gene can be structurally modified in a highly specific manner. Site-directed mutagenesis has both theoretical and practical applications: it can be used to understand the importance of a particular amino acid in a given position in a protein, e.g. the active site of an enzyme, and changes in the amino acid sequence of a protein may lead to it performing more efficiently.

As far as the domestic fowl is concerned there are two developments of importance to date: (i) as a result of sequencing those genes that it has been possible to clone, a better understanding of gene and chromosome organisation has been gained, and (ii) by incorporating particular genes into the domestic fowl it may be possible in future to improve its growth rate and resistance to disease and other desirable qualities. The latter procedure is usually referred to as **transgenics**.

In this chapter the technology of gene transfer is described first, followed by a consideration of the genes that have been cloned and their sequences determined. Finally the potential for 'improved performance' by transgenics is discussed.

## 12.2 Methods for gene manipulation in the domestic fowl

A brief outline of the methods used for gene manipulation is given here, but for further details the monographs by Brown (1986), Old & Primrose (1989) or Kingsman & Kingsman (1988) are strongly recommended. Gene manipulations in the domestic fowl can be divided into three categories:
(a) Cloning genes for DNA sequencing.
Genes or fragments of DNA from the domestic fowl can be cloned in a bacterial host such as *E. coli*, and then, because the latter multiplies so much faster than the fowl, relatively large numbers of copies of the DNA are available for sequence analysis in a short time.
(b) Cloning cDNA.
Messenger RNA from a particular tissue of the domestic fowl can be used to generate a cDNA copy, and the latter cloned in a bacterial host as in (a). The mRNA used in this technique may be the total mRNA from the tissue concerned, or it may be a particular purified mRNA. Using this procedure only genes that are expressed in the tissue under investigation will be cloned.
(c) Transgenics.
DNA fragments isolated from a particular strain of domestic fowl, or potentially useful genes from another species, may be inserted into a different strain of domestic fowl with the aim of improving the latter.

The process of **gene cloning** can be divided into the following stages:

1. Obtaining the gene or piece of DNA to be cloned.
2. Ligating (joining) the gene or piece of DNA to a suitable vector such as a plasmid or viral DNA to form a chimaera.
3. Inserting the chimaera into the host cell and allowing it to replicate.
4. Selection of host cells containing the cloned DNA.

Each will now be considered in turn.

### *12.2a DNA extraction and fragmentation*

DNA is extracted from the appropriate cells. It is fragmented into suitably sized pieces using restriction endonucleases. These enzymes occur naturally in bacteria and generally recognise a particular sequence of 4–6 bases in DNA (Fig. 12.1); over 1000 enzymes of known specificities have now been isolated (Roberts, 1989). Each is given an abbreviated name after the species of bacteria from which they have been isolated, e.g. *EcoR1* from ***Escherichia coli***, and *Hin* from ***Haemophilus influenzae***.

The frequency with which a particular sequence of six bases, e.g. GCATAG, might be expected to recur in a DNA molecule, on the basis of random distribution of the bases, is 1 in $4^6$ (= 1 in 4096). A sequence of 4096 nucleotides, if it were all part of a coding sequence, would be the gene for a protein having 1365 amino acids (triplet genetic code, 4096/3 = 1365.3). This would be a large protein. Of course it should be realised that the fragmentation is very unlikely to occur exactly at the beginning of a coding sequence, but the sizes of DNA fragments formed are of the same order of magnitude as genes. These DNA fragments are then joined either to a suitable viral DNA, or to a plasmid, and the chimaera so formed is used for cloning in the host bacterium (Fig. 12.2). Plasmids are small circular pieces of DNA which have the capability of replicating independently of the main chromosome. They are most well characterised in bacteria such as *E. coli*, but suitable ones for cloning purposes have also been found in lower eukaryotes such as yeasts. The larger plasmids usually promote conjugation in the bacterium, and by this means they are able to spread throughout the culture. The smaller non-conjugative plasmids do not promote conjugation, but can nevertheless become dispersed through the culture if they are present together with a conjugative plasmid.

### *12.2b Choice of plasmid or virus as vector*

The choice of a particular plasmid or viral vector depends on a number of factors. One of these is the size of the DNA fragment to be cloned. It is easier to isolate small DNA molecules, e.g. < 10–20 kilobases (kb) than large ones. The latter are much more liable to degrade during isolation.

**Fig. 12.1** The mode of action of restriction endonucleases. The example given here is the use of a particular restriction endonuclease known as EcoR1 that recognises the sequence within DNA, of six bases GAATTC and cuts between G and A.

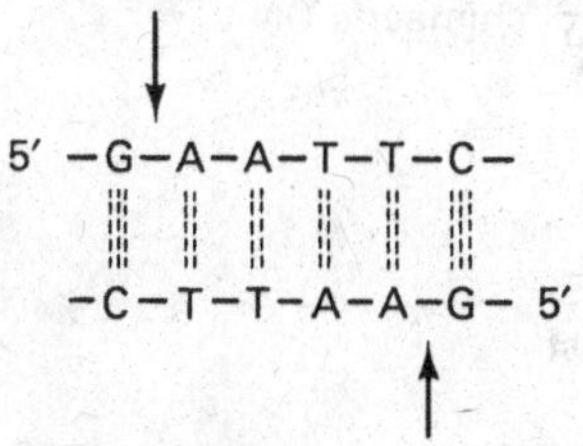

```
   ↓
5′ −G−A−A−T−T−C−
    ||| || || || || |||
   −C−T−T−A−A−G− 5′
                ↑
```

Thus, if a plasmid is used, it is almost invariably a small one, usually 5–10 kb. With such a plasmid it is feasible to insert up to 5 kb of the DNA to be cloned. The efficiency falls off if larger pieces of DNA are cloned. The desirable properties of a plasmid are: (i) its small size which is usually preferable, (ii) possession of markers, usually conferring antibiotic resistance, that enable cells containing the plasmid to be distinguished from those that lack it, and (iii) ability to replicate so as to generate a large number of copies of the plasmid per cell.

The plasmid most frequently used is **pBR322** (p = plasmid, BR = Bolivar & Rodriguez, the developers of this plasmid). This plasmid contains many desirable features, which were incorporated into it during its construction from other plasmids by DNA manipulation. It is small and easily isolated. Under normal conditions up to 50 copies are produced per bacterium. However, addition of chloramphenicol inhibits the bacterial replication without affecting the replication of the plasmid, so permitting the number of copies per bacterium to rise to 1000–3000. Plasmid pBR322 has two markers for antibiotic resistance, one for ampicillin and one for tetracycline. Thus bacteria containing this plasmid will be resistant to both ampicillin and tetracycline. Both genes for resistance to these antibiotics contain unique sites for fragmentation by specific restriction endonuc-

**Fig. 12.2** Steps involved in cloning a DNA fragment.

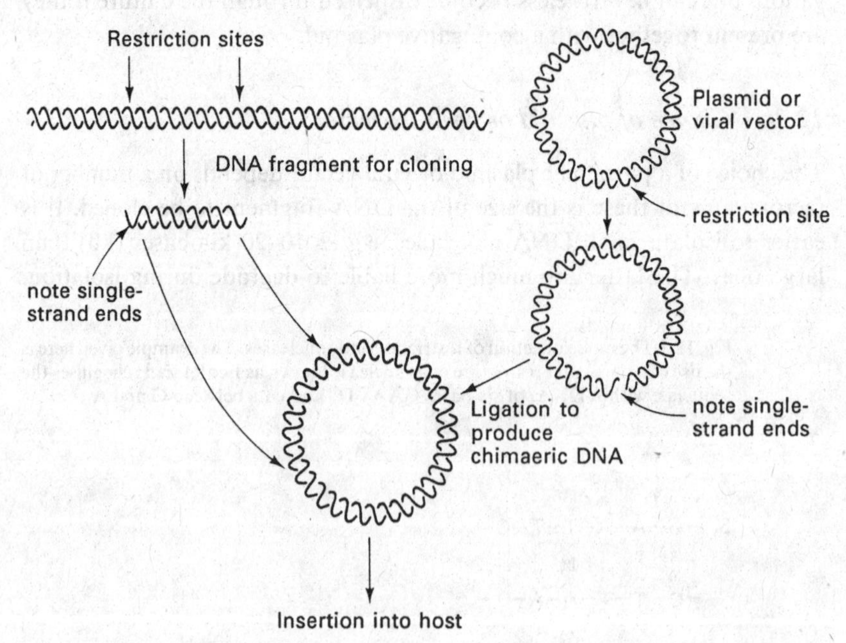

leases. Thus, if a piece of new DNA is incorporated into pBR322 at one of these sites, it will lose its resistance to one of the antibiotics and therefore can be distinguished from the parent pBR322 (Fig. 12.3).

Other plasmids referred to in Table 12.1 below are pMB9 and pUC8. pMB9 is one of the parent types which were used to construct pBR322. It has the tetracycline resistance gene but not the ampicillin resistance gene. pUC8 is a plasmid derived from pBR322 which still contains the gene for ampicillin resistance but also contains a part of the enzyme β-galactosidase. The latter gene contains a number of restriction endonuclease sites suitable for inserting DNA. Bacteria containing normal pUC8 will have ampicillin resistance and β-galactosidase activity, whereas those having pUC8 containing a new DNA insert will lack the β-galactosidase activity. β-galactosidase activity is easily measured by a colour reaction and thus the different bacterial colonies can be readily distinguished experimentally.

The viral vectors used with *E. coli* as host are usually derivatives of the bacteriophages λ or M13. The size of the λ genome is 49 kb. This can be increased by insertion of new DNA up to 52 kb, but beyond this size the genome is too large for packaging into its protein coat. However, it is possible to excise some of the λ genes which are not essential for its lytic growth cycle. In this way space can be made for up to a further 18 kb of new DNA. The normal wild type λ has a large number of restriction sites for endonucleases. This means that if the genome is incubated with a particular

**Fig. 12.3** Two plasmids frequently used in DNA cloning. pBR322 has two sites for antibiotic resistance. If DNA is cloned using one of the three restriction sites (Sca1, Pvu1, or Pst1) in the ampicillin resistance gene (*ampR*), then resistance to ampicillin is lost. pUC8, which is derived from pBR322, still has the ampicillin resistance gene, but also contains part of the β-galactosidase gene (*lacZ*). The latter contains a variety of restriction sites (HindIII, Pst1, Sal1, Ace1, HincII, BamH1, Sma1, Xma1 & EcoR1) that can be used to clone DNA. Insertion of DNA then results in lack of β-galactosidase activity, enabling new chimaerae to be detected.

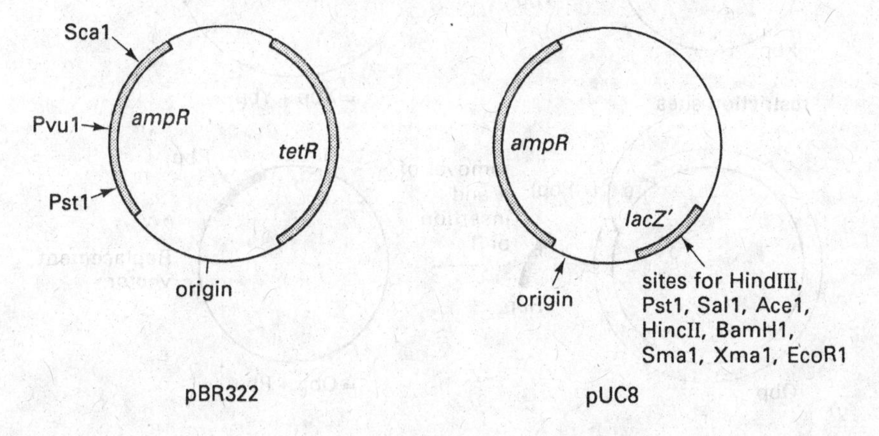

restriction endonuclease it will fragment into five or six pieces, which would be unsatisfactory for cloning. What is required is a λ having either one or two restriction sites in the non-essential regions. By a process of selection, strains of λ have been obtained with only one or two restriction sites for the most frequently used endonucleases. The λ vectors are categorised as either **insertion vectors** or **replacement vectors** (Fig. 12.4).

Insertion vectors have a large non-essential region which is deleted, and a unique restriction site that can be used for insertion of new DNA. The size limit of the DNA to be inserted will depend on how much of the viral DNA has initially been deleted. A replacement vector has two endonuclease recognition sites that flank the DNA to be replaced. Replacement vectors can be used for insertion of larger pieces of DNA. The λ derivatives given in Table 12.1 below are *λgt10*, *λgt11*, λCharon 4A and *λL47*. **λgt10** and **λgt11** are insertional vectors (gt = generalised transducer). Both have genomes too small to be packaged, and this ensures that only when new DNA has been inserted can they be packaged and thus form plaques. *λgt11* also carries the gene for the α-peptide for β-galactosidase. This has a restriction

**Fig. 12.4** Examples of insertion and replacement vectors. Insertion vectors are smaller than the original parent vector and have a single restriction site that is used to incorporate the DNA to be cloned, hence increasing its size (to X bp + Y bp). Replacement vectors typically have two restriction sites which enable a non-essential region (= Pbp) of the virus DNA to be removed and replaced by the DNA (= Rbp) to be cloned; the size of the vector becomes Q bp + R bp.

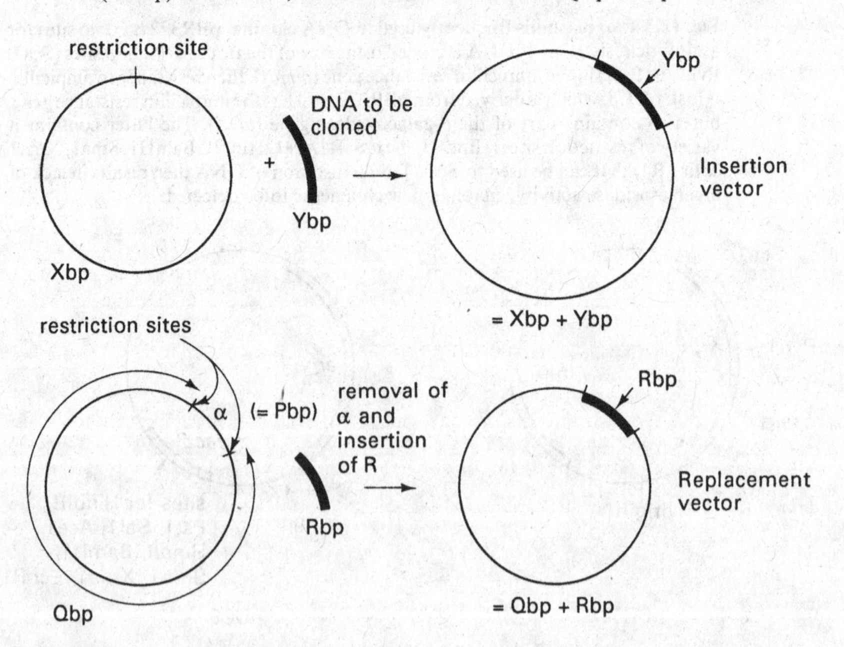

site, as was mentioned previously in connection with pUC8, and thus insertion of a DNA fragment splits the region coding for the $\beta$-galactosidase gene. $\lambda$**Charon 4A** is one of a class of $\lambda$Charon vectors, that have been designed with biological containment in mind. They were named after Charon, the ferryman in Greek mythology who conveyed the spirits of the dead across the River Styx. $\lambda$Charon 4A has two chain termination mutations (amber mutations). This means that it can replicate only in a particular strain of *E. coli* that contains suppressors of these two mutations, thereby overriding the chain termination. It would be unable to replicate in a wild type *E. coli*. $\lambda$**L47** is a high capacity $\lambda$ vector which can accommodate between 6 and 24 kb DNA.

Another vector that can be used if larger pieces of DNA are to be cloned is a **cosmid**. A cosmid is a hybrid constructed from a plasmid and parts of $\lambda$ phage. It is basically a plasmid containing a marker, together with the genes essential for packaging the DNA (*cos* sites) from the phage. Cosmids can allow packaging of up to 40 kb of DNA. The example of the rRNA genes listed in Table 12.1 entails using the cosmid **pHC79** (HC = Hohn & Collins, who constructed the cosmid) to clone a 40 kb DNA fragment.

Another phage sometimes used is M13. It is a small bacteriophage less than 10 kb. Its main advantage is that it is single-stranded. Single-stranded DNA is required for determination of nucleotide sequences, and also for *in vitro* mutagenesis. If double-stranded DNA has been cloned, the strands must be separated before sequencing and thus use of M13 or or its derivatives is often more convenient.

### *12.2c Clone selection*

Having made the clones, the next problem is to select the one containing the gene required. Consider again the initial fragmentation of DNA. The amount of DNA in the haploid cells of the domestic fowl is *c.* $1.3 \times 10^6$ kb. Assuming the average size of fragment from a restriction endonuclease digest is *c.* 4000 bp, there would be *c.* $3 \times 10^5$ different fragments. If each of these were successfully combined with a vector there could theoretically be $3 \times 10^5$ clones from which to select the one required. This is a rather large number to screen, and also it is possible that cleavage may have occurred through the middle of the gene required. A frequent solution to this problem is to make a partial digest using two endonucleases having different specificities. The digest is continued until the average size of fragment is about 20 kb and these large fragments are separated from those which differ substantially from that size. The advantages of using this type of digest are (i) there are few clones to screen, (ii) each clone is likely to contain more than one gene, (iii) there will be overlapping fragments since two different enzymes have been used, (iv) having large and overlapping fragments make it possible to find out which genes, regulatory sequences, and introns lie 'upstream' and 'downstream' of the coding sequence. This complete set of fragments, each ligated to a vector, constitutes a **gene library**.

Another method that also produces fewer clones to screen is to create a **cDNA library**. If the mRNA from a particular tissue is isolated, it will contain the messages for all the genes being expressed in that tissue. This will be considerably less than the total number of genes constituting the genome, since during differentiation only certain genes are normally expressed in given tissues. Using the enzyme reverse transcriptase it is possible to make a DNA copy complementary to the mRNA **(cDNA)**. The cDNAs may be linked to a suitable vector and these collectively constitute a cDNA library. This will differ from a gene library in that it will lack introns (see Chapter 2, section 2.1).

Having obtained either a gene library or cDNA library, it is then necessary to identify the clone required. Three types of probe can be used for this purpose. The first is to use a polynucleotide probe such as the mRNA for the gene to be selected. This method will be possible only where a particular mRNA is present in great abundance in a particular cell type. For example, immature erythrocytes produce a high proportion of globin mRNA since they are specialised cells able to produce very large quantities of haemoglobin. Similarly, the hen oviduct produces a high proportion of mRNA for the egg-white proteins, ovalbumin and ovomucoid. The clones are plated out on agar and then blotted out on to nitrocellulose filters. Radioactively labelled mRNA complementary to the required gene, then mixed on the filter under suitable conditions, will hybridise with any clones containing those gene sequences. By matching the nitrocellulose filter with the agar plate it is possible to isolate the appropriate clone. This method was devised by Southern and is known as Southern blotting (see Fig. 12.5). Using a mRNA probe is feasible only in selected cases. However, if the amino acid sequence or part of the sequence is known for the gene product, it is then possible, applying the genetic code, to synthesise an oligonucleotide probe and use it in the Southern blotting procedure. The third possibility, if the sequence is not known, but that of a related protein or homologous protein from a different species is known, is to use the latter as a probe as it may show sufficient similarity to hybridise with the required clone. All three methods entail using a polynucleotide or oligonucleotide probe to hybridise to the DNA present in the clone of interest.

Another method of screening is immunological screening. In this case, if the gene product, i.e. the protein, has been purified but no sequence data are available, it is possible to raise antibodies to the protein in question. The antibodies may then be used to screen the clones provided they have been ligated to an **expression vector**, e.g. *λgt11*. Very often genes from one organism, when successfully transferred to a different organism, are not generally expressed, although they are replicated along with the host's DNA. The reason is that different promoters are used in different species and there is no interspecies recognition if they are totally unrelated organisms. Thus, in order to ensure that foreign DNA is expressed, it is necessary that the viral or plasmid vector contains the promoter sequences for the host organism. Such vectors are known as expression vectors.

Having outlined the methods used for gene cloning, Table 12.1 illustrates the genes from the domestic fowl that have been cloned in *E. coli*. For a more extensive list of cloned genes, see Bulfield (1990).

## 12.3 Genes cloned and sequenced in the domestic fowl

Estimates of the number of genes in the domestic fowl are in the region of 50 000–100 000 and by 1988 there were over 350 DNA sequences in the GENBANK system and over 100 genes had been cloned (Gavora, 1988). The number of genes cloned can be expected to increase steadily now that the technology is well developed. With so many genes to choose from, what has determined which genes have been sequenced so far? A number of factors are involved, the first of which is practicability. To select a gene from a library requires a suitable probe. From the previous section it will be clear that this requires a suitable polynucleotide, which could be the mRNA, the gene from a related organism, or a synthetic oligonucleotide arrived at from a knowledge of the protein sequence. Thus, for the proteins from the oviduct and the erythrocyte where particular mRNAs predominate,

**Fig. 12.5** The technique of Southern blotting.

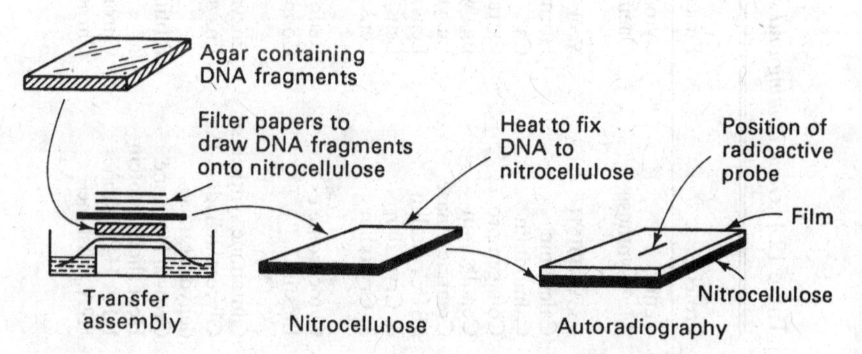

Table 12.1. *Gene cloning and sequencing in the domestic fowl* (Gallus domesticus)

| Protein | Function | Cloning vector | Sequence | Number of introns | Reference |
|---|---|---|---|---|---|
| Actin | Cytoskeleton | λCharon 4A | P | | Chang *et al.*, 1984 |
| ALA synthase | Haem biosynthesis | λgt | C | 10 & 11 | Yamamoto *et al.*, 1985; Riddle *et al.* 1989 |
| apoVLDL II | Yolk protein | | | 3 | Kok *et al.*, 1985 |
| Calcitonin | Hormone | pBR322 | C | | Lasmoles *et al.*, 1985 |
| Calmodulin | $Ca^{++}$ binding | λCharon 4A | P | 7 | Simmen *et al.*, 1985 |
| Collagen pro $\alpha 2$ | Extracellular matrix | λ Charon 4A | C | 51 | Aho *et al.*, 1984 |
| Conalbumin | Egg white | pBR322 | | 16 | Cochet *et al.*, 1979 |
| Creatine kinase | Enzyme | pBR322 | C | | Kwiatkowski *et al.*, 1985 |
| $\beta$B1 Crystallin | Lens protein | λCharon 4A | C | | Fielding Hejtmancik *et al.*, 1986 |
| $\delta_1$ Crystallin | Lens protein | λ Charon 4A | C | 16 | Hawkins *et al.*, 1984 |
| $\delta_2$ Crystallin | Lens protein | λ Charon 4A | C | 16 | Ohno *et al.*, 1985; Nickerson *et al.*, 1986 |
| Cytochrome *c* | Redox protein | λ Charon 4A | C | 1 | Limbach & Wu, 1983 |
| Elastin | Connective tissue | pEX1 | C | | Raju & Anwar, 1987; Bressan *et al.*, 1987 |
| Glutamine synthetase | Enzyme | λgt10 | C | 7 | Pu & Young, 1989 |
| Glyceraldehyde-3-P dehydrogenase | Enzyme | λCharon 4A | C | 11 | Stone *et al.*, 1985 |
| Growth hormone | Pituitary hormone | λgt10 | C | | Lamb *et al.*, 1988 |
| Heat shock protein | ? | pUC8 | P | 1 | Catelli *et al.*, 1985 |
| H1 & H4 Histones | Chromosomes | λ Charon 4A | C | 0 | Sugarman *et al.*, 1983 |
| $H2A._f$ Histone | Chromosome | pUC8 | C | 4 | Dalton *et al.*, 1989 |

| | | | | | |
|---|---|---|---|---|---|
| Ig λ light chain | Antibody | pBR322 | C | | Reynaud *et al.*, 1983 |
| Lysozyme | Egg white | λCharon 4A | C | 3 | Jung *et al.*, 1980 |
| Nicotinic acetyl choline receptor (δ & γ subunits) | Nerve transmission | λL47 | C | 11 | Nef *et al.*, 1984 |
| Ovalbumin | Egg white | pMB9 | C | 7 | McReynolds *et al.*, 1978 |
| Protamine | Sperm chromosome | pWE-15 | C | 0 | Oliva & Dixon, 1989 |
| Proteoglycan core | Cartilage | λEMBL3 | C | 5 | Tanaka *et al.*, 1988 |
| Ovomucoid | Egg white | | P | 7 | Stein *et al.*, 1980 |
| Ribosomal RNA (18S & 28S) | Protein synthesis | pHC79 | | 0 | Mattaj *et al.*, 1982 |
| α Tropomyosin | Muscle | pUC8 | C | | Helfman *et al.*, 1984 |
| Troponin T | Muscle | pBR322 | C | | Cooper & Ordahl, 1984 |
| Vimentin | Cytoskeleton | λCharon | C | 8 | Zehner *et al.*, 1987 |
| Vinculin | Microfilament junction protein | λgt11 | C | | Coutu & Craig, 1988 |
| Vitellogenin II | Egg yolk | λ Charon 4A | C | 34 | Schip *et al.*, 1987 |
| Vitellogenin III | Egg yolk | | C | | Silva *et al.*, 1989 |

C, complete sequence; P, partial sequence.

obvious probes are available. Other factors in the choice are: (i) genes for proteins important in developmental processes, both normal or neoplastic, (ii) genes for regulatory proteins, e.g. insulin, calmodulin, and (iii) genes for groups of proteins having close evolutionary relationships. So far those that have been sequenced fall into the categories of enzymes, oncogenes, endogenous proviruses and tissue specific proteins. Examples of genes that have been sequenced are considered below.

### *12.3a Crystallins*

Crystallins are the principal proteins present in the lens of the eye, making up between 80 and 90% of the total soluble protein. They account almost entirely for the refractive properties of the lens. The ability of the lens to focus incoming light on the retina depends on the concentration and position of the different crystallins within the fibre cells (Fig. 12.6). Sun *et al.* (1984) have shown that there is a smoothly increasing concentration of crystallins from the periphery to the centre of the lens. The protein concentration in a lens may range from 3 to 510 mg/ml (Delaye & Tardieu, 1983). There are four major families or types of crystallins, designated $\alpha$, $\beta$, $\gamma$ and $\delta$, and several less common, e.g. $\epsilon$, $\rho$ and $\tau$. All vertebrates have $\alpha$- and $\beta$-crystallins, but $\gamma$-crystallin, which is present in mammals, is replaced by $\delta$-crystallin in birds and reptiles. The $\epsilon$-crystallin is a major component of the duck lens, $\rho$-crystallin occurs in frog lens, and $\tau$-crystallin has been found in turtle and lamprey. Each has distinct immunological properties although sequence studies indicate homology between $\beta$ and $\gamma$ and these are therefore grouped as a superfamily (Piatigorsky, 1984a). There are at least two types of $\alpha$, seven types of $\beta$ and $\gamma$ together, and two $\delta$.

**Fig. 12.6** The arrangement of cells in the lens.

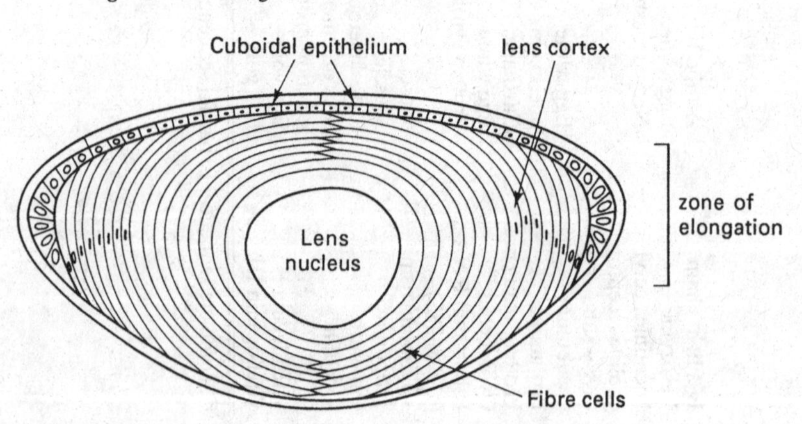

There are several reasons for studying the crystallin proteins. The lens is a good system for studying development, since it is composed almost entirely of fibre cells derived from a single type of epithelial cell (Bloemendal, 1977). Since the crystallins are abundant within the lens the predominant mRNAs in the epithelial cells will be those for crystallins. The structure of the crystallins is of interest since they account for the highly refractive properties of the cells, and do not cause appreciable light scattering as many highly concentrated protein solutions do (Delaye & Tardieu, 1983). This has imposed a constraint on their evolution and it is found that they are highly conserved. The possession of δ-crystallins in birds may be related to their visual acuity which requires extensive accommodation. This in turn requires a soft, more readily deformed lens, which must be highly hydrated. For details of the structure of crystallin in relation to function, see Slingsby (1985) and Wistow & Piatigorsky (1988) and for their genetics, see Yasuda & Okada (1986).

In avian species the domestic fowl has been studied in most detail, although comparative studies have been made of the electrophoretic properties of crystallins from many other species (Gysels, 1964). In the Galliformes δ-crystallin makes up more than 50% of the total crystallin of the lens. It is quite distinct from the other crystallins in having no cysteine residues, but having abundant leucine and a high α-helical content (Nickerson & Piatigorsky, 1984). δ-Crystallin is the crystallin that has been most studied in the domestic fowl. It is the first one to be detected in the embryonic lens at 4 days incubation. Its concentration increases to a plateau at 19 days incubation. After hatching its rate of synthesis declines and Treton, Shinohara & Piatigorsky (1982) have shown that mRNA for δ-crystallin disappears from the fibre cells during the third to the fifth month. Because of the temporal sequence of its synthesis in relation to the development of the lens most of the δ-crystallin is concentrated in the lens nucleus.

It has been shown that δ-crystallins occur as two non-allelic forms, and both have been sequenced from a domestic fowl gene library (Hawkins *et al.*, 1984; Ohno *et al.*, 1985). The sequences of the two genes, $\delta_1$ and $\delta_2$, are highly homologous (*c.* 90%) (Nickerson *et al.*, 1986), and are arranged in tandem on the chromosome separated by a 4 kb intergenic sequence. As can be seen in Fig. 12.7, both comprise 17 exons and 16 introns. Having determined the complete sequences of the genes and their arrangement on the chromosome, it is possible to examine in more detail how their expression is controlled.

Messenger RNA isolated from the lens of the domestic fowl seems to contain mRNA for $\delta_1$ but not for $\delta_2$, suggesting that the $\delta_1$ gene is much more active than that of $\delta_2$. Recently, regulatory sequences for the $\delta_1$ gene have been studied (Das & Piatigorsky, 1986). They have shown that the $\delta_1$ gene can be expressed *in vitro* in Hela cell extracts (certain cultured cells of human cancerous origin) but the expression requires the presence of region $-121$ to $-38$ (i.e. the nucleotides 121 to 38) before the coding region on the gene for RNA polymerase can recognise it. Kondoh *et al.* (1986) have also shown that δ-crystallin genes are expressed when microinjected into mouse cells. They found that expression was much more efficient if they used murine lens tissue than any other type of murine cells. This is particularly interesting since δ-crystallin genes are not present in mouse, yet the basis for recognition of δ genes is *c.* 100 times greater in murine lens than in the other murine tissues. They have shown that a 150 bp region encompassing the transcription initiation site is responsible for the high expression characteristic of lens cells. The duck also has two similar δ crystallin genes suggesting duplication before the divergence of the avian species (Piatigorsky, 1984b).

Four cDNAs have been isolated for β-crystallins in the domestic fowl (Fielding *et al.*, 1986). One of these is of interest from the developmental standpoint. βB1 (previously designated β-35) is specifically expressed in the elongated fibrous cells and not in the cuboidal epithelial cells. It has been cloned and sequenced and found to be highly homologous with the β genes from mammalian tissues. The amino acid sequence is consistent with it having a two domain structure including four Greek key supersecondary structural motifs (for details of secondary structure of proteins, see Creighton, 1983). A β-A crystallin has also been sequenced (Peterson & Piatigorsky, 1986) and the amino acid sequence deduced from the nucleotide sequence is also consistent with the two domain structure, each having two Greek key motifs.

The α-crystallin is the last to be synthesised in the lens during embryogenesis in the domestic fowl (Zwaan & Ikeda, 1968). The complete nucleotide sequence has been determined, together with the 5′ flanking region which is responsible for the control of its expression in the lens

**Fig. 12.7** Organisation of the δ-crystallin genes in the domestic fowl. The exons are numbered and shown as black rectangles with the introns as the lines in between (Nickerson *et al.*, 1986).

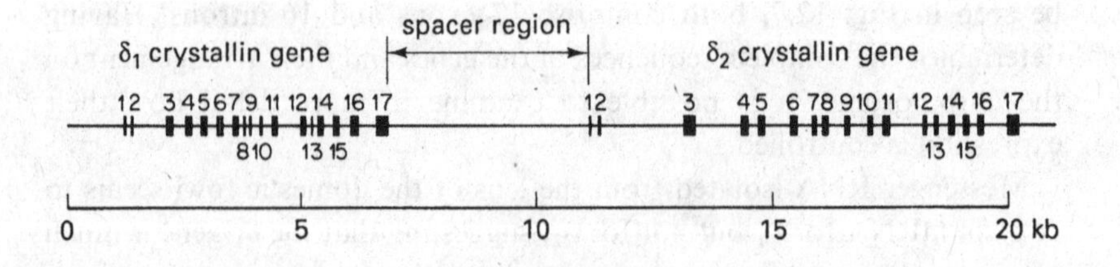

(Thompson, Hawkins & Piatigorsky, 1987). It comprises three exons and two introns.

An unexpected finding in connection with the structure of the crystallins is their relationship to other proteins. The $\alpha$-crystallins closely resemble a group of proteins known as heat shock proteins (Wistow, 1985). The latter occur in both prokaryotes and eukaryotes and are synthesised in response to heat shock, and may have a number of cell functions (see Lindquist, 1986). The $\beta$- and $\gamma$-crystallin family resembles a calcium binding bacterial spore coat protein, but the most surprising relationship is perhaps that of crystallins to certain metabolic enzymes. These latter include argininosuccinate lyase (ASL), lactate dehydrogenase, enolase, glutathione *S*-transferase, and various NADPH-dependent reductases (Piatigorsky *et al.*, 1988). These similarities suggest that the same gene may encode both functions, i.e. the lens function and the metabolic function.

The $\delta$-crystallin from the domestic fowl and the enzyme ASL, important in the urea cycle, are very similar. The $\delta$-crystallin from duck has high ASL activity, and that from the domestic fowl moderate ASL activity. Piatigorsky *et al.* (1988) have shown that cDNA from the domestic fowl $\delta_1$ crystallin and human genomic DNA both hybridise with the human ASL gene (N.B. mammals, including humans lack $\delta$-crystallin as mentioned above). Further evidence for the relationship between $\delta$-crystallin and ASL comes from the structure of the ASL gene from rat liver (Matsubasa *et al.*, 1989). The latter is found also to be split into 16 exons, and all the introns interrupt the coding sequence at the same positions as in $\delta$-crystallins. Collectively, these data on the similarity of crystallin to other functionally unrelated proteins suggest that they have arisen by divergent evolution. Crystallins have evolved very slowly (only 3% change in sequence in $10^8$ years) and this may be in part due to high selective pressure, i.e. selection for two distinct lens and enzymic functions. It is also of interest in relation the the neutralist theory of evolution and random genetic drift (see Chapter 1, section 1.2); random drift would seem less probable if genes are evolving for multiple functions.

### *12.3b Contractile proteins*

A number of proteins are involved, either directly or indirectly in the contractile processes. These may be found in muscle fibres or in other cells of the body (see Chapter 7, section 7.1). In the latter they are involved in the maintainence of, or in changing the shape of, the cell, in spindle formation during mitosis, or in other processes involving cell motility. It is believed that many of the contractile proteins in muscle cells may have evolved from the same ancestral types as those present in the cytoskeleton of other cell types. One of the most studied of these is the actins. $\alpha$-Actins occur only in myogenic cells, whereas $\beta$- and $\gamma$-actins occur in all cells of the body (see Chapter 7, section 7.1). All three types have been sequenced and the $\alpha$-actin differs from the others in its N-terminal sequence and also in that it has a lower isoelectric point. There are also, however, heterogeneities within the three basic types. In the domestic fowl there are at least six types of actin and the possibility of further minor components cannot be excluded. The six forms include three types of $\alpha$-actin and two of $\gamma$-actin. The $\alpha$-actins have been sequenced and there are small differences among the $\alpha$-actins from skeletal muscle, cardiac muscle and smooth muscle (see Chang *et al.*, 1984). In addition, Saborio *et al.* (1979) have shown that the $\gamma$-actin of the gizzard of domestic fowl differs from that of non-muscle cells.

Thus, in analysing the DNA for actin genes from domestic fowl, at least six might be expected. Schwartz & Rothblum (1980) isolated mRNA from the muscle of 3-week-old chicks. This was purified and shown in an *in vitro* translation system to be that for actin. The mRNA was used to generate cDNA and the latter was used as a hybridisation probe to determine the gene dosage. The results indicated that there are 10 or 11 copies per genome. Chang *et al.* (1984) were able to isolate clones from a domestic fowl gene library corresponding to each of the six types of actin. They carried out partial nucleotide sequencing on each of these, which enabled them to identify each.

Having cloned the actin genes (Chang *et al.*, 1984) examination of their regulatory sequences became feasible. This has been done both for both $\alpha$-actin (Bergsma *et al.*, 1986) and $\beta$-actin (Fregien & Davidson, 1986). In order to study the regulatory region or promoter region 'upstream' from the structural genes for actins, the approach used was to insert fragments of the upstream region into an expression vector. The expression vector was derived from an animal virus linked to the gene for the enzyme chloramphenicol acetyltransferase (CAT). If the promoter region is effective then CAT activity can be detected in the host. Using this type of expression vector it is possible to determine (i) how much of the sequence upstream from the structural gene is required for the expression of the latter, and (ii) how 'strong' the promoter is. A strong promoter would enable the structural gene to be expressed more frequently and thus more copies of the gene product would be present in the cell. For the $\alpha$-actin, *c.* 200 base pairs upstream from the structural gene comprise the regulatory element. This region has been sequenced and a number of regulatory signals identified within it. Fregien & Davidson (1986) found that the $\beta$-actin promoter is a very strong promoter and this is consistent with the high abundance of mRNA for β-actin found in many cells.

The most abundant protein in skeletal muscle, myosin is made of two heavy chains ($M_r$ 200 000) and two pairs of light chains ($M_r$ *c*. 20 000). Two of the light chains, $LC_1$ and $LC_3$, have been cloned and sequenced. The arrangement of their genes is somewhat unusual. A single gene comprising nine exons and eight introns codes for both proteins (Fig. 12.8). Exons 1, 4, 5, 6, 7, 8 and 9 code for the mRNA for $LC_1$ and exons 2, 3, 5, 6, 7, 8 and 9 code for the mRNA for $LC_3$ (Nabeshima *et al*., 1984). This has been discussed in more detail in Chapter 7, section 7.1b.

A third gene to be cloned and sequenced from skeletal muscle of the domestic fowl is troponin I (Nikovits, Kuncio & Ordahl, 1986). Troponin is part of the regulatory system for skeletal muscle. It comprises three subunits, troponin C, troponin I and troponin T. Troponin I has been cloned from a domestic fowl gene library; it is *c*. 4.5 kb in length and contains eight exons. The organisation of the gene is shown on Fig. 12.8. The first exon and part of the second contain untranslated sequences and include the promoter region. The coding sequence begins at the triplet ATG

**Fig. 12.8** Organisation of the genes for the light chains of myosin ($LC_1$ and $LC_3$) and for troponin I in the domestic fowl. There is a single gene for myosin $LC_1$ and $LC_3$, which is processed to give two mRNAs transcribed from exons 1, 4, 5, 6, 7, 8 & 9 and from exons 2, 3, 5, 6, 7, 8 & 9 respectively (Nabashima *et al*., 1984). The troponin I gene is about 5 kb (including introns and exons). Only those exons shaded are translated, starting from the ATG sequence on exon 2 (Nikovits, Kuncio & Ordahl, 1986).

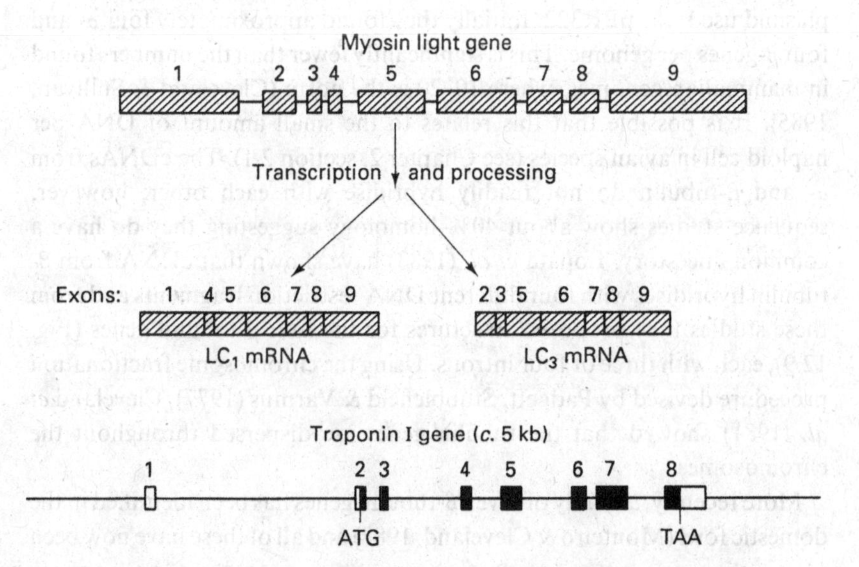

and terminates with TAA. The 3′ untranslated region corresponds to the polyA tail, a feature of most eukaryote mRNAs enabling them to be transported out of the nucleus.

The third exon is one of the shortest identified, being only seven nucleotides long. A comparison has been made of the 5′ flanking sequences of genes for skeletal muscle protein genes such as other troponins, $\alpha$-, $\beta$- and $\gamma$-actins and light chain myosins. All show extensive homologies suggesting a coordinate control in the developmental expression of muscle proteins.

A second group of cytoskeletal proteins are the **tubulins**. These comprise the units that make up the spindle fibres in mitotic and meiotic cells, and they are important elements in the elongation processes characteristic of neurones. Together with actins and a third group of fibres, the intermediate filaments, the tubulins, control the overall shape of a eukaryote cell and also the changes in its shape. About 20 years ago it was shown that the principal protein component of microtubules was tubulin. Two distinct types of monomer, $\alpha$-tubulin and $\beta$-tubulin, form a heterodimer that in turn polymerises to form microtubules. These microtubules occur throughout eukaryotic cells. When more detailed examinations of the $\alpha$- and $\beta$-tubulins were carried out and the proteins were separated, it became apparent that there were a number of different $\alpha$- and $\beta$-tubulins. Furthermore, even if a single cell type was examined the heterogeneity remained. It is therefore not a case of differences between cell types but of heterogeneity within cells. Much of the work on tubulins has been carried out using the chick embryo.

Cleveland *et al.* (1980) constructed plasmids using cDNA copies of the mRNA for $\alpha$- and $\beta$-tubulin isolated from embryonic chick brain. The plasmid used was pBR322. Initially they found approximately four $\alpha$- and four $\beta$-genes per genome. This is significantly fewer than the numbers found in mammalian genomes, where 10–20 is the norm (Cleveland & Sullivan, 1985). It is possible that this relates to the small amount of DNA per haploid cell in avian species (see Chapter 2, section 2.1). The cDNAs from $\alpha$- and $\beta$-tubulin do not readily hybridise with each other; however, sequence studies show about 40% homology suggesting they do have a common ancestory. Lopata *et al.* (1983) have shown that cDNA from $\beta$-tubulin hybridises with four different DNA restriction fragments and from these studies they proposed structures for the four $\beta$-tubulin genes (Fig. 12.9), each with three or four introns. Using the chromosome fractionation procedure devised by Padgett, Stubblefield & Varmus (1977), Cleveland *et al.* (1981) showed that the tubulin genes are dispersed throughout the chromosomes.

More recently, a family of seven $\beta$-tubulin genes has been identified in the domestic fowl (Monteiro & Cleveland, 1988) and all of these have now been sequenced. Some are preferentially expressed in certain tissues, as follows: β-1, in skeletal muscle; β-2, in brain; β-3, mainly in testes, but also in other tissues; β-4, in neurons; β-5, at low levels in most cells, but absent from neurons; β-6, in the microtubules of haemopoietic cells: β-7, in most cells. Each of these β-tubulin genes has a unique pattern of expression during development.

The α-tubulin genes are dispersed on four different chromosomes, including one on chromosome 1 and one on chromosome 8. At least two β-tubulin genes are on chromosome 2. Sullivan & Wilson (1984) separated the α- and β-tubulins from late embryonic and early postnatal chick brain by two dimensional gel electrophoresis, and resolved at least seven α- and ten β-tubulins. Since the number of isoforms appears to be greater than the number of genes, it is possible that some result from post-translational modification.

## *12.3c Collagens*

Collagen is the most abundant protein in the body occurring in virtually all tissues. Fibrils of collagen form the basic material of tendons, bones, cartilage, the walls of blood vessels, basement membrane and skin. In each instance collagen forms part of the extracellular matrix. It is synthesised within cells in a precursor form and then secreted and modified to form a matrix. There are a number of different types of collagen (see Martin *et al.*, 1985), differing in the size of unit that becomes polymerised to form the matrix, and the spatial arrangement of each of the units in the matrix.

**Fig. 12.9** Proposed structure of four of the β-tubulin genes in the domestic fowl (Lopata *et al.*, 1983). The different genes for β-tubulins are dispersed throughout the chromosomes (Cleveland *et al.*, 1981). Exons are shown in black.

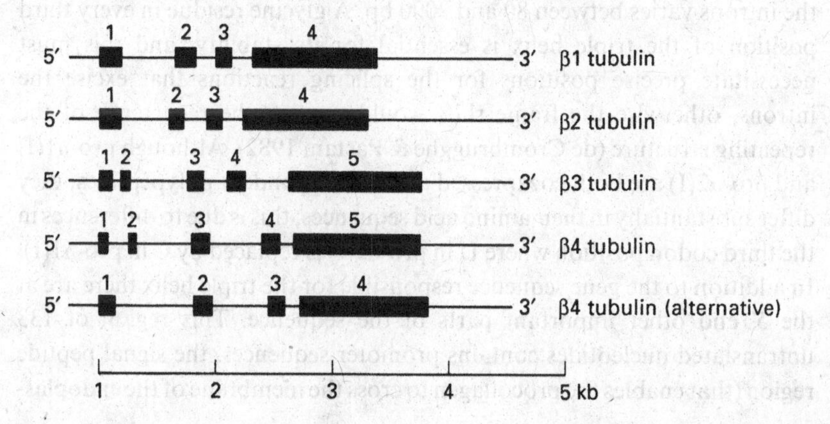

The common features of collagens are the triple helical arrangement of the polypeptide chains that usually comprise about 90% of the total structure, and also large stretches of amino acid sequences of the form (gly-X-Y)$_n$ present in the triple helical structure. Glycine is present as every third amino acid residue, and proline, alanine and serine are common amino acids present. The different types of collagen are numbered I to X; type I is the most abundant, making up about 90% of all collagen. The triple helix of type I collagen is made up of two strands of $\alpha1$ and one strand of $\alpha2$ polypeptide. Collagen is synthesised in the form of a soluble precursor procollagen, that undergoes cleavage before being assembled into the insoluble product, collagen. There are thus genes coding for procollagen $\alpha1$(I) and $\alpha2$(I), the two precursors of collagen type I. The type I collagen has been the one most studied in the domestic fowl where both $\alpha1$(I) and $\alpha2$(I) have been cloned; type VI has also been cloned in the domestic fowl (Bonaldi, Bucciotti & Colombatti, 1987). Fuller & Boedtker (1981) determined the 3′ end of the sequence of pro-$\alpha1$(I) and pro-$\alpha2$(I) using the mRNA to make cDNA, but the entire sequence of pro-$\alpha2$(I) has also been determined (Wozney *et al.*, 1981). The structures of the genes for pro-$\alpha1$(II) (Upholt & Sandell, 1986; Deák *et al.*, 1985) and pro-$\alpha1$(III) (Yamada *et al.*, 1984) in the domestic fowl have been determined.

There are a number of interesting features of the collagen genes in the domestic fowl. The pro-$\alpha2$(I) collagen spans about 40 kb and contains about 50 introns (Wozney *et al.*, 1981). The exons range in size from 45 to 108 bp, 54 bp being the most frequently recurring size. Benveniste-Schrode *et al.* (1985) proposed that collagen arose from a 9 bp primordial sequence which triplicated to 27 bp and then condensed to 54 bp. The 9 bp sequence would correspond to the amino acid repeat sequence in the triple helix: gly-X-Y. They suggest that the nucleotide sequence of the 9 bp for pro-$\alpha1$(I) was GGTCCCCCC and that for pro-$\alpha2$(I) was GGTCCTCCT. The size of the introns varies between 80 and 2000 bp. A glycine residue in every third position of the triple helix is essential for its stability, and this must necessitate precise positions for the splicing reactions that excise the introns, otherwise the frame shift would prevent the recurrence of the repeating structure (de Crombrugghe & Pastan, 1982). Although pro-$\alpha1$(I) and pro-$\alpha2$(I) are both coexpressed and interdependent polypeptides, they differ substantially in their amino acid sequences; this is due to differences in the third codon position where U in pro-$\alpha2$(I) is replaced by C in pro-$\alpha1$(I). In addition to the gene sequence responsible for the triple helix there are at the 5′ end other important parts of the sequence. This region of 133 untranslated nucleotides contains promoter sequences, the signal peptide region (that enables the procollagen to cross the membrane of the endoplasmic reticulum) and also an aminopropeptide. The regions on the gene for type II procollagen in the domestic fowl are shown in Fig. 12.10.

### 12.3d Keratin

Keratin is a major gene product synthesised in terminally differentiating epidermal tissues. It is the predominant protein of hair, wool, feathers, claws, and beaks and so is of obvious importance in birds. The β-keratins present in feathers comprise at least 20 different proteins which are the products of a multigene family. Presland *et al.* (1989a) have cloned and sequenced four of these genes. They all show a high degree of similarity, and all four are expressed in feather tissue from 14-day-old chick embryos. Further, a cluster of 18 feather keratin genes has been cloned and it is flanked on either side by other keratin genes including those coding for scale keratins (Presland, Whitbread & Rogers, 1989b).

### 12.3e Egg proteins

This section is concerned mainly with egg-white proteins, but also includes the egg-yolk protein, lipovitellin. Aspects of particular interest in cloning and sequencing the egg-white proteins are that: (i) most are synthesised in

**Fig. 12.10** Organisation of the gene for type II procollagen from the domestic fowl (Upholt & Sandell, 1986). The gene is approximately 13 kb, comprising 46 exons which are numbered from the 3′ terminus. The exons 5 to 46, and part of exon 4, code for a strand of the triple helix present in mature collagen. Exons 3 and 2, and parts of exons 1 and 4, code for the region of procollagen which is excised before its secretion to form the collagen matrix. The propeptide region codes for the part of the procollagen that is excised on maturation to form collagen.

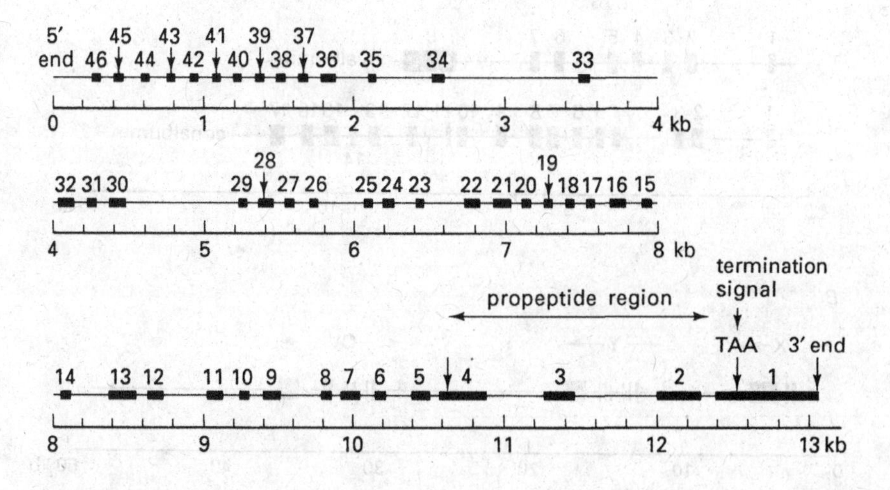

the tubular gland cells of the oviduct and are under hormonal control, and (ii) the structures of the principal proteins have been thoroughly investigated (see Chapter 10, section 10.5). The genes already studied are those for ovalbumin, conalbumin (≡ovotransferrin), ovomucoid, lysozyme and avidin. Although their synthesis is under hormonal control, the response varies between proteins; oestradiol causes a rapid increase in the mRNA for conalbumin whereas the increase for ovalbumin is somewhat delayed (Cochet *et al*., 1979). This may be due to different flanking sequences in the genes which may affect promoter efficiency. In each case the genes contain introns, the numbers being roughly proportional to the size of the protein, e.g. lysozyme ($M_r$ 14 300) and conalbumin ($M_r$ 80 000) have genes containing 4 and 17 exons respectively. The cDNA and gene-containing fragments have been cloned and sequenced or partly sequenced for five egg-white proteins, conalbumin, ovalbumin, ovomucoid, avidin and lysozyme. The distribution of each of the coding sequences is shown on Fig. 12.11. Clones containing the avidin gene have been isolated (Gope *et al*., 1987), but the complete organisation of the gene has not yet been resolved.

**Fig. 12.11** Gene organisation of egg-white proteins from the domestic fowl: *A*, those for lysozyme (Quasba & Safaya, 1984), ovomucoid (Stein *et al*., 1980), ovalbumin (Woo *et al*., 1981), conalbumin (Cochet *et al*., 1979), and *B*, on a smaller scale, the region containing the genes for ovalbumin (ov) together with pseudogenes X and Y (Royal *et al*., 1979).

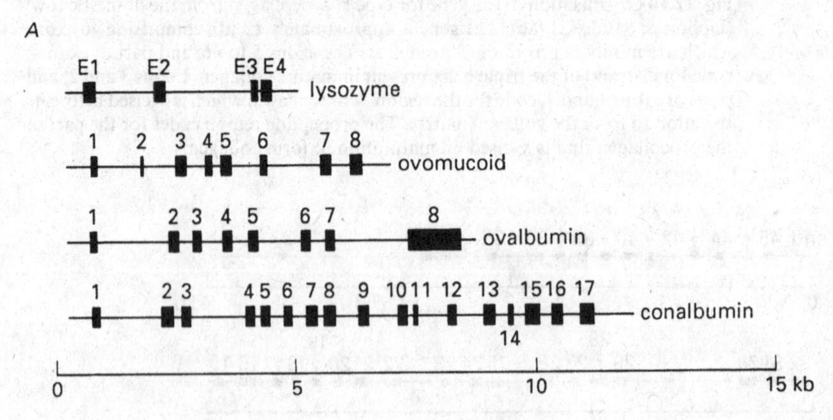


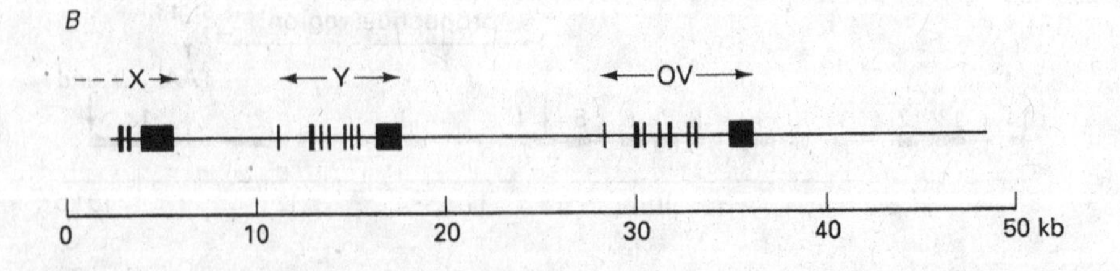

Two genes X and Y, closely related to ovalbumin were discovered (Royal *et al.*, 1979) and they are also under steroid hormonal control. Genes X and Y, together with the ovalbumin gene are contained within a 46 kb fragment of DNA. The genes X and Y are homologous with that of ovalbumin and are considered to have arisen by gene duplication. Each has eight exons, that differ only in the size of the untranslated region. Both the genes X and Y are expressed in the laying hen (LeMeur *et al.*, 1981), but at much lower levels than that of the ovalbumin gene. The mRNAs for X and Y are present at about 2% of the level of ovalbumin mRNA. The mRNA for X and Y has been translated *in vitro* and the proteins are distinct from ovalbumin.

The gene for lysozyme and the cDNA for lysozyme mRNA have been cloned using a λCharon 4A (Jung *et al.* 1980). Of the four exons, exon I codes for the translational signals, the signal peptide (18 residues) and the first 28 residues of lysozyme. There is one lysozyme gene per haploid genome (Grez *et al.*, 1981). The 5′ end of the lysozyme gene sequence has been compared with the equivalent sequence in conalbumin or ovalbumin (Grez *et al.*, 1981) and it reveals partial homology. Lysozyme can be induced by different steroid hormones, e.g. oestrogens, progesterone or glucocorticoids. In order to study the mechanism of these inductions, Von der Ahe *et al.* (1986) examined the binding of purified glucocorticoid receptors and progesterone receptors to the lysozyme gene. The region of the lysozyme gene to which the receptors bind is protected from the action of the enzyme, deoxyribonuclease, and this can be detected by the method known as '**footprinting**'. It involves digesting away the adjacent region with deoxyribonuclease and analysing the region protected, i.e. the 'footprint'. It was found that different, but overlapping footprints upstream of the coding sequence were produced by the progesterone receptor and the glucocorticoid receptor (Fig. 12.12). A detailed map has been made of the deoxyribonuclease-sensitive sites of the chromatin region containing the lysozyme gene from various adult tissues and from the developing chick embryo (Gross & Garrard, 1988).

The complete sequencing of the ovalbumin gene (Woo *et al.*, 1981) reveals how large the introns are compared with the coding region. The complete gene sequence is a polynucleotide stretch of 7564 b (bases), but the coding sequence accounts for only 1872 b. The ovomucoid gene is 5.6 kb and the coding region is 812 b (Lai *et al.*, 1979). Ovomucoid is thought to have arisen by a process of gene duplication, since it is made up of three similar domains (see Chapter 10 section 10.5c). The coding region for each of these domains is separated by an intron (Stein *et al.*, 1980). Altogether there are seven introns and the shortest exon is only 20 b. From the genes that have been sequenced so far it is found that there is a much wider range

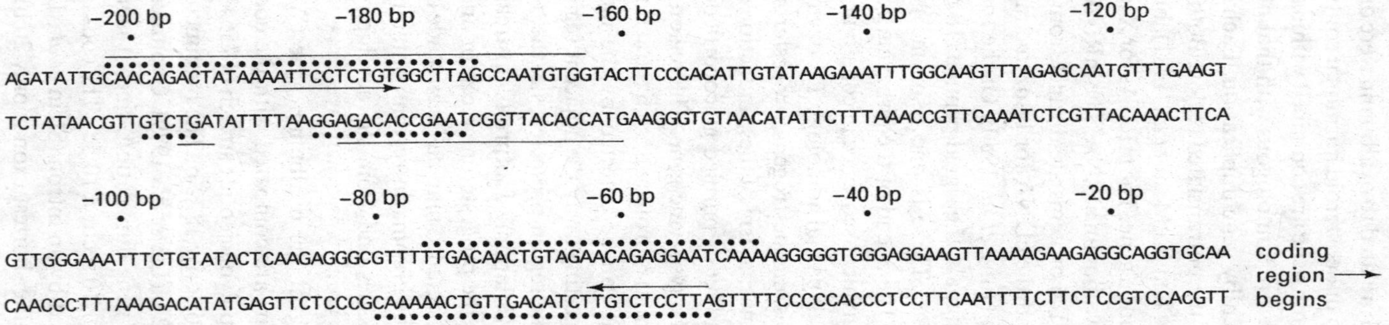

−200 bp −180 bp −160 bp −140 bp −120 bp

AGATATTGCAACAGACTATAAAATTCCTCTGTGGCTTAGCCAATGTGGTACTTCCCACATTGTATAAGAAATTTGGCAAGTTTAGAGCAATGTTTGAAGT

TCTATAACGTTGTCTGATATTTTAAGGAGACACCGAATCGGTTACACCATGAAGGGTGTAACATATTCTTTAAACCGTTCAAATCTCGTTACAAACTTCA

−100 bp −80 bp −60 bp −40 bp −20 bp

GTTGGGAAATTTCTGTATACTCAAGAGGGCGTTTTTGACAACTGTAGAACAGAGGAATCAAAAGGGGGTGGGAGGAAGTTAAAAGAAGAGGCAGGTGCAA

CAACCCTTTAAAGACATATGAGTTCTCCCGCAAAAACTGTTGACATCTTGTCTCCTTAGTTTTCCCCCACCCTCCTTCAATTTTCTTCTCCGTCCACGTT

coding region begins →

**Fig. 12.12** 'Footprinting' the progesterone and glucocorticoid receptors on the lysozyme promoter region. The continous horizontal lines represent the 'footprint' for the progesterone receptor, and the dotted lines that for the glucocorticoid receptor. The region shown (− 210 bp to − 10 bp) is 'upstream' from the coding region (which starts at 0 bp) and represents most of the promoter region consisting of about 200 bp, i.e. from − 200 to 0 (von der Ahe *et al.*, 1986).

of intron sizes (50–20 000 b) than exons (Gilbert, 1985). Exons most commonly code for between 40 and 50 amino acid residues, i.e. 120–150 nucleotides.

**Vitellogenins** are a major family of yolk precursor proteins that are synthesised in the liver in response to oestrogen stimulation. They are large proteins ($M_r$ *c.* 200 000), that are taken up selectively by the developing hen oocytes and then cleaved into lipovitellin and phosvitin. The proteins are stored in granules in the yolk and serve as nutrient for the developing embryo. There are three separate genes coding for vitellogenins and three distinct polypeptides have been isolated. They are referred to as VtgI, VtgII and VtgIII, VtgII being the most abundant. The complete sequence for the VtgII gene has been determined by Schip *et al.* (1987), and in all there are 20 342 bp containing 35 exons (Fig. 12.13*A*). More recently the gene for the minor VtgIII has been cloned, and in addition it has been shown that both genes VtgII and VtgIII are flanked by pseudogenes, ΨVtgIII adjacent to VtgII, and ΨVtgIII adjacent to VtgII (Silva, Fischer & Burch, 1989). A second yolk protein synthesised in the liver in response to oestradiol, is apo-very low density lipoprotein (apoVLDL). The gene encoding it has been sequenced (see Fig. 12.13*B* and Schip *et al.*, 1983). For a detailed review of the evolution of egg-yolk proteins see Byrne *et al.* (1989).

**Fig. 12.13** Organisation of the gene for (*A*) the egg-yolk protein precursor, vitellogenin (Schip *et al.*, 1987), and (*B*) ApoVLDL II gene (Schip *et al.*, 1983). Transcription of the vitellogenin gene starts at position 0. The exons are shaded in black.

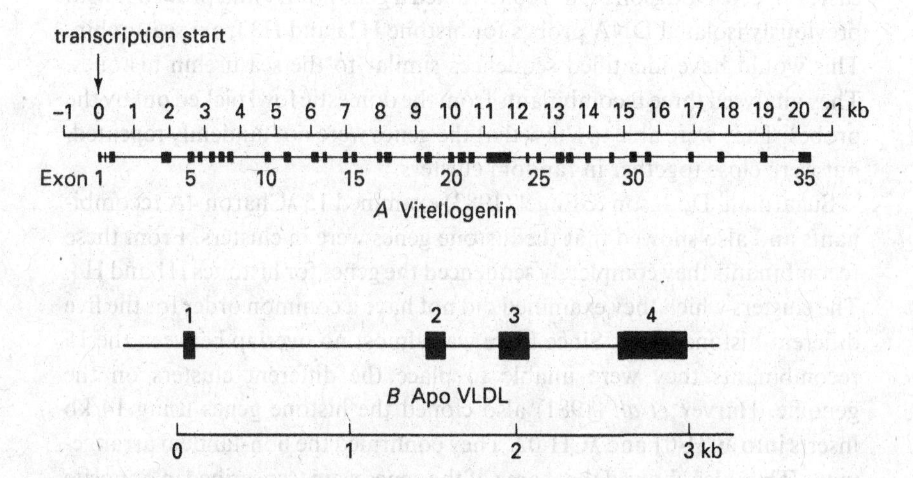

### 12.3f Histones and protamine

Histones are the proteins responsible for maintaining the compact arrangement of DNA in the chromosomes. Their role in the structure of the nucleosome was described in Chapter 2, section 2.1 and Fig. 2.3. In rapidly dividing cells DNA synthesis occurs during the *S* phase and this is matched by a parallel synthesis of histones. Histones are highly conserved proteins.

The genes for histones were amongst the earliest to be studied at a molecular level. One reason for this is that there are many copies of the histone genes per nucleus in certain organisms and this has made analysis easier. The number of copies may be as high as 1600 in the axolotl or as low as only two copies in yeast (Adams, Knowler & Leader, 1986). Those present in the sea-urchin *Psammechinus miliaris*, in which there are hundreds of copies, were the first to be analysed. This amplification of the number of histone genes is believed to allow for rapid cell division such as occurs in the development of the embryo, although the multiple copies remain in the adult. In many animals, although multiple copies occur, the numbers are much more modest, and this may be associated with a slower rate of cell division in the embryo. In the domestic fowl there are about 10 copies of each of the histone genes (Crawford *et al.*, 1979; Harvey *et al.*, 1981).

The multiple copies of the histone genes may be arranged in several ways: present in tandemly repeated clusters, dispersed widely throughout the chromosomes, or present in small clusters. In the domestic fowl, although the early evidence suggested tandem repeated arrangements (Crawford *et al.*, 1979), subsequent more detailed studies have shown this not to be the case. Engel & Dodgson (1981) constructed a gene library and probed it with previously isolated DNA probes for histone H2a and H3 from sea-urchin. This would have identified sequences similar to the sea-urchin histones. They analysed three recombinants from the domestic fowl picked out by the probes. They were able to show that the genes were not tandemly repeated, but were close together in random clusters.

Sugarman, Dodgson & Engel (1983) examined 15 λCharon 4A recombinants and also showed that the histone genes were in clusters. From these recombinants they completely sequenced the genes for histones H1 and H4. The clusters which they examined did not have a common order for the five different histone genes. Since there was almost no overlap between the 15 recombinants they were unable to place the different clusters on the genome. Harvey *et al.* (1981) also cloned the histone genes using 14 kb inserts into λCH-01 and λCH-02. They confirmed the non-tandem arrangement. They also showed that some of the genes were transcribed in opposite directions; this means that they are coded for on opposite strands of the DNA. D'Andrea *et al.* (1985) examined overlapping $\lambda$ clones and also a cosmid clone that contained a larger piece of the genome, and from these they were able to position the clusters in relation to one another; their map of a region of the genome is shown in Fig. 12.14.

A striking feature of the organisation of the histone genes in comparison with other eukaryote genes is that they generally lack introns. However, an H2A variant known as $H2A_F$ has been sequenced in the domestic fowl (Dalton *et al.*, 1989) and found to contain four introns (Fig. 12.15), as has another variant H3.3 (Engel, Sugarman & Dodgson, 1982) from the domestic fowl.

Another feature which these two variants ($H2A_F$ and H3.3) share with H5 is that their expression is independent of the cell cycle, whereas most histone synthesis is tightly coupled to DNA synthesis which occurs in the S

**Fig. 12.14** Organisation of the histone genes in the domestic fowl in random clusters (D'Andrea *et al.*, 1985). The arrows indicate the direction of transcription. Note (i) that histones are often transcribed in opposite directions, and are hence encoded on opposite or complementary strands of DNA, and (ii) the absence of introns within the histone genes.

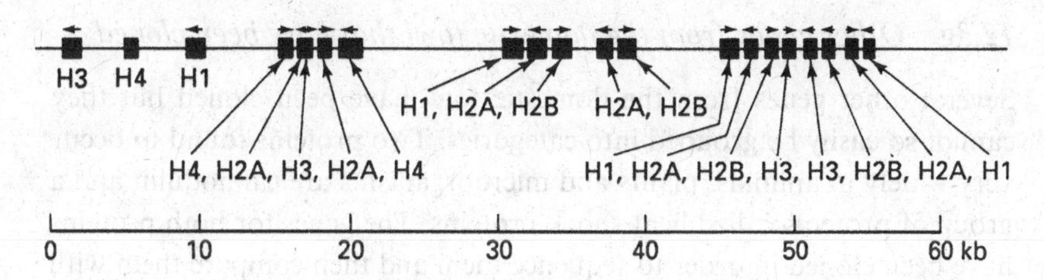


**Fig. 12.15** Histones genes containing introns. Most histone genes lack introns (see Fig. 12.14) but the two variants H3.3 and $H2A_f$, found in the domestic fowl, are exceptions (Engel, Sugarman & Dodgson, 1982; Dalton *et al.*, 1989). Exons are shown in black.

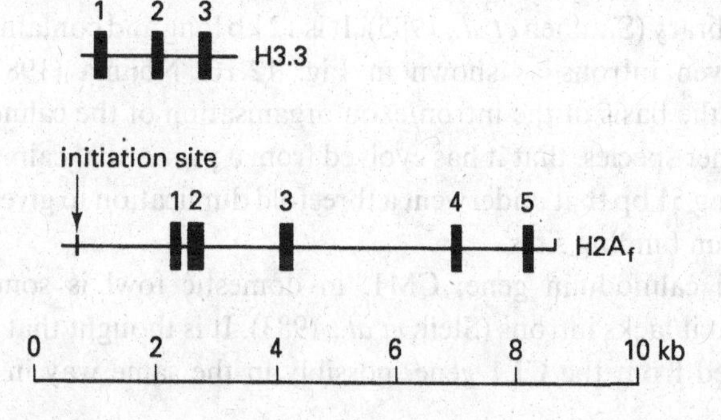

phase of the cell cycle (see Chapter 2, section 2.3). Their importance in chromatin structure is not clear.

During spermatogenesis there is a dramatic change in chromatin structure: the histones present in the nucleosome core (see Fig. 2.3) are completely replaced by another group of basic proteins known as protamines. **Protamines** act by compacting the DNA in the chromatin of the nuclei. They are smaller proteins than histones, having $M_r$ generally between 4000 and 7000, and containing a very high proportion of the basic amino acid, arginine (Ando, Yamasaki & Suzuki, 1973). The protamine genes from a number of different species have been cloned (see Oliva & Dixon, 1989). In the domestic fowl there are two loci for the protamine, galline (each protamine is generally named after the species from which it is derived, e.g. *salmine* from salmon and *iridine* from rainbow trout (*Salmo irideus*)). Both loci have identical coding regions (Oliva & Dixon, 1989), and the number of copies is two per haploid genome, i.e. one at each locus. Neither locus contains introns, in contrast to the mammalian protamines which each contain a single intron, but in common with the protamines from salmonid fishes. The evolutionary significance of this is still a matter for speculation.

### *12.3g Other genes from the domestic fowl that have been cloned*

Several other genes from the domestic fowl have been cloned but they cannot so easily be grouped into categories. Two proteins found to occur very widely in animals, plants and microorganisms are calmodulin and a group of proteins called heat shock proteins. The genes for both proteins have been cloned in order to sequence them and then compare them with homologous proteins in widely different organisms. The protein **calmodulin** has four strong binding sites for $Ca^{2+}$; which is important as an intracellular messenger, and calmodulin is involved in modulating its activity. Two distinct calmodulin genes have been isolated from the domestic fowl, CL1 and CM1 (Putkey, Carroll & Means, 1987). CL1 has been cloned from a λCharon 4A library (Simmen *et al.*, 1985). It is 12 kb long and contains eight exons and seven introns as shown in Fig. 12.16. Nojima (1987) has suggested, on the basis of the intron/exon organisation of the calmodulin genes from other species, that it has evolved from a primordial calmodulin gene comprising 51 bp that underwent a threefold duplication to give rise to the four calcium binding sites.

The second calmodulin gene, CM1, in domestic fowl is somewhat unusual in that it lacks introns (Stein *et al.*, 1983). It is thought that it may have originated from the CL1 gene, possibly in the same way in which pseudogenes are believed to have evolved, i.e. from mRNA transcribed to cDNA, the latter then being incorporated into the genome. Unlike a pseudogene, however, it is expressed, and is part of a fully functional multigene family. It also has been cloned from a λCharon 4A genome library (Stein *et al.*, 1983).

The **calcium-binding protein** that is involved in the uptake of $Ca^{2+}$ in the small intestine, and which is actively synthesised in response to vitamin D, has been cloned and shows a high degree of homology with calmodulin (Wilson, Harding & Lawson, 1985).

The **heat shock proteins** are a group of proteins synthesised in response to heat stress. They are part of an organism's reaction to heat stress, but it is not yet clear what function the proteins serve. They have been detected in such diverse organisms as yeast, *Drosophila* and mammals. In the domestic fowl, four have been detected, designated hsp108, hsp90, hsp70 and hsp23 (Morimoto *et al.*, 1986; Kulomaa *et al.*, 1986), their names corresponding to their $M_r$. The complete nucleotide sequence for hsp108 from hen oviduct and from bursal lymphoma has been determined (Kulomaa *et al.*, 1986). The gene for the most abundant heat shock protein, hsp70, has been cloned and sequenced, and has been found to be a single contiguous reading frame of 1905 nucleotides (Morimoto *et al.*, 1986). The small heat shock proteins, e.g. hsp23, are closely related in structure to α-crystallins (see section 12.3a and Ingolia & Craig, 1982).

Other genes or cDNAs from the domestic fowl which have been cloned include calcitonin (Lasmoles *et al.*, 1985), the progesterone receptor (Jeltsch *et al.*, 1986; Huckaby *et al.*, 1987), nerve growth factor (Meier *et al.*, 1986), globin genes (discussed in Chapter 10, section 10.2), viral genes and oncogenes (Watson, McWilliams & Papas, 1988), erythrocyte anion transport proteins (Kim *et al.*, 1988), ribosomal protein S17 (Trueb *et al.*, 1988), metallothionein (Wei & Andrews, 1988), fatty acid synthase (Back *et al.*, 1986), phospho*enol*pyruvate carboxykinase (Cook *et al.*, 1986), cytochrome *c* (Limbach & Wu, 1983), cytochrome $b_5$ (Zhang & Sommerville, 1988), creatine kinase (Kwiatkowski *et al.*, 1985), adenylate kinase (Suminami *et al.*, 1988), thymidine kinase (Merrill *et al.*, 1984) and proteoglycan (Sai *et al.*, 1986).

**Fig. 12.16** Calmodulin CL1 gene organisation in the domestic fowl (Simmen *et al.*, 1985). The exons are numbered and shown in black.

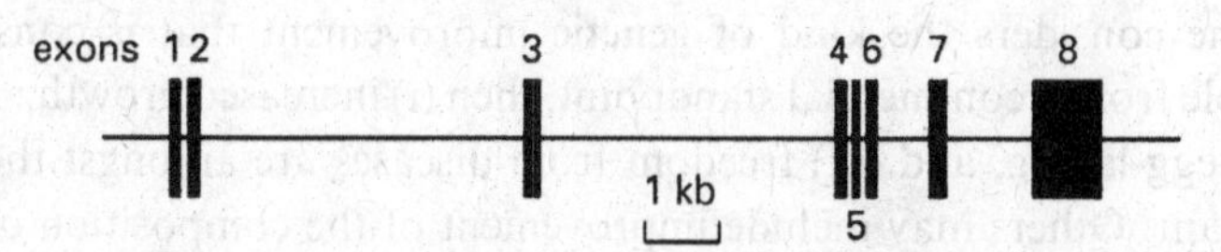

As a result of the sequence data now available the following trends are emerging.

1. The introns are generally large, often occupying more than half of the gene.
2. There is greater conservation of the nucleotide sequence of exons than of introns.
3. Genes for some small proteins, e.g. histones, lack introns.
4. Related genes are often, though not always, clustered on the chromosomes. Examples of clusters are ovalbumins, histones, δ-crystallins, immunoglobulins.
5. Promoter sequences are important in the developmental regulation of gene expression.

## 12.4 Gene transfer into the germline of the domestic fowl

In the previous section the genes from the domestic fowl that have been cloned, using plasmids or viral vectors that propagate in bacteria such as *E. coli*, were described. Generally, the purpose of cloning these genes has been to obtain them in sufficient quantities in order that they can be studied in much greater molecular detail than would otherwise be possible. Another kind of gene manipulation, that will be directed primarily towards improvement of the domestic fowl for commercial purposes, is that of introducing genes into the domestic fowl, either from other species or from different strains of the domestic fowl. This technology involving cloning in higher eukaryotes is at present in its infancy. It is generally known as **transgenics**. The general aim of transgenics applied to the domestic fowl is to improve the breed for commercial purposes. This poses a number of questions. Which genes might it be useful to transfer to the domestic fowl? How can such transfers be brought about? How will the genes behave once transferred, since the expression of particular genes often depends on: (i) where they are inserted into the genome, (ii) whether they have promoter sequences transferred with them, and (iii) the particular genome to which they are added. It is also important to be able to detect whether particular genes have, in fact, been transferred. This may not be obvious from the phenotype, since the transferred gene either may not be expressed at all, or its expression may not be easily recognised against the new genetic background.

If one considers the kind of genetic improvement that is considered desirable from a commercial standpoint, then (i) increased growth rate, (ii) higher egg-laying, and (iii) freedom from diseases are amongst the most important. Others may include improvement of the composition of meat and eggs by, for example, lowering the cholesterol or saturated fatty acid content. One of the difficulties in these cases is to identify which genes are involved. At present only a minute proportion of the total number of genes in the domestic fowl has been identified. The growth rate and ultimate size are regulated only in part by the growth hormone secreted by the anterior pituitary. Disease resistance is determined by a number of components of the immunological system including the major histocompatibility complex, T-cell receptors, immunoglobulins and lymphokines. The genetic control of egg production is much less clearly understood. However, the introduction of improved genes for any of these factors could eventually be advantageous for commercial breeders.

So far the work on the domestic fowl has lagged behind that of mammals used in agriculture. There are two main reasons for this: (i) it is technically more difficult to microinject DNA into the domestic fowl than into mammals, and (ii) an individual fowl is much less valuable than, for example, cattle or sheep, so it is not economically feasible to attempt improvement of an individual fowl because of its low commercial value. The high cost would thus prohibit any manipulation that would improve only an individual domestic fowl, so any improvement in the genotype would have to be transmitted by breeding and giving rise to new strains.

This molecular approach to poultry breeding has been reviewed by Shuman & Shoffner (1986), Freeman & Messer (1985), Freeman & Bumstead (1987) and Shuman (1990). Having identified the desirable genes, the question is then how to transfer them into the host fowl. Gene transfer in higher animals can be effected in a number of ways. Methods have been developed for transfer into cultured cells, and also into whole animals; only the latter will be considered here. The whole animals which have been most studied are *Xenopus*, *Drosophila* and the mouse (Kingsman & Kingsman, 1988), but transfers to domestic animals such as rabbits, pigs and sheep have also been carried out (Hammer *et al.*, 1985). Of the mammals, the mouse has been most extensively studied, where transfer of genes is at present much easier than in the domestic fowl. Microinjection of DNA, which has been used with the mouse, is much more difficult with the domestic fowl, partly due to the fragile nature of the embryo and also because it is optically more opaque. In the mouse one of the methods favoured is to microinject DNA at the pronucleus stage or G-0. This is after fertilisation has occurred, but when both pronuclei from the sperm and ovum are still visible, i.e. before fusion. If foreign DNA is stably integrated at this stage it will be incorporated into all the cells of the embryo. To do this the pronuclei have to be identified under a dissecting microscope. Microinjection is usually made into the male nucleus, since it is larger. If microinjection is performed at a later stage, e.g. blastocyst (4–30 cell stage) it is likely to produce a mosaic, i.e an embryo having cells of different chromosome complements. In the case of the domestic fowl, by the time oviposition occurs approximately 24 h after fertilisation, the embryo contains in the region of 60000 cells, and it is thus far too late for microinjection. Fertilised ova would have to be removed at a much earlier stage for microinjection.

An alternative strategy, that is less traumatic, is to use a retroviral vector, which can be mixed with the developing cells without the need for microinjection into individual cells. The retroviral vector (see Appendix II) can be defective or non-defective. A defective virus is one that has certain genetic elements removed so that it can no longer replicate. This has the advantage of being non-pathogenic, but requires the use of a larger titre than when using a non-defective one, which can replicate within the host cells. So far defective reticuloendothelial virus (REV) vectors have been tested in the domestic fowl, and shown to be transferred to the host cells (see Shuman, 1990). To facilitate embryonic development after gene transfer, Perry (1988) has developed an *in vitro* system that enables microinjection at the one celled stage and, subsequently, development to hatching.

A dramatic demonstration of the power of transgenics was the transfer of rat genomic DNA for growth hormone into mouse embryo by microinjection. The resultant transgenic mice had high numbers of copies (20–40 per cell) of the growth hormone gene and serum levels of growth hormone 100–800 times normal, and the mice grew 2–3 times as fast as the controls, reaching twice their weight at 74 days (see Old & Primrose, 1989). Souza *et al.* (1984) transferred the growth hormone gene into chick embryos using an avian leukosis virus vector. They were able to demonstrate an elevated serum level of circulating growth hormone, but no increase in growth rate.

Clearly transgenic fowl have potential for the future, but at present further improvement in the transfer technology is required, and also further knowledge of the genes that it will be useful to transfer. It seems probable that the genes important for growth, egg production and disease resistance may occur in multigene families (Crittenden, 1986), and thus it is unlikely that they will be able to be cloned as single fragments. At present gene transfer is very costly, and this is a further reason that it is likely to be some time before it becomes a commercial proposition. In the poultry industry the rate of genetic change by selection for growth rate has been estimated as 6.5% per year, and for egg production 1.7% per year. For transgenic fowl to compete, the higher cost will have to be offset by a more rapid rate of improvement than that achieved by the methods already in use.

## References

Adams, R. L. P., Knowler, J. T. & Leader, D. P. (1986). *Biochemistry of the Nucleic Acids*, 10th edn. London: Chapman and Hall.

Aho, S., Tate, V. & Boedtker, H. (1984). Location of the 11bp exon in the chicken pro $\alpha$2(I) collagen gene. *Nucleic Acids Research*, **12**, 6117–25.

Ando, T., Yamasaki, M. & Suzuki, K. (1973). *Protamines*. London: Chapman and Hall.

Back, D. W., Goldman, M. J., Fisch, J. E., Ochs, R. S. & Goodridge, A. G. (1986). The fatty acid synthase gene in avian liver. *Journal of Biological Chemistry*, **261**, 4190–7.

Benveniste-Schrode, K., Doering, J. L., Hauk, W. W., Schrode, J., Kendra, K. L. & Drexler, B. K. (1985). Evolution of chick type I procollagen genes. *Journal of Molecular Evolution*, **22**, 209–19.

Bergsma, D. J., Grichnik, J. M., Gossett, L. M. A. & Schwartz, R. J. (1986). Delimitation and characterization of *cis*-acting DNA sequences required for the regulated expression and transcriptional control of the chicken skeletal $\alpha$-actin gene. *Molecular and Cellular Biology*, **6**, 2462–75.

Bloemendal, H. (1977). The vertebrate eye lens. *Science*, **197**, 127–38.

Bonaldi, P., Bucciotti, F. & Colombatti, A. (1987). Isolation of cDNA clones corresponding to the $M_r = 150\,000$ subunit of chick type VI collagen. *Biochemical and Biophysical Research Communications*, **149**, 347–54.

Bressan, G. M., Argos, P. & Stanley, K. K. (1987). Repeating structure of chick tropoelastin revealed by complementary DNA cloning. *Biochemistry*, **26**, 1497–503.

Brown, T. A. (1986). *Gene Cloning: an introduction*. London: Van Nostrand.

Bulfield, G. (1990). Molecular Genetics, In *Poultry Breeding and Genetics*, ed. R. D. Crawford, pp. 543–84. Amsterdam: Elsevier.

Byrne, B. M., Gruber, M. & AB, G. (1989). The evolution of egg yolk proteins. *Progress in Biophysics and Molecular Biology*, **53**, 33–69.

Catelli, M. G., Binart, N., Feramisco, J. R. & Helfman, D. M. (1985). Cloning of the chick hsp 90 cDNA in expression vector. *Nucleic Acids Research*, **13**, 6035–47.

Chang, K. S., Zimmer, W. E., Bergsma, D. J., Dodgson, J. B. & Schwartz, R.J. (1984). Isolation and characterization of six different actin genes. *Molecular and Cellular Biology*, **4**, 2498–508.

Cleveland, D. W., Hughes, S. H., Stubblefield, E., Kirschner, M. W. & Varmus, H. E. (1981). Multiple $\alpha$ and $\beta$ tubulin genes represent unlinked and dispersed gene families. *Journal of Biological Chemistry*, **256**, 3130–4.

Cleveland, D. W., Lopata, M. A., Macdonald, R. J., Cowan, N. J., Rutter, W. & Kirschner, M. W. (1980). Number and evolutionary conservation of $\alpha$-tubulin and cytoplasmic $\beta$ and $\alpha$ actin genes using specific cloned cDNA probes. *Cell*, **20**, 95–105.

Cleveland, D. W. & Sullivan, K. F. (1985). Molecular biology of tubulin. *Annual Review of Biochemistry*, **54**, 331–65.

Cochet, M., Gannon, F., Hen, R., Marteaux, L., Perrin, F. & Chambon, P. (1979). Organization and sequence studies of the 17-piece chicken conalbumin gene. *Nature*, **282**, 567–74.

Cook, J. S., Weldon, S. L., Garcia-Ruiz, J. P., Hod, Y. & Hanson, R. W. (1986).

Nucleotide sequence of the mRNA encoding the cytosolic form of phospho*enol*-pyruvate carboxykinase(GTP) from the chicken. *Proceedings of the National Academy of Sciences of USA*, **83**, 7583–7.

Cooper, T. A. & Ordahl, C. P. (1984). A single troponin T gene regulated by different programs in cardiac and skeletal muscle development. *Science*, **226**, 979–82.

Coutu, M. D. & Craig, S. W. (1988). cDNA-derived sequence of chicken embryo vinculin. *Proceedings of the National Academy of Sciences of USA*, **85**, 8535–9.

Crawford, R. J., Krieg, P., Harvey, R. P., Hewish, D. A. & Wells, J. R. E. (1979). Histone genes are clustered with a 15-kilobase repeat in the chicken genome. *Nature*, **279**, 132–7.

Creighton, T. E. (1983). *Proteins*. New York: W. H. Freeman.

Crittenden, L. B. (1986). Identification and cloning of genes for insertion. *Poultry Science*, **65**, 1468–73.

Dalton, S., Robins, A. J., Hervey, R. P. & Wells, J. R. E. (1989). Transcription from the intron-containing chicken histone $H2A_F$ gene is not S-phase regulated. *Nucleic Acids Research*, **17**, 1745–56.

Das, G. C. & Piatigorsky, J. (1986). The chicken $\delta_1$-crystallin gene promoter: Binding of transcription factor(s) to the upstream G + C-rich region is necessary for promoter function *in vitro*. *Proceedings of the National Academy of Sciences of USA*, **83**, 3131–5.

D'Andrea, R. J., Coles, L. S., Lesnikowski, C., Tabe, L. & Wells, J. R. E. (1985). Chromosomal organization of chicken histone genes: preferred associations and inverted duplications. *Molecular and Cellular Biology*, **5**, 3108–15.

Deák, F., Argraves, W. S., Kiss, I., Sparkes, K. J. & Goetinck, P. F. (1985). Primary structure of the telopeptide and a portion of the helical domain of chicken type II procollagen as determined by DNA sequence analysis. *Biochemical Journal*, **229**, 189–96.

de Crombrugghe, B. & Pastan, I. (1982). Structure and regulation of a collagen gene. *Trends in Biochemical Sciences*, **7**, 11–13.

Delaye, M. & Tardieu, A. (1983). Short range order of crystallin proteins accounts for eye lens transparency. *Nature*, **302**, 415–7.

Engel, J. D. & Dodgson, J. B. (1981). Histone genes are clustered but not tandemly repeated in the chicken genome. *Proceedings of the National Academy of Sciences of USA*, **78**, 2856–60.

Engel, J. D., Sugarman, B. J. & Dodgson, J. B. (1982). A chicken histone H3 gene contains intervening sequences. *Nature*, **297**, 434–6.

Fielding Hejtmancik, J., Thompson, M., Wiastow, G. & Piatigorsky, J. (1986). cDNA deduced protein sequence for βB-1 crystallin polypeptide of the chicken lens. *Journal of Biological Chemistry*, **261**, 982–7.

Freeman, B. M. & Bumstead, N. (1987). Transgenic poultry: theory and practice. *World's Poultry Science Journal*, **43**, 180–9.

Freeman, B. M. & Messer, L. I. (1985). Gene manipulation in the domestic fowl. *World's Poultry Science Journal*, **41**, 124–32.

Fregien, N. & Davidson, N. (1986). Activating elements in the promoter region of the chicken β-actin gene. *Gene*, **48**, 1–11.

Fuller, F. & Boedtker, H. (1981). Sequence determination and analysis of the 3′ region of the chicken pro-α1(I) and pro-α-2 (I) collagen messenger ribonucleic acids including the carboxy-terminal propeptide sequence. *Biochemistry*, **20**, 996–1006.

Gavora, J. S. (1988). Molecular genetics of birds: a survey. *Proceedings of XVIII World's Poultry Congress*, pp. 502–4. Nagoya, Japan: Japanese Poultry Science Association.

Gilbert, W. (1985). Genes-in-pieces revisited. *Science*, **228**, 823–4.

Gope, M. L., Keinanem, R. A., Kristo, P. A., Coneely, O. M., Beattie, W., Zarucki, G., Schulz, T., O'Malley, B. & Kulomaa, M. S. (1987). Molecular cloning of chicken avidin gene. *Nucleic Acids Research*, **15**, 3595–606.

Grez, M., Land, H., Giesecke, K., Schütz, G., Jung, A. & Sippel, A. E. (1981). Multiple mRNAs are generated from the chicken lysozyme gene. *Cell*, **25**, 743–52.

Gross, D. S. & Garrard, W. T. (1988). Nuclease hypersensitive sites in chromatin. *Annual Review of Biochemistry*, **57**, 159–97.

Gysels, H. (1964). Immunoelectrophoresis of avian lens proteins. *Experientia*, **20**, 145–6.

Hammer, R. E., Pursel, V. G., Rexroad, C. E., Jr, Wall, R. J., Bolt, D. J., Ebert, K. M., Palmiter, R. D. & Brinster, R. L. (1985). Production of transgenic rabbits, sheep and pigs, by microinjection. *Nature*, **315**, 680–3.

Harvey, R. P., Krieg, P. A., Robins, A. J., Coles, L. S. & Wells, J. R. E. (1981). Non-tandem arrangement and divergent transcription of chicken histone genes. *Nature*, **294**, 49–53.

Hawkins, J. W., Nickerson, J. M., Sullivan, M. A. & Piatigorsky, J. (1984). The chicken δ-crystallin gene family. Two genes of similar structure in close chromosomal approximation. *Journal of Biological Chemistry*, **259**, 9821–25.

Helfman, D. M., Feramisco, J. R., Ricci, W. H. & Hughes, S. H. (1984).Isolation and sequence of a cDNA clone that contains the entire coding region for chicken smooth-muscle α-tropomyosin. *Journal of Biological Chemistry*, **259**, 14136–43.

Huckaby, C. S., Conneely, O. M., Beattie, W. G., Dobson, A. D. W., Tsai, M.-J. & O'Malley, B. (1987). Structure of the chromosomal chicken progesterone receptor gene. *Proceedings of the National Academy of Sciences of USA*, **84**, 8380–4.

Hughes, S. N., Kosik, E., Fadly, A. M., Salter, D. W. & Crittenden, L. B. (1986). Design of retroviral vectors for the insertion of foreign deoxyribonucleic acid sequences into the avian germ line. *Poultry Science*, **65**, 1459–67.

Hughes, S. H., Stubblefield, E., Payvar, F., Engel, J. D. Dodgson, J. B., Spector, D., Cordell, B., Schimke, R. T. & Varmus, H. E. (1979). Gene localization by chromosome fractionation: Globin genes are on at least two chromosomes and three estrogen-inducible genes are on three chromosomes. *Proceedings of the National Academy of Sciences of USA*, **76**, 1348–52.

Ingolia, T. D. & Craig, E. A. (1982). Four small *Drosophila* heat shock proteins are related to each other and to mammalian α-crystallin. *Proceedings of the National academy of Sciences of USA*, **79**, 2360–4.

Jeltsch, J. M., Krozowski, Z., Quirin-Stricker, C., Gronemeyer, H., Simpson, R. J., Garnier, J. M., Krust, A., Jacob, F. & Chambon, P. (1986). Cloning of the chicken progesterone receptor. *Proceedings of the National Academy of Sciences of USA*, **83**, 5424–8.

Jung, A., Sippel, A. E., Grez, M. & Schutz, G. (1980). Exons encode functional and structural units of chicken lysozyme. *Proceedings of the National Academy of*

*Sciences of USA*, **77**, 5759–63.
Kim, H.-R. C., Yew, N. S., Ansorge, W., Voss, H., Schwager, C., Vennstrom, B., Zenke, M. & Engel, D. (1988). Two different mRNAs are transcribed from a single genomic locus encoding the chicken erythrocyte anion transport proteins (Band 3). *Molecular and Cellular Biology*, **8**, 4416–24.
Kingsman, S. M. & Kingsman, A. J. (1988). *Genetic Engineering*. Oxford: Blackwell Scientific Publications.
Kok, K., Snippe, L., Geert, A.B. & Gruber, M. (1985). Nuclease- hypersensitive sites in chromatin of the estrogen-inducible apoVLDL II gene of chicken. *Nucleic Acids Research*, **13**, 5189–202.
Kondoh, H., Hayashi, S., Takahashi, Y. & Okada, T. S. (1986). Regulation of δ-crystallin expression: an investigation by transfer of the chicken gene into mouse cells. *Cell Differentiation*, **19**, 151–60.
Kulomaa, M. S., Weigel, W. L., Kleinsek, D. A., Beattie, W. G., Schrader, W. T. & O'Malley, B. W. (1986). Amino acid sequence of a chicken heat shock protein derived from the complementary DNA nucleotide sequence. *Biochemistry*, **25**, 6244–51.
Kwiatkowski, R. W., Ehrismann, R., Scheinfest, C. W. & Dottin, R. P. (1985). Accumulation of creatine kinase mRNA during myogenesis: Molecular cloning of a B-creatine kinase cDNA. *Developmental Biology*, **112**, 84–8.
Lai, E. C., Stein, J. P., Catterrall, J. F., Woo, S. L. C., Mace, M. L., Means, A. R. & O'Malley, B. W. (1979). Molecular structure and flanking nucleotide sequence of the natural chicken ovomucoid gene. *Cell*, **18**, 829–42.
Lamb, I. C., Galehouse, D. M. & Foster, D. N. (1988). Chicken growth hormone cDNA sequence. *Nucleic Acids Research*, **16**, 9339.
Lasmoles, F., Jullienne, A., Day, F., Minvielle, S., Milhaud, G. & Koukhtar, M. S. (1985). Elucidation of the nucleotide sequence of chicken calcitonin mRNA: direct evidence for the expression of a lower vertebrate calcitonin-like gene in man and rat. *European Molecular Biology Organization Journal*, **4**, 2603–7.
LeMeur, M., Glanville, N., Mandel, J. L., Gerlinger, P., Palmiter, R. & Chambon, P. (1981). The ovalbumin gene family: hormonal control of the X and Y gene transcription and mRNA accumulation. *Cell*, **23**, 561–71.
Limbach, K. J. & Wu, R. (1983). Isolation and characterization of two alleles of the chicken cytochrome *c* gene. *Nucleic Acids Research*, **11**, 8931–50.
Lindquist, S. (1986). The heat-shock proteins. *Annual Review of Biochemistry*, **55**, 1151–91.
Lopata, M. A., Havercroft, J. C., Chow, L. T. & Cleveland, D. W. (1983). Four unique genes required for β-tubulin expression in vertebrates. *Cell*, **32**, 713–24.
Lozano, G., Ninomiya, Y., Thompson, H. & Olson, B. R. (1985). A distinct class of vertebrate collagen genes encodes chicken type IX collagen polypeptides. *Proceedings of the National Academy of Sciences of USA*, **82**, 4050–4.
Martin, G. R., Timpl, R., Müller, P. K. & Kühn, K.(1985). The genetically distinct collagens. *Trends in Biochemical Sciences*, **10**, 285–7.
Matsubasa, T., Takiguchi, M., Amaya, Y., Matsuda, I. & Mori, M. (1989). Structure of the rat argininosuccinate lyase gene: Close similarity to chicken δ-crystallin genes. *Proceedings of the National Academy of Sciences of USA*, **86**, 592–6.
Mattaj, I. W., Mackenzie, A. & Jost, J.-P. (1982). Cloning in a cosmid vector of

complete 37kb and 25kb ribosomal DNA repeat units from the chicken. *Biochimica et Biophysica Acta*, **698**, 204–10.

McReynolds, L., O'Malley, B. W., Nisbet, A. D., Fothergill, J. E., Givol, D., Fields, S., Robertson, M. & Brownlee, G. G. (1978). Sequence of chicken ovalbumin mRNA. *Nature*, **273**, 723–8.

Meier, R., Becker-André, M., Götz, R., Heumann, R., Shaw, A. & Thoenen, H. (1986). Molecular cloning of bovine and chick nerve growth factor (NGF): delineation of conserved and unconserved domains and their relationship to the biological activity and antigenicity of NGF. *European Molecular Biology Organization Journal*, **5**, 1489–93.

Meijlink, F. C. P. W., Schip, A. D., Arnberg, A. C., Wieringa, B., Geert, A.B. & Gruber, M. (1981). Structure of the chicken apo very low density lipoprotein II gene. *Journal of Biological Chemistry*, **256**, 9668–71.

Merrill, G. F., Harland, R. M., Groudine, M. and McKnight, S. L. (1984). Physical analysis of the chicken *tk* gene. *Molecular and Cellular Biology*, **4**, 1769–76.

Monteiro, M. J. & Cleveland, D. W. (1988). Sequence of chicken c$\beta$7 tubulin. Analysis of a complete set of vertebrate $\beta$-tubulin isotypes. *Journal of Molecular Biology*, **199**, 439–46.

Morimoto, R. I., Hunt, C., Huang, S.-Y., Berb, K. L. & Banerji, S. S. (1986). Organization, nucleotide sequence, and transcription of the chicken HSP70 gene. *Journal of Biological Chemistry*, **261**, 12692–9.

Nabeshima, Y., Fujii-Kuriyama, Y., Muramatsu, M. & Ogata, K. (1984). Alternative transcription and two modes of splicing result in two myosin light chains from one gene. *Nature*, **308**, 333–8.

Nef, P., Mauron, A., Stalder, R., Alliod, C. & Ballivet, M. (1984). Structure, linkage and sequence of the two genes encoding the $\delta$ and $\gamma$ subunits of the nicotinic acetylcholine receptor. *Proceedings of the National Academy of Sciences of USA*, **81**, 7975–9.

Nickerson, J. M. & Piatigorsky, J. (1984). Sequence of the complete $\delta$-crystallin cDNA. *Proceedings of the National Academy of Sciences of USA*, **81**, 2611–5.

Nickerson, J. M., Wawrousek, E. F., Hawkins, J. W., Wakil, A. S., Wiastow, G., Thomas, G., Norman, B. L. & Piatigorsky, J. (1985). The complete of the $\delta$1 crystallin gene and its 5′ flanking region. *Journal of Biological Chemistry*, **260**, 9100–5.

Nickerson, J. M., Wawroussek, E. F., Borras, T., Hawkins, J. W., Norman, B. L., Filpula, D. R., Nagle, J. W., Ally, A. H. & Piatigorsky, J. (1986). Sequence of the $\delta$2 crystallin gene and its intergenic spacer. Extreme homology with the $\delta$1 crystallin gene. *Journal of Biological Chemistry*, **261**, 552–7.

Nikovits, W., Kuncio, G. & Ordahl, C. P. (1986). The chicken fast skeletal troponin I gene: exon organization and sequence. *Nucleic Acids Research*, **14**, 3377–90.

Nojima, H. (1987). Molecular evolution of the calmodulin gene. *Federation of European Biochemical Societies Letters*, **217**, 187–90.

Ohno, M., Sakamoto, H., Yasuda, K., Okada, T. S. & Shimura, Y. (1985). Nucleotide sequence of a chicken $\delta$-crystallin gene. *Nucleic Acids Research*, **13**, 1593–606.

Old, R. W. & Primrose, S. B. (1989). *Principles of Gene Manipulation*, 4th edn. Oxford: Blackwell Scientific Publications.

Oliva, R. & Dixon, G. H. (1989). Chicken protamine genes are intronless. The complete genomic sequence and organization of the two loci. *Journal of Biological Chemistry*, **264**, 12472–81.

Padgett, T. G., Stubblefield, E. & Varmus, H. E. (1977). Chicken macrochromosomes contain an endogenous provirus and microchromosomes contain sequences related to the transforming gene of ASV. *Cell*, **10**, 649–57.

Perry, M. M. (1988). A complete culture system for the chick embryo. *Nature*, **331**, 70–2.

Peterson, C. A. & Piatigorsky, J. (1986). Preferential conservation of the globular domains of the βA3/A1-crystallin polypeptide of the chicken eye lens. *Gene*, **45**, 139–47.

Piatigorsky, J. (1984a). Lens crystallins and their gene families. *Cell*, **38**, 620–1.

Piatigorsky, J. (1984b). Delta crystallins and their nucleic acids. *Molecular and Cell Biochemistry*, **59**, 33–56.

Piatigorsky, J., O'Brien, W. E., Norman, B. L., Kalumuck, K., Wistow, G. J., Borras, T., Nickerson, J. M. & Wawrousek, E. F. (1988). Gene sharing by δ-crystallin and arginosuccinate lyase. *Proceedings of the National Academy of Sciences of USA*, **85**, 3479–83.

Presland, R. B., Gregg, K., Molloy, P. L., Morris, C. P., Crocker, L. A. & Rogers, G. E. (1989a) Avian keratin genes: I. A molecular analysis of the structure and expression of a group of feather keratin genes. *Journal of Molecular Biology*, **209**, 549–59.

Presland, R. B., Whitbread, L. A. & Rogers, G. E. (1989b) Avian keratin genes.II. Chromosomal arrangement and close linkage of three gene families. *Journal of Molecular Biology*, **209**, 561–76.

Pu, H. & Young, A. P. (1989). The structure of the chicken glutamine synthetase-encoding gene. *Gene*, **81**, 169–75.

Putkey, J. A., Carroll, S. L. & Means, A. R. (1987). The nontranscribed chicken calmodulin pseudogene cross-hybridizes with mRNA from the slow-muscle troponin C gene. *Molecular and Cellular Biology*, **7**, 1549–53.

Quasba, P. K. & Safaya, S. K. (1984). Similarity of the nucleotide sequences of rat α-lactalbumin and chicken lysozyme genes. *Nature*, **308**, 377–80.

Raju, K. & Anwar, R. A. (1987). A comparative analysis of the amino acid and cDNA sequences of bovine elastin a and chick elastin. *Biochemistry and Cellular Biology*, **65**, 842–5.

Reynaud, C.-A., Dahan, A. & Weill, J.-C. (1983). Complete sequence of a chicken λ light chain immunoglobulin derived from the nucleotide sequence of its mRNA. *Proceedings of the National Academy of Sciences of USA*, **80**, 4099–103.

Riddle, R. D., Yamamoto, M. & Engel, J. D. (1989). Expression of δ-aminolevulinate synthase in avian cells: Separate genes encode erythroid-specific and nonspecific isozymes. *Proceedings of the National Academy of Sciences of USA*, **86**, 792–6.

Roberts, R. J. (1989). Restriction enzymes and the isoschizomers. *Nucleic Acids Research*, **17**, r347–87.

Royal, A., Garapin, A., Cami, B., Perrin, F., Mandel, J. L., LeMeur, M., Brégégègre, F., Gannon, F., LePennec, J.P., Chambon, J. & Kourilsky, P. (1979). The ovalbumin gene region: common features in the organization of three genes expressed in chicken oviduct under hormonal control. *Nature*, **279**, 125–32.

Saborio, J. S., Segura, M., Flores, M., Garcia, F. R. & Palmer, E. (1979).

Differential expression of gizzard actin genes during chick embryogenesis. *Journal of Biological Chemistry*, **254**, 11119–25.
Sai, S., Tanaka, T., Kosher, R. A. & Tanzer, M. L. (1986). Cloning sequence analysis of a partial cDNA for chicken cartilage proteoglycan core protein. *Proceedings of the National Academy of Sciences of USA*, **83**, 5081–5.
Schip, A. D., Meijlink, F. C. P. W., Strijker, R., Gruber, M., Vliet, A. J., Klundert, J. A. M. & AB, G. (1983). The nucleotide sequence of the chicken apo very low density lipoprotein II gene. *Nucleic Acids Research*, **11**, 2529–40.
Schip, F. D., Samallo, J., Broos, J., Ophuis, J., Mojet, M., Gruber, M. & AB, G. (1987). Nucleotide sequence of a chicken vitellogenin gene and derived amino acid sequence of the encoded yolk precursor protein. *Journal of Molecular Biology*, **196**, 245–60.
Schwartz, R. J. & Rothblum, K. (1980). Regulation of muscle differentiation: Isolation and purification of chick actin messenger ribonucleic acid. *Biochemistry*, **19**, 2506–14.
Shuman, R. M. (1990). Genetic engineering. In *Poultry Breeding and Genetics*, ed. R. D. Crawford, pp. 585–98. Amsterdam: Elsevier.
Shuman, R. M. & Shoffner, R. N. (1986). Gene transfer by avian retroviruses. *Poultry Science*, **65**, 1437–44.
Shoffner, R. N. (1986). Perspectives for molecular genetics research and application in poultry. *Poultry Science*, **65**, 1489–96.
Sibley, C.G. & Brush, A.H. (1967). An electrophoretic study of avian eye-lens proteins. *Auk*, **84**, 203–19.
Silva, R., Fischer, A. H. & Burch, J. B. E. (1989). The major and minor chicken vitellogenin genes are each adjacent to partially deleted pseudogene copies of the other. *Molecular and Cellular Biology*, **9**, 3557–62.
Simmen, R. C. M., Tanaka, T., Ts'ui, K. F., Putkey, J. A., Scott, M. J., Lai, E. C. & Means, A. R. (1985). The structural organization of the chicken calmodulin gene. *Journal of Biological Chemistry*, **260**, 907–12.
Slingsby, C. (1985). Structural variation in lens crystallins. *Trends in Biochemical Sciences*, **10**, 281–4.
Souza, L. M., Boone, T. C., Murdoch, D., Langley, K., Wypych, J., Fenton, D., Johnson, S., Lai, P. H., Everett, R., Hsu, R.-Y. & Bosselman, R. (1984). Application of recombinant DNA technologies to studies on chicken growth hormone. *Journal of experimental Zoology*, **232**, 465–73.
Stein, J. P., Catterall, J. F., Kristo, P., Means, A. R. & O'Malley, B. W. (1980). Molecular cloning of the ovomucoid gene sequences from partially purified ovomucoid messenger RNA. *Cell*, **21**, 681–7.
Stein, J. P., Munjaal, R. P., Lagage, L., Lai, E. C., O'Malley, B. & Means, A. R. (1983). Tissue-specific expression of a chicken calmodulin pseudogene lacking intervening sequences. *Proceedings of the National Academy of Sciences of USA*, **80**, 6485–9.
Stone, E. M., Rothblum, K. N., Alevy, M. C., Kuo, T. M. & Schwartz, R. J. (1985). Complete sequence of chicken glyceraldehyde-3-phosphate dehydrogenase gene. *Proceedings of the National Academy of Sciences of USA*, **82**, 1628–32.
Sugarman, B. J., Dodgson, J. B. & Engel, J. D. (1983). Genomic organization, DNA sequence, and expression of chicken embryonic histone genes. *Journal of Biological Chemistry*, **258**, 9005–16.

Sullivan, K. F. & Wilson, L. (1984). Developmental and biochemical analysis of chick brain tubulin heterogeneity. *Journal of Neurochemistry*, **42**, 1363–71.

Suminami, Y., Kishi, F., Torigoe, T. & Nakazawa, A. (1988). Structure and complete nucleotide sequence of the gene encoding chicken cytosolic adenylate kinase. *Journal of Biochemistry*, **103**, 611–17.

Sun, S.-T., Tanaka, T., Nishio, I., Peetermans, J., Maizel, J. V. & Piatigorsky, J. (1984). Direct observation of δ-crystallin accumulation by laser light-scattering spectroscopy in the chicken embryo lens. *Proceedings of the National Academy of Sciences of USA*, **81**, 785–7.

Tanaka, T., Har-El, R. & Tanzer, M. L. (1988). Partial structure of the gene for chicken cartilage proteoglycan core protein. *Journal of Biological Chemistry*, **263**, 15831–5.

Thompson, M. A., Hawkins, J. W. & Piatigorsky, J. (1987). Complete nucleotide of the chicken αA-crystallin gene and its 5′ flanking region. *Gene*, **56**, 173–84.

Treton, J. A., Shinohara, T. & Piatigorsky, J. (1982). Degradation of δ-crystallin mRNA in the lens fiber cells of the chicken. *Developmental Biology*, **92**, 60–5.

Trüeb, B., Schreier, T., Winterhalter, K. H. & Strehler, E. E. (1988). Sequence of a cDNA clone encoding chicken ribosomal protein S17. *Nucleic Acids Research*, **16**, 4723.

Upholt, W. B. & Sandell, L. J. (1986). Exon/intron organization of the chicken type II procollagen gene: Intron size distribution suggests a minimal intron size. *Proceedings of the National Academy of Sciences of USA*, **83**, 2325–9.

Von der Ahe, D., Renoir, J.-M., Buchou, T., Baulieu, E.-E. & Beato, M. (1986). Receptors for glucocorticoid steroid and progesterone recognise distinct features of a DNA regulatory element. *Proceedings of the National Academy of Sciences of USA*, **83**, 2817–21.

Watson, D. K., McWilliams, M. J. & Papas, T. S. (1988). Molecular organization of the chicken *ets* locus. *Virology*, **164**, 99–105.

Wei, D. & Andrews, G. K. (1988). Molecular cloning of chicken metalothionein. Deduction of the complete amino acid sequence and analysis of expression using cloned cDNA. *Nucleic Acids Research*, **16**, 537–53.

Wilson, P. W., Harding, M. & Lawson, D. E. M. (1985). Putative amino acid sequence of chick calcium-binding protein deduced from a complementary DNA sequence. *Nucleic Acids Research*, **13**, 8867–81.

Wistow, G. (1985). Domain structure and evolution in α-crystallins and small heat shock proteins. *Federation of European Biochemical Societies Letters*, **181**, 1–6.

Wistow, G. J. & Piatigorsky, J. (1988). Lens crystallins: The evolution and expression of proteins for a highly specialised tissue. *Annual Review of Biochemistry*, **57**, 479–504.

Woo, S. L. C., Beattie, W. G., Catterall, J. F., Dugaiczyk, A., Staden, R., Brownlee, G. G. & O'Malley, B. W. (1981). Complete nucleotide sequence of the chicken chromosomal ovalbumin gene and its biological significance. *Biochemistry*, **20**, 6437–46.

Wozney, J., Hanahan, D., Morimoto, R., Boedtker, H. & Doty, P. (1981). Fine structural analysis of the chicken pro α2 collagen gene. *Proceedings of the National Academy of Sciences of USA*, **78**, 712–16.

Yamada, Y., Liau, G., Mudryj, M., Obici, S. & de Crombrugghe, B. (1984). Conservation of the sizes for one but not another class of exons in two chick collagen genes. *Nature*, **310**, 333–7.

Yamamoto, M., Yew, N. S., Federspiel, M., Dodgson, J. B., Hayashi, N. & Engel, J. D. (1985). Isolation of recombination cDNAs encoding chicken erythroid δ-aminolevulinate synthase. *Proceedings of the National Academy of Sciences of USA*, **82**, 3702–6.

Yasuda, K. & Okada, T. S. (1986). Structure and expression of the chicken crystallin genes. *Oxford Surveys on Eukaryotic Genes*, **3**, 183–209.

Zehner, Z. E., Li, Y., Roe, B. A., Paterson, B. M. & Sax, C. M. (1987). The chicken vimentin gene. Nucleotide sequence, regulation elements, and comparison to the hamster gene. *Journal of Biological Chemistry*, **262**, 8112–20.

Zhang, H. & Sommerville, C. (1988). The primary structure of chicken liver cytochrome $b_5$ deduced from the DNA sequence of a cDNA clone. *Archives of Biochemistry and Biophysics*, **264**, 343–7.

Zwaan, J. & Ikeda, A. (1968). Macromolecular events during differentiation of chicken lens. *Experimental Eye Research*, **7**, 301–11.

APPENDIX I

# *Linkage groups and the chromosome map in the domestic fowl*

The first linkage map for the domestic fowl was published by Hutt (1936); it consisted of 18 loci in five linkage groups. It has since been revised several times as new data became available, and the map given below is that published by Somes in 1988. For the complete details and references, see Somes (1988) and Bitgood & Somes (1990). All the earliest mapping was carried out by determining recombinant frequencies (see Chapter 5, section 5.2), thereby establishing the linkage groups that were given Roman numerals (I to X). Sex-linked characters can be assigned to the sex chromosomes, and apart from the H-W antigen that has been assigned to the small W chromosome, all other sex-linked characters have been assigned to the larger Z chromosome. Apart from the sex chromosomes it was not possible until 1973 to equate a particular linkage group with a particular chromosome. After the first chromosome translocations were studied by Zartman (1973), it became possible in some instances to link particular characters with particular translocations and hence identify the chromosome carrying a particular gene (Chapter 5, section 5.3). Once one gene has been located on a particular chromosome, linkage relations enable others to be linked. Chromosomes are given Arabic numerals in descending size order. The newer methods of mapping genes, using somatic cell hybridisation, *in situ* hybridisation, or chromosome fractionation, establish the particular chromosome, rather than the linkage group used in earlier studies. Until recently it was not possible to distinguish any of the microchromosomes, but with a sequential counterstain-enhanced fluorescence technique it is now possible to identify up to chromosome 18 out of the 39 haploid chromosomes (Auer *et al.*, 1987).

Figures I.1 and I.2 show the mapped linkage relationships, and Table I.1 lists those genes that have been identified on particular chromosomes, by translocations somatic cell hybridization, *in situ* hybridisation, or chromosome fractionation, but their position on the chromosome may not yet be clear.

**Fig. I.1** Linkage map of chromosome Z (Group V) and chromosome 1 (Group III). The dotted lines are for questionable assignments. (From Bitgood & Somes, 1990).

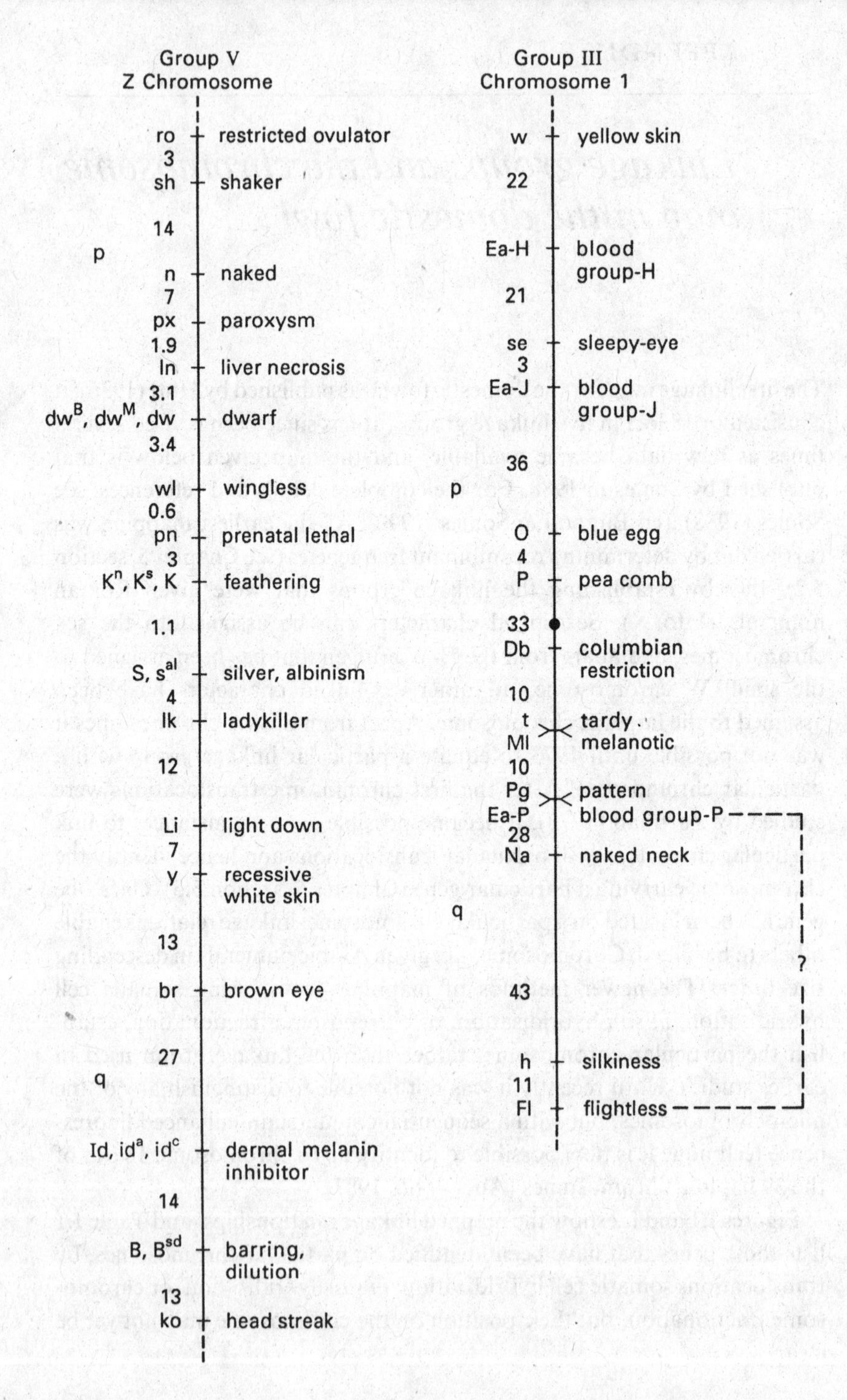

**Fig. I.2** Maps for linkage groups I, II, IV, VIII (chromosome 7) and X (microchromosome 17). The suborder within the MHC (major histocompatibility complex), i.e. B-G, B-F, and B-L, relative to the nucleolar organiser region (NOR) is not known. (From Bitgood & Somes, 1990).

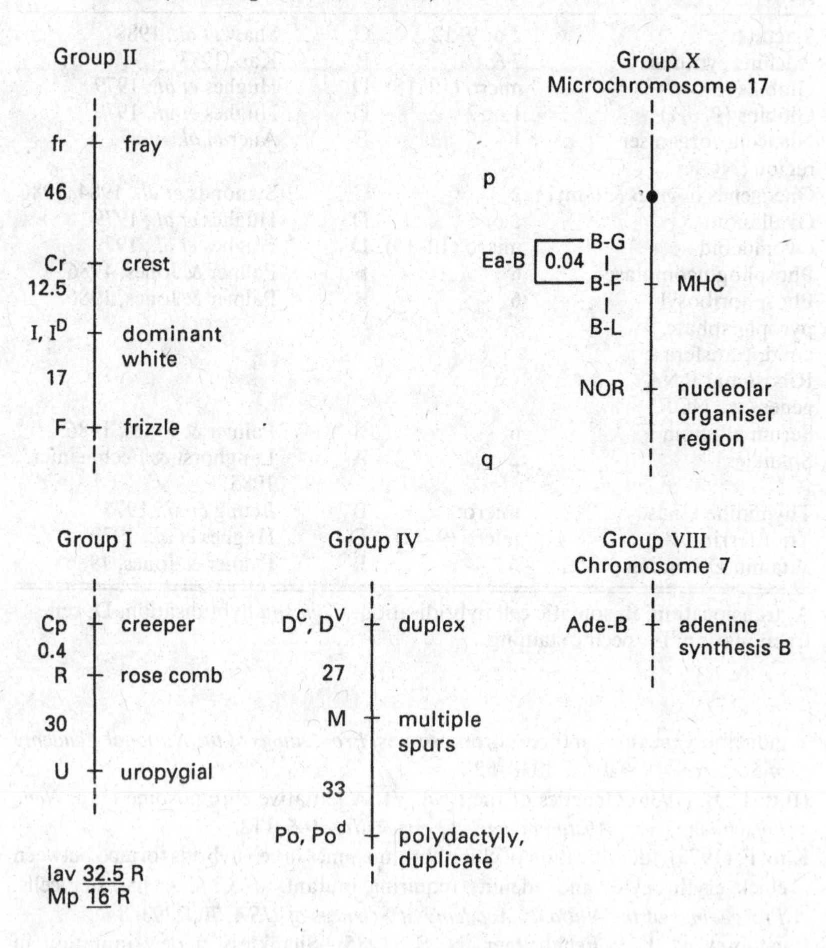


## References


Auer, H., Mayr, B., Lambrou, M. & Schleger, W. (1987). An extended chicken karyotype, including the NOR chromosome. *Cytogenetics and Cell Genetics*, **45**, 218–21.

Bitgood, J. J. & Somes, R. G. (1990) Linkage relationships and gene mapping. In *Poultry Breeding and Genetics*, ed. R. D. Crawford, pp. 469–95. Amsterdam: Elsevier.

Hughes, S. H., Stubblefield, E., Payvar, F., Engel, J. D., Dodgson, J. B., Spector, D., Cordell, B., Schimke, R. T. & Varmus, H. E. (1979). Gene localization by fractionation: globin genes are on at least two chromosomes and three estrogen-

Table I.1. *Chromosome mapping in the domestic fowl*

| Gene | Chromosome | Method | Reference |
|---|---|---|---|
| $\beta$-actin | 2 or 9–12 | C | Shaw *et al.*, 1988 |
| Adenine synthesis | 7 & 8 | B | Kao, 1973 |
| Globins ($\alpha^A$, $\alpha^D$) | micro (10–15) | D | Hughes *et al.*, 1979 |
| Globins ($\beta$, $\rho$, $\epsilon$) | 1 or 2 | D | Hughes *et al.*, 1979 |
| Nucleolar organiser region (NOR) | 17 | E | Auer *et al.*, 1987 |
| Oncogenes (c-*erbB* & c-*myc*) | 2 | C | Symonds *et al.*, 1984, 1986 |
| Ovalbumin | 2 or 3 | D | Hughes *et al.*, 1979 |
| Ovomucoid | micro (10–15) | D | Hughes *et al.*, 1979 |
| Phosphoglucomutase | 6 | B | Palmer & Jones, 1986 |
| Phosphoribosyl pyrophosphate amidotransferase | 6 | B | Palmer & Jones, 1986 |
| Ribosomal RNA genes (see NOR) | | | |
| Serum albumin | 6 | B | Palmer & Jones, 1986 |
| Shankless | 2 | A | Langhorst & Fechheimer, 1985 |
| Thymidine kinase | micro | B | Leung *et al.*, 1975 |
| Transferrin | micro (9–12) | D | Hughes *et al.*, 1979 |
| Vitamin D binding protein | 6 | B | Palmer & Jones, 1986 |

A, translocation; B, somatic cell hybridisation; C, *in situ* hybridisation; D, cell fractionation; E, specific staining.

inducible genes are on three chromosomes. *Proceedings of the National Academy of Sciences of USA*, **76**, 1348–52.

Hutt, F. B. (1936) Genetics of the fowl. VI. A tentative chromosome map. *Neue Forschung. Tierz. Abstammungsl. (Festschrift)*, **105**, 112.

Kao, F. (1973). Identification of chick chromosomes in cell hybrids formed between chick erythrocytes and adenine requiring mutants of Chinese hamster cells. *Proceedings of the National Academy of Sciences of USA*, **70**, 2893–8.

Langhorst, L. J. & Fechheimer, N. S. (1985). Shankless, a new mutation in chromosome 2 in the chicken. *Journal of Heredity*, **76**, 182–6.

Leung, W.-C., Chen, T. R., Dubbs, D. R. & Kit, S. (1975). Identification of chick thymidine kinase determinant in somatic cell hybrids of chick erythrocytes and thymidine kinase-deficient mouse cells. *Experimental Cell Research*, **95**, 320–6.

Palmer, D. K. & Jones, C. (1986). Gene mapping in chicken–Chinese hamster somatic cell hybrids. *Journal of Heredity*, **77**, 106–8.

Somes, R. G. (1988) International Registry of Poultry Genetic Stocks. *Storrs Agricultural Experimental Station Bulletin*, 476.

Shaw, E. M., Guise, K. S. & Shoffner, R. N. (1989). Chromosomal localization of chicken sequences homologous to the $\beta$-actin gene by in situ hybridization. *Journal of Heredity*, **80**, 475–8.

Symonds, G., Stubblefield, E., Guyaux, M. & Bishop, J. M. (1984). Cellular oncogenes (*c-erb-A* and *c-erb-B*) located on different chicken chromosomes can be transduced into the same retroviral genome. *Molecular and Cellular Biology*, **4**, 1627–30.

Symonds, G., Quintrell, N., Stubblefield, E. & Bishop, J. M. (1986) Dispersed chromosomal localization of the proto-oncogenes transduced into the genome of Mill Hill 2 or E26 leukemia virus. *Journal of Virology*, **59**, 172–5.

Zartman, D. L. (1973). Location of the pea comb gene. *Poultry Science*, **52**, 1455–62.

APPENDIX II

# *Oncogenes*

One class of genes not described in this book is the oncogenes. It now seems fairly certain that many cancers are caused either by the malfunctioning of certain cellular genes or by a virus entering a host cell and so introducing additional genes able to transform normal cells into cancer cells. In recent years there have been great advances in understanding and defining these genes, which are termed oncogenes. Many of the oncogenes so far identified occur in the domestic fowl. Thus, in a review of oncogenes Bishop (1983) listed 25 of which nine were from the domestic fowl. At present more than 40 are known (Reddy, Skalka & Curran, 1988), and it is thought that the number may rise to the region of 100 (Watson *et al.*, 1987), although the number of different classes appears to be less than 30. Since a number of the cellular oncogenes are present in normal uninfected cells they constitute part of the normal genome of the domestic fowl. Those oncogenes that are introduced on viral infection are related to their cellular counterparts. In this appendix the different types of oncogenes known to be present in the domestic fowl are listed, together with their protein products.

Two types of virus may cause cancer: DNA viruses and RNA viruses. The oncogenic DNA viruses are diverse in structure and the oncogene that forms part of the viral genome is required for transformation. Generally two or three different oncogenes are necessary to transform host cells into cancer cells. All the known oncogenic RNA viruses are retroviruses and they appear to be related – sufficiently so to have descended from a single ancestral type. Retroviruses are RNA viruses which, when infecting host cells, use the enzyme reverse transcriptase; this copies the RNA genome into a DNA provirus which becomes integrated into the host genome. When the host DNA is transcribed it therefore produces more virus RNA.

The best understood oncogenes in the domestic fowl are those from retroviruses and cellullar oncogenes. The retroviruses can be divided into two types: (i) those which have a long latent period before transforming cells into cancerous cells (often referred to as either non-defective viruses or

Table II.1. *Regions of retroviral genomes*

| Region | Function |
|---|---|
| *gag* (group specific antigen) | Encoding internal structural proteins of the virion |
| *pol* (polymerase) | Encoding reverse transcriptase |
| *env* (envelope) | Encoding surface glycoprotein of the virion envelope |
| *onc* (oncogene) | Encoding oncogene for cell transformation |
| $U_3$ (unique to 3′ end) | 0.2–1 kb region concerned with translational control |
| R (repeat sequence) | 20–80 nucleotides concerned with transfer of DNA chain during synthesis |
| $U_5$ (unique to 5′ end) | 80–100 nucleotides, function ? |
| LTR (long terminal repeat) | Combination of $U_3$-R-$U_5$ at each end of unintegrated proviral DNA |

replication-competent viruses); and (ii) acute transforming viruses, that have a very short latent period (Watson *et al.*, 1987). Many leukaemia viruses, e.g. avian leukaemia viruses (ALV), are in the former category, whereas Rous Sarcoma Virus (RSV) is an acute transforming virus. The main regions of the retroviral genomes are listed in Table II.1. The acute transforming RNA retroviruses generally have a genome which includes one or two of the following genes: *gag*, *pol*, *env*, together with a gene for transforming normal cells to cancerous cells; in some cases only truncated versions of these genes are present in the viral genome. The Rous Sarcoma Virus (Fig. II.1) has all four of these genes and so is able both to replicate and to transform, but most other acute transforming viruses lack one of *gag*, *pol* or *env* and so although able to transform are unable to replicate without a helper virus that contains the missing gene(s). One example of a virus containing each of the different oncogenes is given on Fig. II.1.

The retroviruses which have a long latent period differ from the acute transforming viruses, in that the former lack an oncogene. Their ability to transform is dependent on the position at which the provirus integrates into the host genome. Only when the integration occurs adjacent to a cellular oncogene does transformation occur. Since the position of integration is random the probability of transformation is low.

A number of viral oncogenes have now been identified and also their cellular counterparts. The two types differ in that only the cellular oncogenes have introns; there may also be regions deleted from the viral oncogenes. Viral oncogenes are designated with the prefix ‘*v*’, e.g. *vsrc* and

**Fig. II.1** Genomic structure of transforming viruses in the domestic fowl. For an explanation of the terms, see Table II.1. Oncogenes are shown as hatched blocks.

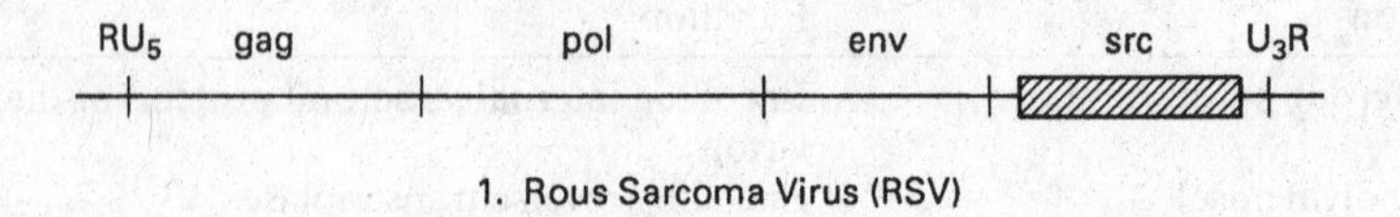


1. Rous Sarcoma Virus (RSV)

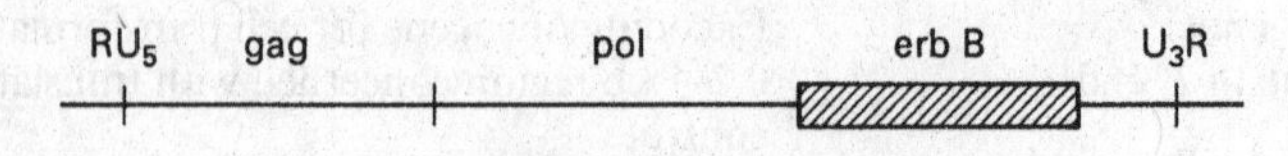


2. Avian Erythroblastosis Virus (AEV-H)

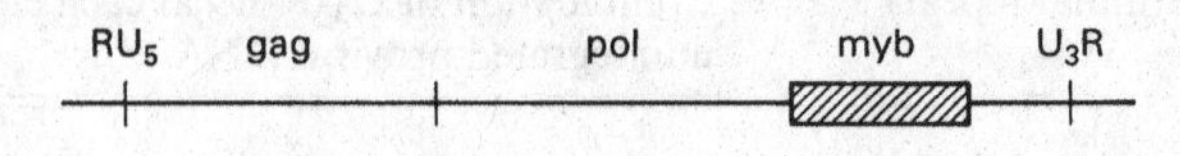


3. Avian Myeloblastosis Virus (AMV)

4. Avian Erythroblastosis Virus: E26 isolate (E26-AEV)

5. Avian Erythroblastosis Virus: S13 isolate (S13-AEV)

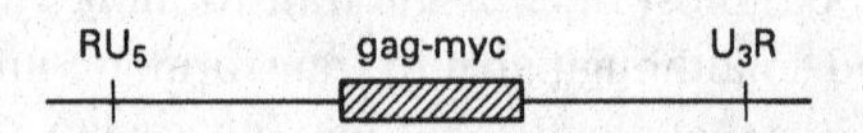


6. Avian Myelocytomatosis Virus : MC29 isolate (MC29)

7. Avian Myelocytomatosis Virus : MH2 isolate (MH2)

8. Avian Sarcoma Virus: Y73 isolate (Y73)

cellular oncogenes '*c*', e.g. *csrc*. For a number of oncogenes the protein product synthesised when the oncogene is expressed has been identified; and so far the products fall into the following five categories (Reddy *et al.*, 1988):

1. Protein kinases which phosphorylate tyrosine residues on proteins.
2. Growth factors or growth factor receptors.
3. Guanine nucleotide binding proteins (GTP-binding proteins).
4. Nuclear proteins.
5. Proteins with multiple membrane-spanning domains.

Of these groups the protein kinases are the most numerous, and they can be subdivided into two categories: (i) integral membrane proteins that span the membrane bilayer, and (ii) extrinsic membrane proteins associated with the inner surface of the plasma membrane. The protein products are often the product of expression of a fused gene.

As can be seen, most of the oncogene products have functions that are related to the control of growth and cell division; their 'abnormal' expression may thus lead to 'uncontrolled' growth and proliferation such as occurs in cancerous tissue. A list of acute transforming retroviruses which infect the domestic fowl and related avian species is given in Table II.2. The viruses are arranged in groups according to their oncogenes. The oncogenes were named according to the virus from which they were first isolated and are therefore trivial names and do not imply target specificity. In each group there may be a number of different isolates, e.g. the oncogene first found in the Fujinami sarcoma virus, *fps*, is also present in the related strains of virus, PRCII, PRCIV, UR1 and 16L.

The commmonest type of oncogene, of those so far characterised from viruses infecting the domestic fowl, is that coding for a protein kinase. These include the Rous sarcoma virus (RSV) and related strains, B77, recovered avian sarcoma virus, and the Prague strain of RSV. The Fujinami sarcoma virus group causes similar fibrosarcomas; however, unlike RSV they require a helper virus for replication since they have incomplete gag, pol and env. The Esh sarcoma virus (ESV- yes) was isolated in 1966 from a tumour on a Leghorn owned by a Pennsylvanian farmer, Mr Esh, and the same oncogene was later found in the Y73 virus. Both strains cause fibrosarcomas.

The avian erythroblastosis viruses (AEV) having oncogenes *erbA* and *erbB* cause erythroleukaemias and sarcomas. The oncogene *v-erbB* has been extensively studied because of its homology to the epidermal growth factor receptor. The epidermal growth factor receptor is a transmembrane protein, the extracellular portion of which contains the extracellular-

Table II.2. *Transforming retroviruses in the domestic fowl*[a]

| Oncogene | *v-onc* | Virus strain | Protein product[b] | Activity of protein[c] | Tumour |
|---|---|---|---|---|---|
| *src* | RSV-*src* | Rous sarcoma virus | $pp60^{src}$ | PK | Sarcoma |
| *src* | B77-*src* | B77 avian (or Rous) sarcoma virus | $pp60^{src}$ | PK | Sarcoma |
| *src* | rASV-*src* | Recovered avian sarcoma virus | $pp60^{src}$ | PK | Sarcoma |
| *src* | PR-RSV-*src* | Prague strain RSV | $pp60^{src}$ | PK | Sarcoma |
| *fps* | FuSV-*fps* | Fujinami sarcoma virus | $P130^{gag\text{-}fps}$ | PK | Sarcoma |
| *fps* | PRCII-*fps* | PRCII sarcoma virus | $P105^{gag\text{-}fps}$ | PK | Sarcoma |
| *fps* | PRCIV-*fps* | PRCIV sarcoma virus | $P170^{gag\text{-}fps}$ | PK | Sarcoma |
| *fps* | UR1-*fps* | Rochester sarcoma virus | $P150^{gag\text{-}fps}$ | PK | Sarcoma |
| *fps* | 16L-*fps* | 16L recovered avian sarcoma virus | $P142^{gag\text{-}fps}$ | PK | Sarcoma |
| *yes* | Y73-*yes* | Y73 avian sarcoma virus | $P90^{gag\text{-}yes}$ | PK | Sarcoma |
| *yes* | ESV-*yes* | Esh sarcoma virus | $P80^{gag\text{-}yes}$ | PK | Sarcoma |
| *sea* | AEV-S13 | Avian erythroblastosis virus | $gp155^{env\text{-}sea}$ | PK | Sarcoma, erythroblastosis |
| *rel* | REV-T-*rel* | Reticuloendotheliosis strain T | $pp59^{rel}$ | PK | Leukaemia |
| *ros* | UR2-*ros* | Avian sarcoma virus UR-2 | $P68^{gag\text{-}ros}$ | PK | |
| *erb*-A | AEV-*erb*-A | Avian erythroblastosis virus | $P75^{gag\text{-}erbA}$ | R | Erythroblastosis |
| *erb*-B | AEV-*erb*-B | Avian erythroblastosis virus | $gp65^{erbB??}$ | R | Erythroblastosis |
| *myc* | MC29-*myc* | Avian myelocytomatosis virus MC29 | $P110^{gag\text{-}myc}$ | NP | Sarcoma |
| *myc* | CMII-*myc* | Avian myelocytomatosis virus CMII | $P90^{gag\text{-}myc}$ | NP | Carcinoma |
| *myc* | MH2-*myc* | Avian myelocytomatosis virus MH2 | $P100^{gag\text{-}myc}$ | NP | Myelocytoma |
| *myc* | OK10-*myc* | Avian myelocytomatosis virus OK10 | $P200^{gag\text{-}pol\text{-}myc}$ | NP | Myelocytoma |
| *myb* | AMV-*myb* | Avian myeloblastosis virus strain BAI-A | $p45^{myb}$ | NP | Myeloblastosis |
| *myb* | E26-*myb* | Avian leukosis virus strain E26 | $P135^{gag\text{-}myb\text{-}ets}$ | NP | Myelo & erythroblastosis |
| *ets* | E26-*ets* | Avian leukosis virus strain E26 | $P135^{gag\text{-}myb\text{-}ets}$ | U | Myelo & erythroblastosis |
| *ski* | ALV-tdB77 | Avian leukosis virus | $P^{gag\text{-}ski}$ | NP | |

[a] For complete details see the references listed below.

[b] The following nomenclature is used: P, protein; pp, phosphoprotein; gp, glycoprotein. The number indicates the size of the protein in kDa. The superscript indicates the gene(s) from which the product is derived. In several cases the product is coded for by two or more gene fragments which have become fused.

binding domain. The *v-erbB* oncogene protein has a large portion of the extracellular region excised, but has the intracellular region intact (Fig. II.2). This latter region contains protein kinase activity. The *c-erbA* gene is located on a microchromosome, and *c-erbB* on one of the macrochromosomes, probably number 2.

The UR2 virus, containing the oncogene *ros*, induces tumours histologically similar to those of RSV. The protein product P68$^{gag\text{-}ros}$ is similar to that produced by *v-erbB* in that it is a transmembrane protein having protein kinase activity (Fig. II.2).

The *v-myc* oncogene is found in the acute transforming virus M29 and three related viruses, CMII, MH2 and OK10, each of which is defective and

**Fig. II.2** Comparison of the structures of the oncogene protein products from *v-erbB* and *v-ros* with that of the epidermal growth factor receptor.

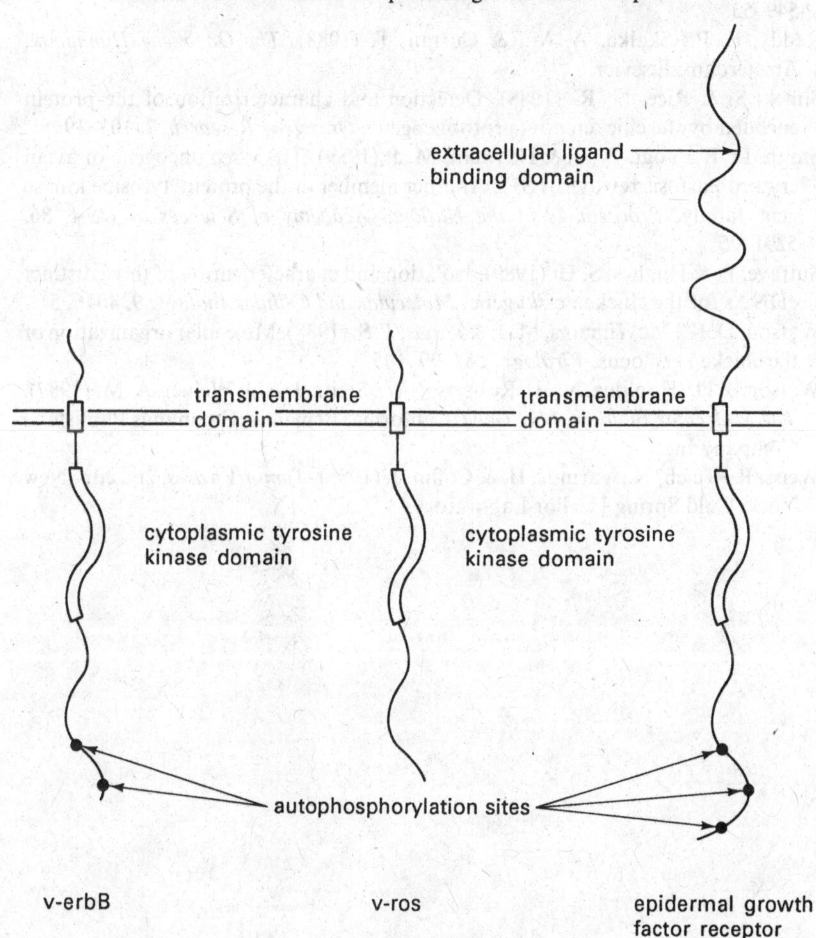

requires helper virus. The oncogene products differ in that MC29 and CMII produce a single fusion protein, whereas OK10 and MH2 each produce two protein products. Proteolytic cleavage does not occur on maturation in M29 or CMII, in contrast to OK10 and MH2. The protein products are located in the nucleus.

## References

Bishop, J. M. (1983). Cellular oncogenes and retroviruses. *Annual Review of Biochemistry*, **52**, 301–54.

Chen, L., Lim, M. Y., Bose, H. & Bishop, J. M. (1988). Rearrangements of chicken immunoglobulin genes in lymphoid cells transformed by the avian retroviral oncogene v-*rel*. *Proceedings of the National Academy of Sciences of USA*, **85**, 549–53.

Reddy, E. P., Skalka, A. M., & Curran, T. (1988). *The Oncogene Handbook*. Amsterdam: Elsevier.

Simek, S. & Rice, N. R. (1988). Detection and characterization of the protein encoded by the chicken *c-rel* protooncogene. *Oncogene Research*, **2**, 103–19.

Smith, D. R., Vogt, P. K. & Hayman, M. J. (1989). The v-*sea* oncogene of avian erythroblastosis retrovirus S13: Another member of the protein–tyrosine kinase gene family. *Proceedings of the National Academy of Sciences of USA*, **86**, 5291–95.

Sutrave, P. & Hughes, S. H. (1989). Isolation and characterization of three distinct cDNAs for the chicken c-*ski* gene. *Molecular and Cellular Biology*, **9**, 4046–51.

Watson, D. K., McWilliams, M. J. & Papas, T. S. (1988). Molecular organization of the chicken *ets* locus. *Virology*, **164**, 99–105.

Watson, J. D., Hopkins, N. H., Roberts, J. W., Steitz, J. A. & Weiner, A. M. (1987). *The Molecular Biology of the Gene*. California: Benjamin/Cummings Publishing Company Inc.

Weiss, R., Teich, N., Varmus, H. & Coffin, J. (1985). *Tumor Viruses*, 2nd edn. New York: Cold Spring Harbor Laboratory.

APPENDIX III

# *The Chi squared ($\chi^2$) test*

It is often necessary in studies of Mendelian genetics to compare the results of particular crosses with those expected on the basis of a particular hypothesis. For example, if a homozygous dominant is crossed with a homozygous recessive and the progeny self-crossed, then we expect a 3:1 ratio of dominant:recessive phenotypes in the $F_2$ generation (see Chapter 3, section 3.2). This ratio is dependent on the random segregation of homologous chromatids during meiosis. Being a chance event, the ratio will not be exactly 3:1 every time such a cross is performed. It is rather like tossing a coin ten times; on average we expect it to fall five times on heads and five times on tails. However, this will not be the case each time 10 tosses are carried out. We therefore need a method of assessing whether the results of a particular cross are close enough to fit our hypothesis. The statistical method used is the Chi squared ($\chi^2$) test. It enables us to assess whether or not a difference between an expected and observed result is significant.

A brief explanation of the test will now be given (for further details of this and other statistical tests, see Mead & Curnow, 1983) and the test will be applied to two sets of data given in Chapters 3 and 5.

In essence, the difference between the observed numbers of a particular phenotype and the expected numbers is determined. This must be done using the actual number used, and not the frequency or percentage of each. The larger the number of observations the less likelihood there is of a chance aberrant result.

The value of $\chi^2 = \sum(O-E)^2/E$ is determined. It is then necessary to know what is called the 'degrees of freedom'. This is given by ($n$–1), where '$n$' is the number of different categories any type can fit into. For example, if the progeny of a cross has either barred or non-barred plumage, there are two types possible and therefore one degree of freedom, i.e. one alternative. The value of $\chi^2$ for a particular number of degrees of freedom is compared in $\chi^2$ distribution tables (Table III.1) that give a probability value (P). Generally $P < 0.05$ is regarded as significant and $P < 0.01$ as highly significant. These

Table III.1. $\chi^2$ *distribution*

| Degrees of freedom | Probability 0.95 | 0.90 | 0.80 | 0.70 | 0.50 | 0.30 | 0.20 | 0.10 | 0.05 | 0.01 | 0.001 |
|---|---|---|---|---|---|---|---|---|---|---|---|
| 1 | 0.004 | 0.02 | 0.06 | 0.15 | 0.46 | 1.07 | 1.64 | 2.71 | 3.84 | 6.64 | 10.83 |
| 2 | 0.10 | 0.21 | 0.45 | 0.71 | 1.39 | 2.41 | 3.22 | 4.60 | 5.99 | 9.21 | 13.82 |
| 3 | 0.35 | 0.58 | 1.01 | 1.42 | 2.37 | 3.66 | 4.64 | 6.25 | 7.82 | 11.34 | 16.27 |
| 4 | 0.71 | 1.06 | 1.65 | 2.20 | 3.36 | 4.88 | 5.99 | 7.78 | 9.49 | 13.28 | 18.47 |
| 5 | 1.14 | 1.61 | 2.34 | 3.00 | 4.35 | 6.06 | 7.29 | 9.24 | 11.07 | 15.09 | 20.52 |
| 6 | 1.63 | 2.20 | 3.07 | 3.83 | 5.35 | 7.23 | 8.56 | 10.64 | 12.59 | 16.81 | 22.46 |
| 7 | 2.17 | 2.83 | 3.82 | 4.67 | 6.35 | 8.38 | 9.80 | 12.02 | 14.07 | 18.48 | 24.32 |
| 8 | 2.73 | 3.49 | 4.59 | 5.53 | 7.34 | 9.52 | 11.03 | 13.36 | 15.51 | 20.09 | 26.12 |
| 9 | 3.32 | 4.17 | 5.38 | 6.39 | 8.34 | 10.66 | 12.24 | 14.68 | 16.92 | 21.67 | 27.88 |
| 10 | 3.94 | 4.86 | 6.18 | 7.27 | 9.34 | 11.78 | 13.44 | 15.99 | 18.31 | 23.21 | 29.59 |
| | Nonsignificant | | | | | | | | Significant | | |

*Source:* R. A. Fisher and F. Yates, *Statistical Tables for Biological, Agricultural and Medical Research* (6th edition), Table IV, Oliver & Boyd, Ltd., Edinburgh, by permission of the authors and publishers.

probabilities mean that a chance difference would only occur in 1 out of 20 (5%) and 1 out of 100 (1%) occasions, respectively. Two examples are given to illustrate the test.

Example 1: Test for linkage of sleepy-eye with other traits.
See Table 5.1 for details.

If the characters given in Table 5.1 segregate independently from sleepy-eye then we would expect a ratio of 1:1 of parental types and recombinant types. The observed ($O_1$ and $O_2$) and expected results ($E_1$ and $E_2$) can be compared and $\chi^2$ determined. In each case there is one degree of freedom, thus *P* values can be found from the appropriate column in Table III.1 as shown below.

| Gene | Observed Parental Types($O_1$) | Expected Parental Types($E_1$) | Observed Parental Types($O_2$) | Expected Parental Types($E_2$) | $(O_1 - E_1)^2/E_1$ | $(O_2 - E_2)2/E_2$ | $\chi^2$ | P |
|---|---|---|---|---|---|---|---|---|
| *I* | 614 | 615 | 616 | 615 | 0.0016 | 0.0016 | 0.0032 | >0.95 |
| *Cr* | 187 | 192.5 | 198 | 192.5 | 0.167 | 0.167 | 0.314 | >0.5 |
| *D* | 194 | 202.5 | 211 | 202.5 | 0.357 | 0.357 | 0.714 | >0.3 |
| *E* | 354 | 353 | 352 | 353 | 0.0028 | 0.0028 | 0.0056 | >0.9 |
| *R* | 89 | 89 | 89 | 89 | 0 | 0 | 0 | =1 |
| *P* | 1063 | 967.5 | 872 | 967.5 | 9.43 | 9.43 | 18.86 | <0.001 |
| *w* | 314 | 277 | 240 | 277 | 4.94 | 4.94 | 9.88 | <0.01 |

The data indicate that the differences from the expected ratio of 1:1 are not significant for *I, Cr, D, E,* and *R*, but are highly significant for *P* and *w*. Therefore linkage between *se* and *P*, and between *se* and *w* can be assumed.

Example 2: Plumage Colour
See Chapter 3, section 3.6.

Morejohn analysed plumage colours by making crosses between the genotypes $e^+$, $e^b$ and $e^y$. His hypothesis was that $e^+$ is dominant to $e^y$, and $e^+$ is incompletely dominant to $e^b$, and $e^y$ is incompletely dominant to $e^b$. The results from self crossing the three heterozygotes ($F_1$ generation) are given together with the expected results, from which the $P$ values can be calculated. Note that the expected ratio for complete dominance is 3:1 but that for incomplete dominance is 1:2:1.

| Crosses | Observed No. of phenotypes | Expected No. of phenotypes | $\sum(O-E)^2/E$ | Degrees of Freedom | P |
|---|---|---|---|---|---|
| $e^+e^y \times e^+e^y$ | 66:19 | 63.75:21.25 | 0.246 | 1 | $>0.5$ |
| $e^+e^b \times e^+e^b$ | 12:11:8 | 7.75:15.5:7.75 | 2.64 | 2 | $>0.1$ |
| $e^ye^b \times e^ye^b$ | 32:50:28 | 27.5:55:27.5 | 1.20 | 2 | $>0.5$ |

All three $P$ values indicate that the differences observed are not significantly from that expected on the basis of the dominance and incomplete relationships mentioned above, and thus Morejohn's hypothesis is acceptable.

## Reference

Mead, R. & Curnow, R. N. (1983). *Statistical Methods in Agriculture and Experimental Biology*. London: Chapman and Hall.

## APPENDIX IV

# *One letter amino acid code*

Twenty different amino acid are commonly found in proteins. They are frequently abbreviated to three letters, most commonly the initial three, but it is convenient for many purposes and especially when expressing long protein sequences to use a single letter abbreviation. These are given below, together with the corresponding three letter abbreviations. Two pairs of amino acids are closely related in structure, aspartic acid/asparagine and glutamic acid/glutamine, the second of each pair being the amide derivative of the first. Amino acid analysis involving acid hydrolysis does not distinguish these pairs, and so it is useful also to have an abbreviation covering both.

| Amino acid | Abbreviations | |
|---|---|---|
| Alanine | Ala | A |
| Arginine | Arg | R |
| Asparagine | Asn | N |
| Aspartic acid | Asp | D |
| Cysteine | Cys | C |
| Glutamic acid | Glu | E |
| Glutamine | Gln | Q |
| Glycine | Gly | G |
| Histidine | His | H |
| Isoleucine | Ile | I |
| Leucine | Leu | L |
| Lysine | Lys | K |
| Methionine | Met | M |
| Phenylalanine | Phe | F |
| Proline | Pro | P |
| Serine | Ser | S |
| Threonine | Thr | T |
| Tryptophan | Trp | W |
| Tyrosine | Tyr | Y |
| Valine | Val | V |
| Aspartic acid or asparagine | Asx | B |
| Glutamic acid or glutamine | Gsx | Z |

APPENDIX V

# *The genetic code*

The genetic code is the consecutive sequence of nucleotide triplets (codons) of DNA and RNA that specify the sequence of amino acids for protein synthesis. The four bases present in DNA and RNA give rise to $4^3$ possible combinations of triplets or codons; the amino acid to which they correspond is given in the table below

| First base | Second base U | C | A | G | Third base |
|---|---|---|---|---|---|
| U | phe | ser | tyr | cys | U |
| U | phe | ser | tyr | cys | C |
| U | leu | ser | Ter | Ter | A |
| U | leu | ser | Ter | trp | G |
| C | leu | pro | his | arg | U |
| C | leu | pro | his | arg | C |
| C | leu | pro | gln | arg | A |
| C | leu | pro | gln | arg | G |
| A | ile | thr | asn | ser | U |
| A | ile | thr | asn | ser | C |
| A | ile | thr | lys | arg | A |
| A | met and fmet | thr | lys | arg | G |
| G | val | ala | asp | gly | U |
| G | val | ala | asp | gly | C |
| G | val | ala | glu | gly | A |
| G | val | ala | glu | gly | G |

Ter, terminator codon, fmet, N-formylmethionine, the initiator codon.

# *Glossary*

**$\alpha$-helix** A helical arrangement of a protein chain having 3.6 amino acid residues per turn and stabilised by hydrogen bonds.
**allele** One of a series of possible alternative forms of a given gene.
**amber mutation** A mutation which leads to the premature termination of a polypeptide chain. It results from the change of a codon for an amino acid to the termination codon UAG.
**antigenic determinant** Particular site on an antigen molecule eliciting the formation of a specific antibody.
**antiserum** Serum containing antibodies against a particular antigen.
**autosome** A chromosome other than a sex chromosome.

**$\beta$-sheet** A sheet-like arrangement of a protein chain in which stretches of the chain lie either parallel or antiparallel to one another, linked by hydrogen bonds.
**Barr body** A condensed single X-chromosome normally found in the somatic cells of a female.
**basal metabolic rate** Resting rate of metabolism, expressed in terms of rate of heat production.
**base** A nitrogenous base (as opposed to acid), of which there are four major types in DNA and in RNA.
**bursectomy** The surgical removal of the bursa.

**C-terminus** The end of a protein chain, which terminates with a carboxyl group.
**cDNA library** A library of complementary DNA inserted into a vector, derived from the total cell mRNAs by the action of reverse transcriptase before insertion into a vector. It contains the gene sequences for all the genes that are being expressed in that particular cell, but lacks the introns.
**centromere** A region of the chromosome to which spindle fibres attach during mitosis.
**chiasma** The points of junction of non-sister chromatids at which crossing over occurs.
**chimaera** An individual composed of a mixture of genetically different cells, or, in the context of gene cloning, the vector (plasmid or virus) containing DNA from a genetically different source.
**chloramphenicol** An antibiotic, which is a potent inhibitor of protein synthesis.
**chymotrypsin** A proteolytic enzyme which catalyses the cleavage of specific peptide

bonds in a protein on the carboxyl side of aromatic amino acids.
**chromatid** The two daughter strands of a duplicated chromosome which are joined by the single centromere.
**circadian rhythm** A rhythm with a cycle of about 24 h.
**cladistic method** A method of classification that attempts to reconstruct phylogenies in terms of branching sequences of ancestor–descendant lineages.
**clone** 1. A group of genetically identical cells or organisms all descended from a single ancestral cell or organism. 2. Genetically engineered replicas of DNA sequences.
**complement system** Part of the defence system in higher organisms, comprising a group of proteins present on the blood which combine with antibody/antigen complexes, causing lysis of bacterial and other cells.
**complementation** The appearence of wild type phenotype in an organism or cell containing two different mutations combined in a hybrid diploid or a heterokaryon.
**complexity** The total length of different sequences of DNA present in a given preparation, usually expressed in base pairs or in daltons. Some organisms may have a large amount of DNA per cell because they have multiple copies of particular sequences. Complexity enables comparison of the amount of different sequences.
**cytoskeleton** The internal proteinaceous framework of a eukaryote cell.

**diploid** Each chromosome represented twice, except the sex chromosome in the heterogametic sex.
**distal** Furthest from the point of attachment.
**diverticulum** A blind ending pouch or tube.
**domain** A lobe-like area of supersecondary structure, when used with reference to the three dimensional structure of a globular protein.

**electrophoresis** A method used for the separation of charged molecules or ions by application of an electric field.
**endoplasmic reticulum** A membranous network within the cytoplasm of cells separating the aqueous phase into two compartments; also a major site of protein synthesis.
**enhancer** A type of control site on DNA that enhances transcription.
**epistasis** The interaction of non-allelic genes in which one gene masks the expression of another.
**erythrocyte** Red blood cell.
**eukaryote** All organisms, the cells of which contain a true nucleus bounded by a nuclear membrane. This group includes all animals, plants, fungi and protozoa.
**exon** A portion of a split gene, the transcript of which is present on mRNA. The initial transcript is processed before forming the mature mRNA. It exits the nucleus.
**expression vector** A vector containing the necessary regulatory sequences to enable expression (transcription and translation) to occur.
**expressivity** The range of phenotypes expressed by a given genotype under any given set of environmental conditions.

**fibula** One of two long bones in the lower limb.
**frame-shift** A mutation that causes a change in the reading frame, by addition or deletion of nucleotides in a DNA sequence.

**genome** All the genes carried by a single gamete, i.e. the genes present on the complete set of haploid chromosomes.
**genomic library** A collection of recombinant DNA molecules (cloned fragments) representative of the individual's genome.
**genotype** The genetic constitution of an organism
**Giemsa stain** A stain containing azur and eosin in glycerol and methanol which can be used to stain regions of the chromosomes.
**glycosylated** When used with reference to protein, it means that the protein contains a carbohydrate moiety that is linked to the protein either by an amide or ether link.

**haemopoiesis** The formation and development of the various types of blood cells.
**haploid** A nucleus, cell or organism possessing a single set of unpaired chromosomes.
**haplotype** A set of closely linked genes that tend to be inherited together.
**hemizygous** A genetic locus present in only one copy.
**heterogametic sex** The sex in which the two sex chromosomes differ, e.g. male (XY) in mammals; female in birds (WZ).
**heterokaryon** A cell containing two or more genetically different nuclei.
**heterosis** The greater vigour, in terms of growth, survival and fertility of hybrids, usually achieved from crosses between highly inbred strains.
**heterozygote** An individual that has inherited different alleles at one or more loci, and therefore does not breed true.
**homology** Structures or processes in different organisms that show a fundamental similarity because they have descended from a common ancestor. The structures may refer to organs or to macromolecules such as proteins.
**homozygous** An organism having identical alleles in the corresponding loci of homologous chromosomes and therefore breeding true.
**hypervariable regions** Regions in the sequence of genes, mRNA or proteins that show considerable variation in the nucleotide sequence or amino acid sequence within their particular class, e.g. the sequence of amino acids in immunoglobulins.

**intervening sequence** See intron.
**intron** A segment of transcribed nuclear RNA that is excised during the processing of mRNA, generally before it exits the nucleus.
**isoelectric focusing** A method used to separate proteins based on differences in their isoelectric point.
**isoform** A family of functionally related proteins that differ slightly in their amino acid sequences.

**karyotype** The somatic chromosomal complement of an individual or species. The term is often used for photomicrographs of the metaphase chromosomes arranged in a standard sequence.
**kb kilobase** A sequence of nucleotides (usually in DNA or RNA) containing 1000 bases.
**kDa** kilodalton, a unit equal to the mass of 1000 hydrogen atoms ($1.67 \times 10^{-21}$ g)
**keratoglobus** A condition in which there is a marked increase in intraoccular fluid with enlargement of the eyeball.

**linkage group** A group of genes having their loci on the same chromosome.
**locus** The position of a gene on a chromosome.
**luteinising hormone** A hormone of the anterior pituitary acting to stimulate oocyte maturation.
**lymphokine** A group of glycoproteins released from T lymphocytes after contact with a cognate antigen.

**macrophage** A type of white blood cell that ingests invading microorganisms.
**mast cell** Cells found in connective tissue that secrete heparin and histamine, and are involved in some hypersensitivity reactions.
**melanocyte** A pigment cell containing melanin granules.
**melanosome** A subcellular organelle producing melanin.
**mesenchyme** An embryonic type of connective tissue.
**metastasis** The spread of malignant neoplastic cells from the original site to another part of the body.
**micromelia** The condition of having disproportionately small or short limbs.
**microtubule** Hollow tube of protein, made up of the protein, tubulin, a component of the cytoskeleton
**mitogen** A compound that stimulates cells to undergo mitosis.
**monoclonal antibodies** Antibodies derived from a single clone of plasma cells, which are therefore all structurally identical.
**mosaic** An individual composed of two or more cell lines of different genetic constitution, both cells being derived from the same zygote; in contrast to those in a chimaera, in which the cells are from different individuals.
**$M_r$** Relative molecular mass, or molecular weight
**mutagen** A physical or chemical agent that raises the frequency of mutation above the spontaneous rate.
**myelination** The acquisition of of a myelin sheath by a nerve cell.
**myoblasts** Cells that differentiate into striated muscle cells.
**myofibrillogenesis** The production of myofibrils.
**myogenic cells** Cells originating in or starting from muscle.
**myopathy** An abnormal condition or disease of muscle tissues.

**N-terminus** The amino terminus, with reference to the sequence of amino acids in a protein.
**nanomelia** The condition of extreme smallness of the extremities.
**neural crest** The ridge of the ectoderm that forms above the neural tube during early embryogenesis in chordates and provides the cells that develop into the peripheral nervous system.
**nucleolar organiser region** The region on a chromosome controlling the formation of the nucleolus.
**nucleolus** An RNA-rich body within the nucleus associated with a chromosome, and concerned with the production of ribosomes.
**nucleosome** The bead-like structure of eukaryote chromosomes comprising histones and DNA.

**oligosaccharide** A carbohydrate comprising a small number of monosaccharides, e.g. glucose, galactose, linked together by a glycoside link.

**oncogene** A gene that induces uncontrolled cell proliferation.
**oviposition** Egg-laying.
**ovotestis** A gonad in which both testicular and ovarian components are present.
**ovulation** Shedding of an egg from the ovary.

**palaeontology** The study of past life on Earth from fossils and fossil impressions.
**penetrance** The proportion of individuals of a specified genotype that show the expected phenotype under a defined set of environmental conditions.
**peptide bond** The bond linking amino acids together in the formation of proteins.
**pericentric inversion** An inversion of part of a chromosome that includes the centromere.
**phenetic method** A taxonomic classification based on phenotypic characteristics without regard to phylogenetic relationships.
**phenotype** The observable characters of an organism, produced by the genotype in conjunction with the environment.
**phylogeny** The evolutionary history and line of descent of a species or higher taxonomic group.
**plasmid** A small self-replicating circular DNA independent of the chromosome in bacteria and unicellular eukaryotes.
**pleiotropic** The phenomenon of a single gene being responsible for a number of distinct and seemingly unrelated phenotypic effects.
**polyA** Polyadenylic acid, a form of RNA containing a long sequence of nucleotides containing the base adenine (A), which occurs at the end of many types of mRNA.
**polydactyly** A condition of having more than the normal number of digits.
**polygenic character** A quantitatively variable phenotype dependent on the interaction of numerous genes.
**polymorphism** The existence of two or more genetically different classes in the same interbreeding population
**polypeptide** A polymer of amino acids linked by peptide bonds. A protein comprises one or more polypeptide chains.
**premelanosome** The precursor of a melanosome, an intracellular organelle found in melanocytes.
**primordium** The early cells that serve as the progenitors of an organ during development.
**probe** In molecular biology, any biochemical compound used to identify or isolate a gene or gene product. The probe is usually radioactively labelled.
**prokaryote** An organism that lacks a membrane-bound nucleus. The superkingdom that contains all classes of bacteria.
**promoter** A region on a DNA molecule to which RNA polymerase binds and initiates transcription.
**proteinase** An enzyme that catalyses the breakdown of proteins by cleaving peptide bonds.
**protooncogene** A cellular gene that functions in controlling the normal proliferation of cells. It can be converted into an oncogene as a result of (i) a somatic mutational event, or (ii) recombination with a viral genome.
**proximal** Nearer to the point of attachment.
**pseudogene** A gene bearing a close ressemblance to a known gene at a different locus, but rendered non-functional by additions or deletions in its structure that prevent

normal transcription and/or translation.
**Purkinje cell** A nerve cell with many dendritic branches between the molecular and granular layers of the cerebellar cortex.

**restriction endonuclease** One of a group of enzymes that catalyse the cleavage of DNA at specific positions determined by the base sequence.
**reverse transcriptase** An enzyme that catalyses the synthesis of DNA using an RNA template.

**sarcoma** A cancer of connective tissue.
**Schwann cell** The cell that enfolds a myelinated nerve fibre.
**sialic acid** An acidic amino sugar that often occurs in glycoproteins.
**signal peptide** The region on a nascent polypeptide chain, that is critical for it to attach to the endoplasmic reticulum. It occurs on all proteins that are secreted from cells.
**silent replacement** The replacement of a base in DNA, which may occur by mutation, that does not result in any changes in the amino acid sequence of the gene product. This is due to the redundancy built into the genetic code.
**smooth endoplasmic reticulum** Part of the membraneous network that occurs throughout the cytoplasm, and which lacks attached ribosomes and so has a smooth appearance.
**stoichiometry** The proportions of components in a reaction.
**suppression** The restoration of a lost or aberrant genetic function.

**tandem repeat** Two or more adjacent copies of a gene or nucleotide sequence all arranged in the same orientation.
**tetraploid** The condition of having four haploid sets of chromosomes.
**thyroglobulin** An iodine-containing protein in the thyroid gland.
**thyroiditis** The inflammation of the thyroid gland.
**thyroxine** A hormone secreted by the thyroid essential for normal growth and development.
**tibia** The larger of the two medial bones in the leg.
**transduction** The transfer of bacterial genetic material from one bacterium to another using a phage as vector.
**tri-iodothyronine** A hormone secreted by the thyroid gland.
**trisomy** The state of an organism or cell in which the chromosome number is $2n+1$, where $n=$ the haploid number.
**triplet** A unit of three successive bases in DNA or RNA that codes for a specific amino acid.
**trypsin** An enzyme that catalyses the cleavage of polypeptide chains at specific positions adjacent to the basic amino acids arginine or lysine.

**vector** A self-replicating DNA molecule that transfers a DNA segment between host cells.

# Index